Infancy to Early Childhood

INFANCY TO EARLY CHILDHOOD

Genetic and Environmental Influences
on Developmental Change

EDITED BY
Robert N. Emde
John K. Hewitt

SECTION EDITORS
Jerome Kagan
JoAnn L. Robinson
David W. Fulker
Robert N. Emde
John K. Hewitt

OXFORD
UNIVERSITY PRESS

2001

MT

OXFORD
UNIVERSITY PRESS

Oxford   New York
Athens   Auckland   Bangkok   Bogotá   Buenos Aires   Calcutta
Cape Town   Chennai   Dar es Salaam   Delhi   Florence   Hong Kong   Istanbul
Karachi   Kuala Lumpur   Madrid   Melbourne   Mexico City   Mumbai   Nairobi
Paris   São Paulo   Shanghai   Singapore   Taipei   Tokyo   Toronto   Warsaw

and associated companies in
Berlin   Ibadan

Library of Congress Cataloging-in-Publication Data
Infancy to early childhood : genetic and environmental influences on
developmental change / edited by Robert N. Emde and John K. Hewitt;
section editors, Jerome Kagan . . . [et al.].
p.   cm.
Includes bibliographical references.
ISBN 0-19-513012-X
1. Twins—Psychology.   2. Twins—Longitudinal studies.   3. Nature and
nurture—Longitudinal studies.   4. Child psychology.   5. Toddlers—Psychology.
I. Emde, Robert N.   II. Hewitt, John K.
BF723.T9 T73   2000
155.44'4—dc21   00-022514

9 8 7 6 5 4 3 2 1

Printed in the United States of America
on acid-free paper

7/29/03

We dedicate this volume to the memory of
our gifted and much-missed colleague,
David W. Fulker,
who died on July 9, 1998.

Acknowledgments

The work in this book was supported by the John D. and Catherine T. MacArthur Foundation and its Research Network on Early Childhood Transitions. We also wish to acknowledge the assistance of the many investigators of that network, as well as the assistance of John DeFries and his faculty and staff of the Institute for Behavioral Genetics at the University of Colorado. We are grateful to Sally-Ann Rhea for her coordinating efforts with the infant twin study and to Christina Hewitt for her expertise in indexing. We are also most appreciative for the careful processing of text and excellent editorial guidance of our administrative assistant, Shirley M. Speller. Most important, we are enduringly grateful to the twins and their families who participated in our longitudinal study.

Contents

Contributors

Marcie L. Chambers
Institute for Behavioral Genetics,
University of Colorado at
Boulder

Stacey S. Cherny
Wellcome Trust Centre
for Human Genetics, Oxford

Robin P. Corley
Institute for Behavioral Genetics,
University of Colorado at
Boulder

John C. DeFries
Institute for Behavioral Genetics,
University of Colorado at
Boulder

Robert N. Emde
University of Colorado
Health Sciences Center, Denver

David W. Fulker
Institute for Behavioral Genetics,
University of Colorado at
Boulder

Julia D. Grant
Pennsylvania State University

Scott L. Hershberger
University of Kansas

John K. Hewitt
Institute for Behavioral Genetics,
University of Colorado at
Boulder

Jerome Kagan
Harvard University

Lorraine F. Kubicek
University of Colorado
Health Sciences Center, Denver

Beth Manke
University of Houston

Charles Martin
University of Kansas

Jude McGrath
Institute for Behavioral Genetics,
University of Colorado at
Boulder

Shirley McGuire
University of California,
San Diego

Diana Nikkari
University of Colorado
Health Sciences Center, Denver

Robert Plomin
Institute of Psychiatry and
Kings College, London

J. Steven Reznick
Yale University

JoAnn L. Robinson
University of Colorado
Health Sciences Center, Denver

Anne-Catherine Roch-Levecq
University of California,
San Diego

Kimberly J. Saudino
Boston University

Kimberly Schiro
National Institute of
Mental Health

Stephanie Schmitz
Institute for Behavioral Genetics,
University of Colorado at
Boulder

Steven M. Wilson
Institute for Behavioral Genetics,
University of Colorado at
Boulder

Susan Young
Institute for Behavioral Genetics,
University of Colorado at
Boulder

Carolyn Zahn-Waxler
National Institute of
Mental Health

Part I

Introduction

1

The Dynamics of Development in a Unique Multidisciplinary Collaboration

Robert N. Emde

The MacArthur Longitudinal Twin Study began as a unique and exciting venture that joined a multidisciplinary group of scientists in a research network sponsored by the John D. and Catherine T. MacArthur Foundation. The venture was a bold one, considering the diverse views of investigators, the questions we asked, and the scope of the study. The motives for this bold venture were fueled by the research network as a whole, which was entitled "The Transition from Infancy to Early Childhood." The larger network arose in response to a practical and conceptual rationale.

The practical rationale of the network was to initiate a social experiment in science to see if a selected group of committed investigators in a targeted research area could engage in long-term innovative research planning and joint work and generate a collaborative culture. Accordingly, the network sought to overcome traditional competitive modes, socialized by disciplinary training, and establish the productive excitement of a creative, multidisciplinary working group of colleagues, students, and others. Closely related to these aspirations were formulating methods through large-scale collaboration and working within this methodological framework.

The conceptional rationale for the network included the fact that the transition from infancy to early childhood (1–3 years) marked a crucial but understudied developmental period during which many important psychological qualities emerge. These qualities include important changes in the broad domains of cognition and emotion, such as the acquisition of language, the attribution of causality, and the development of self-awareness, as well as play and new social skills, changes in positive and negative emotions, and the first

3

signs of empathy. Moreover, the domain of temperament seemed crucial. Children were known to differ in their temperaments and in the ways they engage their environments, and temperament was known to predispose them, in certain contexts, to be inhibited or exuberant.

The network began in a high state of energy and excitement.[1] Enthusiasm was fueled by a synergism that came from fusing the practical and conceptual rationales of the network. Discussions then led to disagreements about what needed to be done. As we began to learn from each other, there were different views from different disciplines and different views about what needed to be known and about what areas provided opportunities for discovery. The MacArthur Foundation support allowed us to plan ambitiously and to address large questions across disciplines in a programmatic way.

It was in an initial time of ferment that discussions began about the desirability of a large-scale collaborative research project that could harness the energy and expertise of its investigators and serve as a landmark study of development during the second year of life. The idea of the MacArthur Longitudinal Twin Study emerged and was given impetus by John DeFries and Robert Plomin. The idea was to join multiple developmental scientific approaches and to use a behavioral genetics framework for addressing questions pertaining not only to the rationale of the network but pertaining to central processes of early development. Collaborators would contribute their expertise and study variables of interest in a large study in which each investigator would be a stakeholder in the success of the overall project. As we began to design the study, we also realized that it would provide a unique opportunity for multivariate analysis across domains. We realized, in addition, that our large longitudinal twin study would complement our interests in understanding developmental processes.

At this point it is helpful to review our thinking about development, its dynamic processes, and some of the major questions we address in this book. A definition of development, tacitly held by most of us in the early days of the network but articulated later in its history, is presented below. Next, we present questions about developmental processes in the second year that are addressed by the MacArthur Longitudinal Twin Study.

A DEFINITION OF DEVELOPMENT

Development consists of change within individuals over time. It involves increasingly organized complexity. As such, it involves processes of differentiation (divisions into subsystems), integration (articulation of wholeness), and successive ordering of parts and wholes (hierarchialization). Moreover, development necessarily involves dynamic exchanges within the environment.

Human behavioral development must be considered not only in its biological context, wherein adaptive aspects of interactions involving genetic expression, maturation, and cognitive construction are salient, but also in its sociocultural context, wherein adaptive aspects of interactions involving social roles, networks, and varied environments are important. Human development must be considered not only from the perspective of stability (continuity), but also from the perspective of transformational change (discontinuity), and not only from the perspective of successful adaptation (health), but also from the perspective of unsuccessful adaptation (disorder and pathology).

CORE FEATURES OF DEVELOPMENT AND MOTIVATING QUESTIONS OF THE LONGITUDINAL STUDY

The above definition of development highlights a number of features about the dynamics of development. Our study could not deal with all features and their implications, but four core features can be highlighted. These gave rise to a number of unanswered questions about early development. They also led to a fifth set of questions about genetic and environmental influences that motivated our twin design. We sought to address all these questions with a major longitudinal study that combined the approaches of developmental science with those of behavioral genetics.

1. Questions about increasing complexity. The first set of questions emerge from the feature of increasing complexity. In early behavioral development, psychologists had assumed a progression from undifferentiated to differentiated behavioral functioning, with integrative processes following. In other words, early development was characterized by a period of global interrelatedness of behaviors, followed by a period of specialization and less interrelatedness; and then still later with behavioral connectedness at a higher level (Sameroff, 1983; Werner, 1948). Recent discoveries, however, have shown the extent to which many socioemotional and cognitive behaviors are more differentiated in infancy, suggesting that many are to some extent already specialized and pre-programmed by evolution to serve adaptation during this age period (see reviews in Emde, 1988; Stern, 1986). In addition, some contemporary approaches to early development had found it more useful to think about more differentiated components of behavioral systems in infancy which then become configured and interrelated as a result of experience. (See approaches that are referred to as "dynamic systems models," Smith & Thelen, 1993, and "connectionist models," Ellman et al., 1996). We might refer to this as a "components and configurations" view: earlier component behaviors that are relatively independent in their functioning become configured and connected in various ways over the course of development.

Newer questions about increasing complexity have emerged as a result of this new way of thinking. To what extent are behaviors in the child's second year independent of one another (reflecting differentiation) versus to what extent are they correlated with one another (reflecting global interrelatedness or undifferentiation)? Although the principle of differentiation as put forth by Werner (1948) was phrased in terms of normative development, it can be placed in an individual differences framework. Thus, differentiation can be seen as a lesser magnitude of intercorrelations among measures across time, providing that variability exists between individuals for such measures (Plomin et al., 1990). Developmental observations in a longitudinal study involving large numbers of children could therefore provide the opportunity to assess individual differences at multiple time points and, correspondingly, provide the opportunity to make inferences about correlations (or their absence) between behaviors with respect to even low levels of association. But the methods of behavioral genetics would add more, as we will see below.

This brings us to the second feature of development we addressed, namely, the role of context.

2. Questions about context. All development takes place in context, and the context for early development includes strong environmental as well as biological influences. We are immersed in the beginnings of a genetic revolution in the health sciences in which biologic discoveries related to molecular genetics are reconfiguring our knowledge about disorder and risk, as well as our knowledge about the way genes work. Genes work with the environment, and genetic expression is influenced by environmental co-actions at all levels of developmental systems, from cell to society (Gottlieb, 1992; Scarr & Kidd, 1983). Thus, old questions of nature versus nurture are inappropriate, and contemporary questions are formed in terms of mutual influences (i.e., nature *and* nurture; Plomin, 1990). Correspondingly, genetic influences and expression would be expected to change across development as would environmental influences.

Longitudinal studies need to include measured aspects of the environment. Our longitudinal twin study in its planning phases not only outlined detailed developmental assessments of the individual but also outlined detailed assessments of the individual's environment such as observations of family and community. The scale of this already large landmark study, however, did not allow for funding a full-scale study of the environment (see Plomin et al., 1990). The study did include some measures of the family, and the methods of behavioral genetics allowed us to study a variety of questions about genetic and environmental influences—that is, about nature and nurture during this important developmental period (see questions under 5, following page).

3. Questions about continuities and change. Development is characterized by both behavioral continuity and transformational change. Sometimes there are pervasive changes across behavioral domains that are stepwise rather than linear and are referred to as developmental transitions or times of discontinuity. Such times are particularly prominent in early development (Emde et al., 1976; Kagan, 1984), and the need to study these episodes entered into the rationale for the research network. Some of the questions stemming from this feature of development are compelling and obvious.

To what extent is there continuity of individual differences in behavior across time within the domains of temperament, cognition, and emotion? To what extent is there change and transformation? Is there evidence for more pervasive and substantial change across domains in a transition period toward the end of the child's second year, as would be predicted from previous research (e.g. Kagan, 1981)? A study with a large number of children, followed longitudinally and with multiple measures, would give us some confidence in our estimates of continuity and change. But even more exciting to us was the prospect of looking deeper using the twin design and behavioral genetic approaches, a prospect that is discussed below.

4. Questions about extremes. Still another set of questions suggested by our definition of development concerns an analysis of extremes. Individual variability in development is considerable. Are extremes in variation among individuals qualitatively different? Or are extremes of variation on a quantitative continuum? In a longitudinal study with sufficiently large numbers, one has the opportunity to examine different parts of the distribution of children on a given aspect of behavior at multiple points in time. Questions about extremes may be important for considerations of temperament, as well for disorder and risk for disorder. Many contemporary clinicians debate whether disorder is better regarded on a continuum with the normal or as separate in a categorical sense. Although our study of normal development could not be expected to resolve such a question about the separateness of disorder, it might help to frame future research directed toward either broad categories or dimensions that are related to disorder.

5. Questions about genetic and environmental etiology. Development occurs with both genetic and environmental influences. The methods of behavioral genetics, using a longitudinal twin design where comparisons are made between twins that are genetically identical (i.e., monozygotic) and twins that share on average half of their genes (dizygotic) provides an opportunity to go beyond description. In terms of our core features of development, we can go beyond the phenotypic descriptions of complexity, context, continuity/change, and extremes and we can move to the level of explanation. Chapter 2 introduces special opportunities and strategies made possible by a longitudinal twin design. Below we discuss some of the questions we sought to answer concerning additional core features of development.

Concerning increasing complexity, using a view about components and configurations, we can ask important additional questions about development. To what extent would different component behaviors, within the emotion domain, for example, have different genetic and environmental influences? To what extent would new configurations have entirely new genetic and environmental influences?

Concerning questions about context, the twin design allows us to seek answers to questions about genetic and environmental origins of behavior during a toddler's life. What is the extent of such influences at particular ages? To what extent are there environmental influences as compared with genetic influences? To what extent are environmental influences common to siblings (i.e., shared by siblings) and to what extent are environmental influences unique to individuals within a family (i.e., not shared by siblings)? Do patterns of influence vary at different ages? Can we find evidence for genetic–environmental correlations at some ages?

Perhaps most exciting to us were questions about the genetic and environmental etiology of continuity and change. Longitudinal genetic analyses make it possible to chart developmental continuities and change at the level of genetic and environmental influences and to ask time-related dynamic questions about the underpinnings of development. To what extent is continuity accounted for by genetic factors? To what extent is continuity accounted for by shared environmental factors? By nonshared environmental factors?

We now know that development is characterized by changing influences across time, both environmental and genetic (Plomin et al., 1997). How much evidence exists for new genetic influences on given behaviors at later ages? How much evidence exists for new environmental influences, either shared or nonshared? And, finally, to what extent are there patterns of influence that go across domains over time? This last question is similar to our questions about increasing complexity, but it puts the matter in terms of continuity and change. In other words, how much evidence exists for globality during the course of the second year (i.e., with similar patterns of continuity and change across domains) and to what extent is there relative independence or differentiation?

Questions about extremes are also enhanced. The twin methodology and behavioral genetic analytic techniques enable us to explore a key question: do the extremes on any given variable (i.e., either very high or very low) have different genetic and environmental influences than the middle of the distribution? In other words, would there be evidence of a separate genetic and environmental etiology for extremes that might in turn contain distinguishable configurations or syndromes?

THE SITE OF THE STUDY AND OUR APPROACH

The site of the study was the Institute of Behavioral Genetics at the University of Colorado, Boulder. The twins were obtained from birth registries along the front range region of Colorado. The principal investigator for most of the duration of the study was John DeFries, with strong leadership also from Robert Plomin. Early on, the research team decided to focus on the previously unexplored domains of behavioral inhibition, empathy, emotional expressivity, and experimental cognitive tasks that, in addition to Bayley exams, included categorizations, spatial memory, and word comprehension. The approach included multiple methods, employing assessments both in the home and in the laboratory and including eliciting situations, videotaped observations, and ratings by observers and parents. In addition, standard measures were included to assess temperament, mental development, and language development. A longitudinal design was critical to analyze change and continuity. The test ages were strategically chosen. Twenty and 24 months bracketed the major transition period that we expected to encompass language spurts such as the use of two-word phrases, growth of a sense of self, and the development of sophisticated emotional communication. A period of change was bracketed by assessments during periods that we anticipated would be ones of relative consolidation rather than change, namely, at 14 and 36 months.

OVERVIEW OF THE VOLUME

The chapters of this introductory section provide overviews that orient the reader with respect to the detailed reporting of findings in subsequent chapters. Chapter 2 describes the twin method and also contains an introduction to our analytic strategies. This will be useful for the general developmental reader and may be skipped by those familiar with such strategies. Chapter 3 details the sample and procedures of our study. This chapter may be skimmed by the reader and used as a reference for later chapters that report analyses and findings.

Findings of our study are presented in four parts, each with preface that guides the reader to the important issues and questions addressed within that part. Parts II, III, and IV deal with the domains of temperament, emotion, and cognition, respectively, and part V deals with environmental contributions and cross-domain integrations. The major questions posed in this chapter frame the contributions of this volume. As such, they are guiding questions, and all questions are not addressed in every analysis or chapter.

The final part reviews major themes and conclusions. It returns to our major questions and indicates, not surprisingly, that these questions do not have simple answers that apply to all domains or to all ages. The MacArthur Longitudinal Twin Study is only a first step toward addressing those questions. We hope that its successes, as well as its limitations, will stimulate more thinking and research on the dynamics of development during the transition from infancy to early childhood. Above all, we hope the reader will experience the excitement of this special journey in multidisciplinary collaboration.

NOTES

1. Principal investigators of five geographic nodes were joined in the network, which began its operations in late 1982. Each node had 10–20 investigators. The principal investigators were Kathryn Barnard, University of Washington; Elizabeth Bates, University of California at San Diego; Robert Emde, University of Colorado, Denver, who served as network director; Jerome Kagan, Harvard University; and Marian Radke-Yarrow of the Intramural Program of

the National Institute of Mental Health. Other collaborative projects from the research network are documented and summarized in published volumes listed at the end of this chapter.

REFERENCES

Ellman, J. L., Bates, E. A., Johnson, M. H., Karmiloff-Smith, A., Parisi, D., & Plunkett, K. (1996). *Rethinking innateness: A connectionist perspective on development*. Cambridge, MA: MIT Press.

Emde, R. N. (1988). Development terminable and interminable: I. Innate and motivational factors from infancy. *International Journal of Psycho-Analysis, 69,* 23–42.

Emde, R. N., Gaensbauer, T. J., & Harmon, R. J. (1976). Emotional expression in infancy: A biobehavioral study. *Psychological Issues, a Monograph Series, Inc., 10*(37). New York: International Universities Press.

Gottlieb, G. (1992). *Individual development & evolution*. New York: Oxford University Press.

Kagan, J. (1981). *The second year: The emergence of self-awareness*. Cambridge, MA: Harvard University Press.

Kagan, J. (1984). *The nature of the child*. New York: Basic Books.

Plomin, R. (1990). *Nature and nurture—an introduction to human behavioral genetics*. Pacific Grove, CA: Brooks/Cole.

Plomin, R., Campos, J., Corley, R., Emde, R. N., Fulker, D. W., Kagan, J., Reznick, J. S., Robinson, J., Zahn-Waxler, C., & DeFries, J. D. (1990). Individual differences during the second year of life: The MacArthur longitudinal twin study. In J. Colombo & F. Fagen (Eds.), *Individual differences in infancy: Reliability, stability, and predictability* (pp. 431–455). Hillsdale, NJ: Erlbaum.

Plomin, R., DeFries, J. C., McClearn, G. E., & Rutter, M. (1997). *Behavioral genetics* (3rd ed.). New York: Freeman.

Sameroff, A. J. (1983). Developmental systems: Contexts and evolution. In E. M. Hetherington (Ed.) & P. H. Mussen (Series Ed.), *Handbook of child psychology: Vol. 1. Socialization, personality, and social development* (pp. 237–294). New York: Wiley.

Scarr, S., & Kidd, K. K. (1983). Developmental behavior genetics. In E. M. Hetherington (Ed.), *Handbook of child psychology: Vol. 4*. New York: Wiley.

Smith, L. B., & Thelen, E. (Eds.). (1993). *A dynamic systems approach to development; applications*. Cambridge, MA: MIT Press.

Stern, D. (1985). *The interpersonal world of the infant*. New York: Basic Books.

Werner, H. (1948). *Comparative psychology of mental development*. New York: International Universities Press.

Published Volumes from Collaborative Studies

The volumes listed below provide a review and documentation of studies other than the MacArthur Longitudinal Twin Study that resulted from the Research Network that received its support from the John D. and Catherine T. MacArthur Foundation between 1982 and 1994. The Network for its first five years was known as the MacArthur Research Network on the Transition from Infancy to Early Childhood; after that it was known as "The Early Childhood Transitions Network." Robert Emde served as Chair of the Network.

Bretherton, I., & Watson, M. W. (Eds.). (1990). *New directions for child development: Children's perspectives on the family.* San Francisco: Jossey-Bass.

Cicchetti, D., & Beeghly, M. (Eds.). (1990). *The self in transition: Infancy to childhood.* Chicago: University of Chicago Press.

Cummings, E. M., & Davies, P. T. (1994). *Children and marital conflict: The impact of family dispute and resolution.* New York: Guilford Press.

Elman, J. L., Bates, E. A., Johnson, M. H., Karmiloff-Smith, A., Parisi, D., & Plunkett, K. (1996). *Rethinking innateness.* A connectionist perspective on development. Cambridge, MA: MIT Press.

Fenson, L., Dale, P. S., Reznick, J. S., Bates, E., Thal, D. J., & Pethick, S. J. (1994). Variability in early communicative development. *Monographs of the Society for Research in Child Development, 59* (5, Serial No. 242).

Fenson, L., Dale, P. S., Reznick, J. S., Thal, D., Bates, E., Hartung, J. P., Pethick, S., & Reilly, J. S. (1993). *The MacArthur communicative development inventories: User's guide and technical manual.* San Diego: Singular Press.

Fox, N. A. (Ed.) (1994). The development of emotion regulation: Biological and behavioral considerations. *Monographs of the Society for Research in Child Development, 59* (2–3, Serial No. 240).

Greenberg, M., Cicchetti, D., & Cummings, M. (Eds.). (1990). *Attachment in the preschool years.* Chicago: University of Chicago Press.

Haith, M. M., Benson, J. B., Roberts, R. J. Jr., & Pennington, B. F. (Eds.). (1994). *The development of future-oriented processes.* Chicago: University of Chicago Press.

Kagan, J. (1994). *Galen's Prophecy.* New York: Basic Books.

Kagan, J., & Lamb, S. (eds.). (1987). *The emergence of morality in young children.* Chicago: University of Chicago Press.

Plunkett, K., & Elman, J. L. (1997). *Exercises in rethinking innateness. A handbook for connectionist simulations.* Cambridge, MA: MIT Press.

Reiss, D., Richters, J., Radke-Yarrow, M., & Scharff, D. (1993). *Violence in public and private spaces: Its impact on children.* New York: Guilford Press.

Reznick, S. (1989). *Perspectives on behavioral inhibition.* Chicago: University of Chicago Press.

Reznick. J. S., Corley, R., & Robinson, J. (1997). A longitudinal twin study of intelligence in the second year. With commentary by A. P. Matheny, Jr. *Monographs of the Society for Research in Child Development, 62*(1, Serial No. 249).

Rubin, K., & Asendorpf, J. (Eds.). (1993). *Social withdrawal, inhibition, and shyness in children.* Hillsdale, NJ: Erlbaum.

Sameroff, A. J., & Emde, R. N. (Eds.). (1989). *Relationship disturbances in early childhood: A developmental approach.* New York: Basic Books.

Sameroff, A. J., & Haith, M. M. (Eds.). (1996). *The five to seven year shift. The age of reason and responsibility.* Chicago: University of Chicago Press.

Stiles-Davis, J., Kritchevsky, M., & Bellugi, U. (Eds.). (1988). *Spatial cognition: Brain bases and development.* Hillsdale, NJ: Erlbaum.

2

The Twin Method
What We Can Learn from a Longitudinal Study

John K. Hewitt
Robert N. Emde
Robert Plomin

The studies reported in this volume address issues about the organization of behavior and its development in infancy. Of interest to the research team were the roles of genes, shared family environments, and individual experiences in shaping individual differences during early development. Thus, the study was designed to make use of the "natural experiment" of twinning that provides one of the most direct and well-controlled methods for the detection and quantitative assessment of the impact of heritable (genetic) and nonheritable (environmental) influences (Neale and Cardon, 1992).

Although a first step in any behavioral genetic study is to assess the extent of heritable and environmental influences, the most exciting analyses, and the reason for undertaking this study, are those aimed at determining the roles of genes and environment in determining the changes and continuities that are the hallmark of development. One of the important generalizations from the last decade of twin and family research is that the extent of both genetic and environmental influences changes during development (Plomin et al., 1997); moreover, new genetic influences may be switched on at different periods of the life cycle (Fabsitz et al., 1992), just as different environmental influences become important as individuals pass through significant developmental transitions. In other words, we cannot think of genes as affecting only those characteristics that are present at birth and are passive throughout life, while the dynamic actions of the environment mold and shape development. Genes, as well as environmental influences, may contribute to both continuity and change. The combination in the MacArthur Longitudinal Twin Study of the classical twin study with a longitudinal design made it possible to explore this issue.

In this introductory chapter, we first discuss how the twin study design can be used to assess the relative importance of genetic and environmental effects, and then we introduce in a general way some of the methods that allow us to make inferences about developmental change and continuity. Finally, any approach to the study of complex questions about human development will have limitations and will make assumptions that may only hold to a first approximation in actuality. The longitudinal twin study is no exception, and so we conclude this chapter with a discussion of those assumptions and limitations. The details of the statistical modeling methods are presented throughout the book in the context of the studies in which they were applied. We hope that we have managed to avoid unnecessary detail while retaining sufficient information about the basis for the results and conclusions. In most cases, descriptions of models or statistical procedures appear as end notes to the relevant chapters.

ESTIMATING THE EFFECTS OF GENES AND ENVIRONMENTS ON INDIVIDUAL DIFFERENCES IN BEHAVIOR

The twin method compares pairs of twins who are genetically identical (monozygotic, MZ) to pairs of fraternal twins (dizygotic, DZ) who are approximately half as similar genetically. If heredity affects a trait, the twofold greater genetic similarity of MZ twins is expected to make them more similar than DZ twins. Thus, for a particular measure, an MZ correlation that is significantly higher than the DZ correlation suggests genetic influence. The effect size of genetic influence can be estimated by doubling the difference between the MZ and DZ correlations. With some assumptions, which often hold approximately in practice, this value estimates the heritability, the extent to which observed variance in behavior can be attributed to genetic influences. The difference between the MZ and DZ twin correlations is doubled because MZ twins are identical genetically, whereas the genetic correlation for DZ twins is assumed to be .50. Thus, the difference between the MZ and DZ twin correlations is an estimate of half of the heritability and is doubled to estimate heritability. The rest of the variance is attributed to environmental factors, perhaps more properly referred to as "nongenetic factors" because they include all nonheritable influences such as biological influences and accidents as well as systematic psychosocial environmental influences.

The environmental component of variance can be decomposed into two subcomponents. One subcomponent, referred to as shared, family, or common environmental influence, is estimated in the twin method as twin resemblance not explained by hereditary resemblance. The twin estimate of shared environment is twice the DZ correlation minus the MZ correlation, although the twin method by itself does not provide as much power to estimate shared environmental variance as would, say, an appropriate estimate from an adoption study (Martin et al., 1978; Plomin et al., 1997). The other subcomponent is the residual environmental variance, which includes error of measurement and nonshared, individual, or unique environmental influences that make family members different from each other. Differences within pairs of MZ twins are due to nonshared environment and to measurement error. Differences within DZ pairs are due to these influences together with the effects of genetic segregation.

The distinction between shared and nonshared environment is important because in studies later in life, nearly all environmental influences on behavior have been found to be of the nonshared variety (Plomin & Daniels, 1987), suggesting that the environmental impact of what is shared and common to children within the family is less long lasting than has often been supposed. However, there are two caveats. First, many environmental influences that might be thought of as common to the family (e.g., parental behavior) might well be differentially applied to individual children; in this case, the influence on the child's behavior would be detected as a contribution to the nonshared variance. Thus, failure to find evidence of shared influences does not necessarily imply that parents are ineffective in shaping their children's behavior. Rather, it suggests that either they are not exerting a substantial influence or, alternatively, they are adjusting behavior to individual children. Second, the nonshared variance component includes all of the random errors of measurement psychometrists usually call unreliability. For this reason, well-developed standardized assessments, such as many cognitive tests, some personality assessment inventories, and most anthropometric measurements, will have quite high reliabilities and often reveal smaller nonshared environmental variance components than do less well-developed observational, interview, or other ratings of children's behavior. The relative magnitude of the nonshared environmental component should be interpreted with this in mind.

In this volume, two general approaches have been used to capitalize on the basic properties of family resemblance of MZ and DZ twins to yield estimates of genetic and environmental components. The first approach we refer to as the "DeFries-Fulker regression procedure" (DeFries & Fulker, 1985, 1988). This analytic approach capitalizes on the statistical property that if one group of twin pairs (MZ) shows greater family resemblance than another group of twin pairs (DZ), then the linear regression of cotwins on each other's scores will interact with the group membership (zygosity). This basic linear regression procedure was also elegantly exploited by DeFries and Fulker to provide estimates of the genetic and shared environmental variance in the case of samples of twins selected for extreme scores on a given behavioral dimension (e.g., the poorest readers). This approach has the advantage that when extreme groups can be identified (e.g., the most shy or behaviorally inhibited children), there is an opportunity to test whether the heritability in the extremes of the distribution is same as that in the normal range. The DeFries-Fulker regression procedure also has the merit that it is easy to apply and easy to extend to allow the inclusion of covariates that may moderate the the influence of genes or environments.

The second approach uses "path analysis" first developed by geneticist Sewall Wright in 1921 and subsequently adopted widely in the social sciences (Kenny, 1979; Wright, 1921). Specific hypotheses about genetic and environmental influences can be summarized in terms of path diagrams or structural equation models. These in turn lead to precise expectations for the observed individual variances and covariances among different kinds of relatives (MZ and DZ twins in our study). The expectations can be compared to the observed statistics, and a test of the goodness of fit of the model to the data can be made. Given that the model is adequate, estimates of the genetic and environmental parameters can be obtained, and the statistical significance of each parameter can be tested. The use of the these methods has become standard in twin and family studies because of their statistical rigor and their ready generalization from univariate to multivariate and developmental analyses. A detailed account of this approach in the context of twin and family studies can be found in Neale and Cardon (1992). (These methods are also the subject of annual training workshops supported by the National Institute of Mental Health and by NATO. Readers with a special interest in such methodological training are invited to contact one of the editors [J. K. H] for further information about these workshops.)

The chapters that follow vary in the detail of presentation of the statistical and mathematical aspects of these analyses depending on the level of detail appropriate to the argument of each chapter, the questions being addressed, and the data being analyzed. The chapters on cognitive abilities (Part III) typically present more detailed explication of the data analysis and model fitting than do the chapters on temperament and emotion. To a large extent, this is because the argument and conclusions of the chapters on cognitive abilities lean most heavily on the details of the model fitting and comparisons among the genetic and environmental models.

ASSESSING GENETIC CHANGE

Two types of genetically mediated change can be investigated from the perspective of quantitative genetic theory. The first type involves the magnitude of genetic effects, indexed by the heritability statistic, which can change during development. The second type involves genetic influence on age-to-age change. Genetic factors can contribute to change from age to age even if heritability remains the same across ages. Such age-to-age genetic change can occur, for example, because the changing developmental context of the child engages different sets of genes for a particular task at different ages. In other words, developmental changes in the genetic underpinnings can occur because of changes in the nature of a variable which includes changes in its measurement, validity, and reliability.

Both types of genetic change (i.e., in magnitude of heritability from age to age and in genetic influence on age-to-age change) are most likely during a period of rapid developmental change such as infancy. However, little is known about the role of either type of genetic change during infancy. It is usually assumed that the relative magnitude of genetic influence diminishes as the lives of children diverge due to cascading differences in their experiences. This view, which implies decreasing heritability during development, has been suggested by life-span theories of development. Non-normative life events are thought to increase throughout the life span as "significant life events take on a more and more important role in determining the course of human development" (Baltes et al., 1980, p. 78).

The reasonableness of this hypothesis of decreasing heritability contrasts with the meager data on the topic: when heritability changes, it tends to increase rather than decrease (Plomin, 1986). Heritability increases most clearly in the realm of mental development from infancy to middle childhood (Fulker et al., 1988). For personality, changes in heritability are sometimes seen earlier in the life span but seldom in adulthood (Goldsmith, 1983; Loehlin, 1992: McCartney et al., 1990; Plomin & Nesselroade, 1990).

Levels of heritability can be the same at two ages for different genetic reasons. That is, theoretically, the genes that affect a trait at one age could differ entirely from the genes that affect the trait at another age, but the overall magnitude of genetic effects (heritability) could be the same at the two ages.

The second type of genetic change explores the extent to which genetic effects at one age differ from genetic effects at another age. The simplest approach to this type of genetically mediated change is to analyze change scores.

Genetically mediated change from age to age is not merely the flip side of genetic continuity. Although traits can show genetic contributions to both change and continuity, it is also possible that genetic influences on a trait at two ages largely contribute to change but not to continuity or to continuity but not to change. The essence of a twin analysis of cross-age continuity is the cross-twin correlation—that is, the correlation of one twin's score at age 1 and the cotwin's score at age 2. A genetic contribution to continuity is suggested to the extent that such cross-twin correlations are greater for identical twins than for fraternal twins.

These simple approaches to genetic change and continuity are related to more general developmental genetic models that simultaneously analyze genetic contributions to covariance between ages and take into account heritability at both ages as well as phenotypic stability between ages (Boomsma et al., 1989; Eaves et al., 1986; Hewitt et al., 1989; Loehlin et al., 1989; Phillips & Fulker, 1989).

These more general methods, which are used extensively in later chapters, are extensions of the path analytic, or structural equation modeling, approach to the analysis of twin data. In their most simple form, they can be thought of as testing three different models for how each source of variation—genetic, shared family environmental, or nonshared individual environmental—contributes to development. To be concrete, let us consider genetic influences, while recognizing at the same time that what we portray can apply equally to the environmental influences.

The first model says that genetic influences contribute only to continuity across 14, 20, 24, and 36 months. This model posits a single source of genetic influence that, like a general factor in a factor analysis, influences the behavior at each age, and there are no age-specific genetic influences. If this model fits the data, we can conclude that genes contribute to developmental continuity rather than to change. We should note, however, that even with a single general genetic source of influence, the quantitative impact of the influence could increase or decrease, just as in a psychometric general factor model some variables are more greatly influenced than others by that general factor.

The second model says that genetic influences are specific to each age of assessment; the genetic influences on behavior at age 14 months are not the same as those that influence the behavior at 20 months, are different again from those that are effective at 24 months, and differ again from those effective at 36 months. Although this pattern could arise from the statistical independence of genetic effects that nevertheless are controlled by the same genes, this is what would be observed in the extreme case of genes being switched on and switched off as development proceeds. The model is analogous to a factor model with no general factor, only specific factors. If this model were correct, we would observe significant heritability at each age, but no evidence of higher age-to-age cross-correlations for MZ than for DZ twins. Genes would be driving developmental changes (at a furious pace), but not contributing at all to developmental continuity.

The third model allows for both kinds of process. In this model, a genetic factor has loadings (paths) for each age, starting at 14 months in our study. Then, at 20 months, a second genetic factor may be introduced that has influences (paths to) 20, 24, and 36 months. A third genetic factor may influence the observations at 24 and 36 months. Finally, a distinct genetic factor may influence behavior at 36 months, over and above any influences from the earlier factors. In its full form, this model is really just a description of the genetic variation at the four ages and is referred to as the Cholesky or triangular decomposition. Submodels corresponding to various patterns of continuity and change may be assessed, and the end result will be a quantitative description of the pattern and magnitude of genetic contributions to development.

These developmental models can be examined separately for each genetic and environmental source of variation, and this would lead to a full developmental behavior genetic analysis. Many analyses reported in this book involve variants of these development models, with the details and technical aspects of the modeling procedures reported in the relevant chapters.

ASSUMPTIONS ABOUT THE TWIN METHOD AND
SOME CAUTIONARY NOTES

The kinds of analyses and their interpretation that we have outlined depend to a greater or lesser extent on assumptions about the twin method. The first assumption is that of equal environments. The difference between MZ and DZ twin correlations indexes genetic influence if we assume that environmental contributions to twin resemblance are similar for the two types of twins. In the current study, both MZ and DZ twins are born from the same womb, they are the same age and sex, and they are growing up in the same family. However, if environments are more similar for MZ twins (e.g., if MZ twins are treated more similarly than are DZ twins, MZ correlations that exceed DZ correlations could be due to environmental rather than genetic factors). Where this possibility has been studied, it has usually been found that different degrees of twin resemblance cannot be attributed to systematic differences in treatment, based either on actual zygosity or on what twins or their parents believe to be their zygosity (see Kendler et al., 1993, for a recent example and review). In the case of the MacArthur Longitudinal Twin Study, analyses of the data reported by McGuire and Roch-Leveque in this volume suggest that mothers more often perceive themselves as behaving more similarly with MZ twins. Even though this maternal perception may be a response to, rather than a cause of, twin similarity in behavior, in the absence of any definitive test of the equal environments assumption, we should maintain some caution in interpreting the results of a twin study.

A second assumption concerns generalizability. The twin method assumes that analyses involving twins will generalize to nontwin populations. A major concern about generalizability, especially in infancy, is that twins are born nearly a month premature on average and about 30% lighter and 20% shorter at birth, differences that disappear by middle childhood (MacGillivray et al., 1975). Twin deficits in childhood have been noted for verbal ability (Hay & O'Brien, 1983), especially for language (Savic, 1980), although most of these deficits also diminish by the early school years (Wilson, 1983). Our results for the Bayley Mental Development Index, however, show means and variances similar to those of the standardization sample. Still, it is possible that our twins might show deficits relative to nontwin singletons in families matched for parental education. A possible comparison group offering within-family control would be nontwin siblings of the twins. Moreover, the possible effect of perinatal factors on twin results can be broached empirically by using indices such as birth weight as covariates, as is done in this volume.

A third assumption concerns nonadditive genetic variance. Estimating the heritability by doubling the difference of the MZ and DZ correlations assumes that all genetic effects are additive—that is, that effects of substituting one gene for another combine in a linear manner to influence the trait. If nonadditive genetic effects are influential, an estimate of heritability that doubles the difference between MZ and DZ correlations may overestimate genetic influence.

A useful corrective is to remember that heritability cannot exceed the MZ correlation. Throughout this volume, we use model-fitting procedures that test the adequacy of the assumption that genetic effects act additively and make corrections with respect to variables where nonadditive genetic effects are indicated. The assumption that genetic effects operate primarily in an additive manner appears to be reasonable, although evidence for some nonadditive genetic influence is beginning to emerge in the domain of personality (Plomin et al., 1990). For example, for the personality trait extraversion, DZ twin resemblance is consistently less than would be predicted in the absence of nonadditive genetic effects (Loehlin, 1992).

A fourth assumption is that assortative mating, indexed by resemblance between spouses, is negligible. Assortative mating has an effect opposite to that of nonadditive genetic variance: It increases the genetic correlation of DZ twins beyond the .50 genetic correlation expected if genetic effects are additive. Thus, assortative mating is a conservative bias in the sense that it lowers heritability estimates based on the difference between MZ and DZ correlations.

Unlike nonadditive genetic variance, assortative mating might be estimated directly. Resemblance between parents could be assessed, and the expected genetic correlation for DZ twins can be adjusted upward to account for their increased genetic resemblance due to assortative mating. Although there is little information about how assortative mating affects behavior in early development, the literature on adults is informative. Assortative mating is low for most personality traits—about .05 for the two major dimensions of extraversion and neuroticism, for example (Eaves et al., 1989)—and can safely be ignored. However, for cognitive variables, assortative mating is more substantial—.33 in the world's literature on IQ scores (Bouchard & McGue, 1981).

In the MacArthur Longitudinal Twin Study, parents have not been assessed for most variables, which means that the degree of assortative mating cannot be calculated. Indeed, for many traits in infancy, it may not be possible to assess trait-relevant assortative mating of parents because the adult versions of infant traits are not known, for example, for infant hedonic tone and empathy, and perhaps even for infant temperament, mental ability, and language development. Thus we must remember that although the effects are unlikely to be substantial, it is possible that some of our estimates of heritability may be lowered, and estimates of shared environmental effects raised, by the consequences of unassessed assortative mating.

A fifth assumption concerns genotype–environment correlation and interaction. Our estimates of genetic and environmental effects assume that correlations and interactions between those effects are of negligible importance. This may or may not be true. It may be, for example, that a child with a genetic bias for shyness and inhibition will have a role model in a parent who is also extremely shy and inhibited. Possible interactions between the genetic bias and the environmental bias would then be impossible to assess in our current design. Such effects on twin analyses in general are complex and not completely understood (e.g., Plomin et al., 1977), although one example of such effects may be the results reported by McGuire and Roch-Leveque in this volume where the maternal perception of treatment of twins is correlated with the twins' similarity in behavior.

Another consideration providing limitations on interpretation concerns statistical power. The twin method estimate of heritability reflects the difference between MZ and DZ correlations, and differences in correlations have large standard errors of estimation. Sample sizes of approximately 200 MZ pairs and 175 DZ pairs, available for our analyses, are able to detect only substantial heritabilities. Given reasonable patterns of twin correlations, heritabilities will be detected as significant ($p < .05$, one tailed) only if they exceed 20% when MZ and DZ correlations are high (e.g., .8 and .7, respectively). Detecting heritabilities with 80% statistical power (i.e., detecting a particular heritability as significant four out of five times) requires heritabilities in excess of 37% for moderate twin correlations (e.g., MZ and DZ correlations .60 and .42, respectively; calculations based on Neale, 1997).

Despite these cautions, the twin method remains a valuable screening device for genetic influence. For measures substantially influenced by heredity, the approximately twofold difference in genetic similarity for the two types of twins can reasonably be expected to exceed by far the effect of complicating factors such as nonadditive genetic variance and unequal environments for MZ and DZ twins. Analyses of the MacArthur Longitudinal Twin Study data indicate that, as well as the environment, heredity exerts substantial influence on the transition from infancy to early childhood across the domains of temperament, emotion, and cognition.

Acknowledgments: Portions of this chapter are based on material previously published in Emde, R. N., Plomin, R., Robinson, J. A., Corley, R., DeFries, J., Fulker, D. W., Reznick, J. S., Campos, J., Kagan, J., & Zahn-Waxler, C. (1992). Temperament, emotion, and cognition at fourteen months: The MacArthur Longitudinal Twin Study. *Child Development, 63,* 1437–1455; and Plomin, R., Emde, R. N., Braungart, J. M., Campos, J., Corley, R., Fulker, D. W., Kagan, J., Reznick, J. S., Robinson, J., Zahn-Waxler, C., & DeFries, J. C. (1993). Genetic change and continuity from fourteen to twenty months: The MacArthur Longitudinal Twin Study. *Child Development, 64,* 1354–1376.

REFERENCES

Baltes, P. B., Reese, H. W., & Lipsitt, L. P. (1980). Life-span developmental psychology. *Annual Review of Psychology, 31,* 64–110.
Boomsma, D., Martin, N. G., & Molenaar, P. (1989). Factor and simplex models for repeated measures: Application to two psychomotor measures of alcohol sensitivity in twins. *Behavior Genetics, 19,* 79–96.

Bouchard, T. J., & McGue, M. (1981). Familial studies of intelligence: A review. *Science, 212,* 1055–1059.

DeFries, J. C., & Fulker, D. W. (1985). Multiple regression analysis of twin data. *Behavior Genetics, 15,* 467–473.

DeFries, J. C., & Fulker, D. W. (1988). Multiple regression analysis of twin data: Etiology of deviant scores versus individual differences. *Acta Geneticae Medicae and Gemellologiae, 36,* 205–216.

Eaves, L. J., Eysenck, H. J., & Martin, N. G. (1989). *Genes, culture and personality: An empirical approach.* New York: Academic Press.

Eaves, L. J., Long, J., & Heath, A. C. (1986). A theory of developmental change in quantitative phenotypes applied to cognitive development. *Behavior Genetics, 16,* 143–162.

Fabsitz, R., Carmelli, D., & Hewitt, J. K. (1992). Evidence for independent genetic influences on obesity in middle age. *International Journal of Obesity, 16,* 657–666.

Fulker, D. W., DeFries, J. C., & Plomin, R. (1988). Genetic influences on general mental ability increases between infancy and middle childhood. *Nature, 336,* 767–769.

Goldsmith, H. H. (1983). Genetic influences on personality from infancy to middle childhood. *Child Development, 54,* 331–355.

Hay, D. A., & O'Brien, P. J. (1983). The La Trobe Twin Study: A genetic approach to the structure and development of cognition in twin children. *Child Development, 54,* 317–330.

Hewitt, J. K., Eaves, L. J., Neale, M. C., & Meyer, J. M. (1989). Resolving causes of developmental continuity or 'tracking'. I. Longitudinal twin studies during growth. *Behavior Genetics, 18,* 133–151.

Kendler, K. S., Neale, M. C., Kessler, R. C., Heath, A. C., & Eaves, L. J. (1993). A test of the equal-environment assumption in twin studies of psychiatric illness. *Behavior Genetics, 23,* 21–28.

Kenny, D. A. (1979). *Correlation and causality.* New York: Wiley.

Loehlin, J. C. (1992). *Genes and environment in personality development.* Newbury Park, CA: Sage.

Loehlin, J. C., Horn, J. M., & Willerman, L. (1989). Modeling IQ change: Evidence from the Texas Adoption Project. *Child Development, 60,* 993–1004.

MacGillivray, L., Nylander, P. P. S., & Corney, G. (1975). *Human multiple reproduction.* London: Saunders.

Martin, N. G., Eaves, L. J., Kearsey, M. J., & Davies, P. (1978). The power of the classical twin study. *Heredity, 40,* 97–116.

McCartney, K., Harris, M. J., & Bernieri, F. (1990). Growing up and growing apart: A developmental meta-analysis of twin studies. *Psychological Bulletin, 107,* 226–237.

Neale, M. C. (1997). *Mx: Statistical modeling* (4th ed.). Richmond, VA: Department of Psychiatry, Medical College of Virginia.

Neale, M. C., & Cardon, L. R. (1992). *Methodology for genetic studies of twins and families.* Dordrecht: Kluwer.

Phillips, K., & Fulker, D. W. (1989). Quantitative genetic analysis of longitudinal trends in adoption designs with application to IQ in the Colorado Adoption Project. *Behavior Genetics, 19,* 621–658.

Plomin, R. (1986). *Development, genetics, and psychology*. Hillsdale, NJ: Erlbaum.

Plomin, R., Chipuer, H. M., & Loehlin, J. C. (1990). Behavioral genetics and personality. In L. A. Pervin (Ed.), *Handbook of personality theory and research* (pp. 225–243). New York: Guilford.

Plomin, R., & Daniels, D. (1987). Why are children in the same family so different? *Behavioral and Brain Sciences, 10*, 1–16.

Plomin, R., DeFries, J. C., & Loehlin, J. (1977). Genotype-environment interaction and correlation in the analysis of human behavior. *Psychological Bulletin, 84*, 309–322.

Plomin, R., DeFries, J. C., McClearn, G. E., & Rutter, M. (1997). *Behavioral genetics* (3rd ed.). New York: Freeman.

Plomin, R., & Nesselroade, J. R. (1990). Behavioral genetics and personality change. *Journal of Personality, 58*, 191–220.

Savic, S. (1980). *How twins learn to talk: A study of the speech development of twins from 1 to 3*. New York: Academic Press.

Wilson, R. S. (1983). The Louisville Twin Study: Developmental synchronies in behavior. *Child Development, 54*, 298–316.

Wright, S. (1921). Correlation and causation. *Journal of Agricultural Research, 20*, 557–585.

3

The Conduct of the Study

Sample and Procedures

JoAnn L. Robinson
Jude McGrath
Robin P. Corley

Same-sex twin pairs born in the state of Colorado between 1986 and 1990 were recruited as participants. Parents of the twins were initially contacted through a mailing from the Colorado Department of Health. Those parents who expressed a willingness to be contacted were sent additional information and enrolled if their twins met health criteria and if they lived within a 2-hr driving radius of the testing site in Boulder, Colorado. More than 50% of the eligible families agreed to participate. Fifty-six percent of the sample of 408 twin pairs had participated in an earlier infant twin study of development in the first year of life but were initially recruited as described above.

FAMILY CHARACTERISTICS

The majority of participating families were two-parent families when the families were recruited. However, by the time the twins turned three, in 34 families (8.3%) the biological father of the twins was no longer living with the mother of the twins. Based on self-reported information on the twins' birth certificates, the ethnic distribution of the parents of the 408 participating twin pairs was 89.8% European-American, 5.3% Hispanic-American, and less than 1% African-American or Asian-American.[1] The ethnicity of the remaining parents (3.9%) was unreported or was listed as "Other" on the birth certificate. Some other pertinent characteristics of the parents are shown in table 3.1. Using information supplied by the Colorado Department of Health, we estimated the average years of education of parents of all twins born during the

Table 3.1. Demographic characteristics of parents

	Mothers			Fathers		
	Mean	SD	*n*	Mean	SD	*n*
Age at birth (years)	29.62	4.51	402	31.61	5.17	393
Years of education	14.29	2.30	105	14.42	2.56	397
NORC occupation[1]	38.88	16.53	369	48.59	13.59	352
WAIS vocabulary	10.58	2.74	386	10.97	2.30	277

[1] National Opinion Research Council

1986–1990 period to be 13.9 and 14.1 for mothers and fathers, respectively. The age at birth for parents of all twins born during 1986–1990 was 28.3 and 30.9 for mothers and fathers, respectively. Parents of participating twins were thus marginally better educated and somewhat older than the average parent of twins born during the recruitment period. Substantial variability exists within the sample of parents, however. For example, 5% of the participating parents did not complete high school, 29% completed high school without formal post-secondary education, 49% had some post-secondary education, and 17% had some graduate-level education.

TWIN CHARACTERISTICS

Table 3.2 presents birth weights and 5-min Apgar scores for participating twins. Participating twins were evenly divided by gender. For all twins born in Colorado during 1986–1990, the average birth weight was 2370 g for girls and 2446 g for boys. The corresponding average 5 min Apgar scores were 8.58 and 8.54, respectively.

Participating twins were thus heavier, and looked somewhat healthier at 5 min after birth than the average Colorado twin. This results from the exclusion of very low birth weight twins (< 1000 g) or those with very poor perinatal health (requiring > 2 weeks hospitalization) from the study. Only 4% of participating twins weighed < 1700 g at birth.

Table 3.2. Birth weight and Apgar scores for female and male twins

	Female twins			Male twins		
	Mean	SD	*n*	Mean	SD	*n*
Birth weight	2538.38	448.23	408	2607.03	488.13	408
5-min Apgar	8.77	0.60	406	8.65	0.86	405

Table 3.3. Child sex by zygosity

	MZ	DZ	Ambiguous	Total
	n(%)	*n*(%)	*n*(%)	*n*(%)
Female	104 (51.0)	80 (39.2)	20 (9.8)	204 (50.0)
Male	99 (48.5)	95 (46.6)	10 (4.9)	204 (50.0)
Total	203	175	30	408

ZYGOSITY DETERMINATION

Zygosity of the twins was determined through aggregation of independent tester ratings across test locations and test ages. At each in-person test, including earlier year-one tests, testers rated twin similarity on 10 physical characteristics of the head, including eye color, hair color, and shape of the ears. The attributes were selected on the basis of the diagnostic rules developed by Nichols and Bilbro (1966). When the features were rated consistently as highly similar (1 or 2 on a 5-point similarity scale), the twins were rated as monozygotic. If two or more features were rated as only somewhat similar (score of 3), or if one feature was rated as not at all similar between the twins (scores of 4 or 5), the twins were rated as dizygotic. To be assigned a composite zygosity diagnosis, at least four raters had to have rated the twins' similarity. To be regarded as unambiguously monozygotic (MZ), or unambiguously dizygotic (DZ), 85% of the raters had to agree on zygosity. In nine ambiguous families, results of blood testing were used to assign zygosity. Zygosity determination was regarded as an ongoing process; whenever the twins were seen in-person, additional similarity ratings were done. The composite zygosity rating used in this book includes in-person tests through age five. As shown in table 3.3, there were 203 MZ pairs, 175 DZ pairs, and 30 ambiguous pairs in the study (six of whom were ambiguous because too few total ratings were done).

DATA COLLECTION

Data collection for the study can be divided into four main types: home visits, laboratory visits, maternal interviews, and parent questionnaires. Each type of data was gathered at the four child ages targeted for study: 14, 20, 24, and 36 months. Fourteen-month data were gathered when the children were between 13.5 and 15.5 months of age; 20-month data were gathered between 19.5 and 21.5 months of age; 24-month data were gathered between 23.5 and 26.5 months of age, and 36-month data were gathered between 35.5 and 38 months.

Home Visit Procedures

Home visits were planned for times when it was convenient for mothers and when the children were well rested. Each visit lasted 2–2.5 hr. Two female examiners visited each home. The order of procedures was presented consistently and varied only when children were too shy to tolerate either of the emotion challenges (e.g., anger procedures, prohibitions) which occurred early in the visit. The order of the procedures were as follows: (1) home entry, (2) anger procedures, (3) prohibition, (4) Bayley Scales of Infant Development, (5) Sequenced Inventory of Communication Development, (6) examiner distress/child empathy probe, (7) mother–child interaction, (8) mother distress/child empathy probe, and (9) free play. These procedures were carried out at each of the four ages. In addition, a sibling cooperation/competition procedure was introduced at 20 through 36 months, and a story stem narrative procedure was introduced at 36 months.

Home Entry

The entry procedure consisted of five intervals which reflected a gradual engagement process with the children.

Interval 1: minutes 0–2. Mothers were instructed beforehand by telephone not to hold either child as they answered the door so that the children's movements toward and away from the pair of adult strangers could be observed. One examiner carried a video camera (the camera person) and filmed the children's response to our arrival, beginning from the moment the door was opened and continuing for 6 min. Each mother was asked to identify the twins by name, and the camera person then verbally described what each child wore for future identification purposes. Mothers were then given consent forms and expense vouchers to complete. They were also asked at this time not to encourage the children in any way for the next 4 min, as we wanted to see what the children were comfortable doing on their own. However, if a child wanted to be held during this time it was permitted.

Interval 2: minutes 2–3. The second examiner took out two toys from her equipment bag, placed them near her but still easily accessible to the children, and proceeded to play with each toy. She briefly talked out loud about the toy but did not directly reference the children either visually or verbally. If either child approached the toys at this time or any time from min 2–5, he or she was permitted to play freely with the toys until min 5.

Interval 3: minutes 3–4. At min 3, if the children had not approached the toys, the examiner called one child by name and invited him or her to play with the toys. This was then repeated for the second child. To record the twins' responses, the camera person followed each child in turn as the child responded to this direct invitation.

Interval 4: minutes 4–5. At min 4, if a child had not approached the toy, the examiner picked up the toy and offered it to one of the children by extending her hand and asking if he or she would like to play with it. If the toy was not taken, it was placed either in the child's hands or close to the child. This procedure was then repeated for the second twin, if that child had not approached the toys.

Interval 5: minutes 5–6. At min 5, whether or not the children touched or played with the toys, the examiner attempted to place an identifying vest on each child. (Upon entering the study, each twin in the pair was assigned the color red or yellow. At 14–36 months, small vests that fastened in the back were worn by each child.) Mothers were asked to assist and fit the children with the vests if they protested strongly and could not be coaxed by the examiner.

Anger Procedures

After home entry, two anger-eliciting probes, the measurement/restraint and the toy removal, were administered to each twin independently. The measurement/restraint procedure always preceded the toy removal for each child.

Height Measurement/Restraint. One child accompanied the mother and an examiner to a family room or bedroom that contained a door which could be closed so that the children would not be able to hear each other or the other examiner's instructions. Mothers were instructed to speak only to the examiner and to refrain from touching, smiling, or reassuring their children during this procedure. To complete the measurement, the child was laid on his/her back on the floor with mom seated at the child's head. A videocamera, positioned earlier, was 3–4 feet away from the child recording the procedure. Immediately after placing the child on the floor, the examiner, seated next to the child, stretched a measuring tape from the top of the child's head to his or her foot and stated the results of the measurement for the camera. Then the examiner immediately placed a hand on the child's abdomen and said solemnly, "Now, (child's name), lie still." The restraint lasted a maximum of three minutes and was later coded for latency to protest as well as the intensity and manner of any protest. The procedure ended early if the child struggled persistently against the examiner's hand, accompanied by clear distress for several seconds or if the mother requested the procedure be ended.

The 36-month measurement was done with the child standing against a wall with the mother and examiner standing on either side of the child. The restraint immediately followed the measurement with the examiner holding the child's wrist that was closest to her with one of her hands and saying, "Now, (child's name), stand still."

Toy Removal. After the measurement/restraint probe, each child went into the living room with the other examiner to complete the toy removal procedure.

At 14 and 20 months of age, an air pressure manipulative toy was used for this procedure. At 24 months, a push button barn with animal sounds and moving figures was used, and at 36 months, a hexagonal-shaped box with color-coded doors, keys, and animals was used. Each child was seated on the floor facing a video camera 4 feet away. A bag was placed on the floor between the child and the camera. The examiner took the age-appropriate toy out of the bag and allowed 1 min for the child to approach and play with the toy spontaneously. After 1 min, the examiner sat with the child to encourage play and model how to use the toy. When the child was intently involved with the toy for 2 min (as indicated by using both hands to play, vocalizing, or drooling, at 14 months) the examiner abruptly removed the toy and said, "(Child's name), I have to put this away now." The toy was put into the bag, zipped closed, and the bag was placed, half-way between the child and the camera, but well within the child's reach. The examiner then stepped behind the camera to film the child's reaction for the next 2 min. If a child showed great distress or left the room before the 2 min period ended, the procedure ended early.

Prohibition

This procedure was done during the 14-, 20-, 24-, and 36-month visits. At 14, 20, and 24 months it preceded the administration of the Bayley mental scales and Sequenced Inventory of Communication Development (SICD) for which each child was seated in a high chair in different, but usually adjacent rooms. After seating the child in the high chair, the examiner took out a 4-inch acrylic glitter wand. For 30 s, the examiner drew attention to the wand by shaking it back and forth. She made eye contact with the child, and then lowered the wand to the high chair tray table saying, "Now, (child's name), don't touch." The examiner then looked away and after 30 s she looked back at the child and removed the prohibition saying, "It's okay, you can touch it now." The prohibition was removed at this time whether or not the child had touched the wand. Children who still would not touch the wand were encouraged to do so for the next 30 s. All other children were filmed the additional 30 s with no further encouragement. At 36 months, the prohibition followed the Stanford-Binet and SICD administrations (described below).

Bayley Scales of Infant Development and Sequenced Inventory of Communication Development

These two assessments were integrated into one procedure at 14, 20, and 24 months, making a more efficient presentation for children of these ages. The mental development scale from the Bayley Scales of Infant Development (Bayley, 1976) consists of many tasks representative of problem solving, fine motor coordination, and expressive and receptive vocabulary. An additional measure of child language, the Sequenced Inventory of Communication Development (SICD; Hedrick et al., 1975), was used to assess more thoroughly each child's receptive and productive language. Mothers were told that the children's best performance may depend on either her close proximity or, conversely, her absence during this assessment. Their cooperation was sought in creating an optimal situation for each child whenever possible. This may have meant that a child was seated in his or her mother's lap or that the mother was continually observable by the child or that the mother was out of sight in an adjacent room.

Examiners completed a standard administration of the Bayley scales, obtaining a baseline and ceiling of the child's abilities. The SICD items that did not duplicate items already assessed as part of the Bayley scales were administered at the end of Bayley testing. The SICD also included a separate parental report interview in which mothers were asked questions pertaining to each child's expressive and receptive communication. This interview was usually completed at the end of the home visit but was sometimes completed at a later date over the phone.

At 36 months of age, each child's general cognitive abilities were assessed with the Stanford-Binet Intelligence Scale, version L-M (Terman & Merrill, 1973). Administration of the SICD followed the Stanford Binet as a separate measure at this age.

Empathic Responses

Examiner Distress. The child's empathy was assessed through naturalistic enactments of distress by the examiner and mother, which were adapted from earlier work by Zahn-Waxler et al. (1979). At 14, 20, and 24 months, upon completion of the Bayley/SICD assessment, each examiner pretended to pinch her finger in a test materials case. In each simulation which lasted 90 s, the examiner grasped her finger, saying, "Ooh" and "Aah" for 30 s. She was careful not to make eye contact with the child during this time. At 30 s, the examiner expressed that the injury "Feels better now" and for 30 s more talked about how the injury was better. At 60 s, the examiner reengaged the child. The camera continued to record the next 30 s for any delayed responses. At 3 years of age, empathy for the examiner was assessed at the completion of the Stanford-Binet and before administration of the SICD.

Mother Distress. Empathic response to mothers' distress was assessed at 14, 20, and 24 months approximately 15 min after the examiner feigned distress. (During this 15-min interval, a lengthy transition from the cognitive testing occurred, as well as the mother–child teaching task.) The mother sat on the floor with each child separately and played with the toy previously used in the toy removal procedure. After she had engaged the child with the toy, she stood up and pretended to hurt her knee, going back down to the ground on one knee and rubbing the other knee as she said, "Ooh" and "Aah." Again, mothers were instructed not to engage the child through eye contact or to call the child's name. The examiner signaled the mother after 30 s by asking, "(Mother's name), are you okay?" At this point, the mother expressed that the injury was, "Feeling better now," and the examiner and mother discussed her injury for another 30 s. At 60 s, the examiner signaled to the mother that the procedure was completed by stating, "I'm glad that it's feeling better now" and engaged the mother in a brief discussion of her child's response. The camera continued to record the child's behavior for an additional 30 s. At 3 years of age, empathy for the mother was assessed following the mother–child construction procedure. Each mother pretended in this simulation to hurt her knee as she attempted to get up to leave the table.

Mother–Child Interactions

Mother–child interactions were assessed triadically at 14, 20, and 24 months of age using the teaching task procedure; at 36 months, a dyadic procedure (co-construction) was used.

Mother–Child Teaching Task. This procedure was designed to assess how a mother differentially deploys her warmth, attention, and teaching strategies simultaneously to the twins in this triadic situation. This assessment was done at 14, 20, 24, and 36 months at the end of the SICD testing and before simulation of maternal distress described above. Each mother was asked to sit on the floor with one child sitting by each of her knees. Each child was given a peg board with colored shapes at 14 months of age, a square board with colored blocks at 20 and 24 months of age, and a sorting board with colored blocks of varying shapes at 36 months of age. The mother was asked to teach the children to put the shapes on the appropriate pegs or to sort the colored blocks. Once the mother and her children were engaged in the task, the examiner turned on the video camera (6–8 feet in front of the triad) and left the room for 7 min.

Co-Construction. This procedure was administered at 3 years of age and was based on work by Oppenheim et al. (1997). Mothers played/worked with each of their children separately at the kitchen table, building a farm scene using a wooden barn and farm animals. A video camera was placed on a tripod 6–8 feet in front of the dyad. The examiner instructed the mother to assist her child as needed in the construction. The examiner placed an unopened duffel bag containing the toys on the table between mother and child and left the room for 10 min. This procedure formed the basis for rating mother–child emotional availability.

Free Play

The final procedure in the home visit at 14 months was the free play. Examiners set the stage for the free play by arranging an age-specific set of 12 toys in a standardized array on the floor in the family's living room. The mother occupied the children elsewhere during this time. The toy set differed at 14, 20, and 24 months. Toys included two versions of six toys, one intact and one altered in its appearance or function. For example, at 14 months, the toys included two small, stuffed Mickey Mouse dolls with one missing a leg; at 20 months the toy set included two rubber balls although one had a portion removed; at 24 months there were two books but one was obviously torn. Upon completion of the setup, examiners sat on the floor at the edges of the space, holding the video camera. The children and their mothers were called into the room. The mother was instructed to sit on a sofa, if available, or in a comfortable chair that was accessible to the children. Mothers were instructed not to label any of the toys for the children and not to initiate any interaction with the twins. They could respond if approached by the children but were requested to keep interactions to a minimum. Children played freely, without interruption by the examiners, for 15 min. This situation provided an opportunity to observe qualities of the twins' interactions with each other. Following a similar setup, at 36 months the children were given blocks, two vehicles, and several animal and people figures to use in free play.

Sibling Cooperation/Competition

A further opportunity to observe a child's cooperative style with a sibling was introduced at 20 and 24 months of age and continued at all later ages. The cooperative play procedure lasted 10 min. Mothers were told at 20 and 24 months, in particular, to intervene as they normally would in any situation that needed resolution. Mothers were also told they could encourage turn taking and even model how the children could play together with the toy. At 20 months, a Sit-and-Spin was used as the cooperative toy, at 24 months, a wagon was used, and at 36 months, the children were given Play-Doh, one rolling pin, and one cookie cutter.

Story Stem Narratives

This assessment was introduced as part of the 36-month home and laboratory and was based on procedures established by Buchsbaum and Emde (1990). Each series of narratives began with a "warm-up" story depicting the family at a birthday party. This warm-up story was used to encourage the children to become engaged to complete a story by manipulating a set of dolls and props and verbalizing actions using the dolls. If a child did not do these things spontaneously, the examiner used the story to model actions and dialogues among the dolls. Children were instructed that the child and examiner were going to tell some stories together, the examiner would start the stories and the children could finish them any way they liked. A total of six stories were completed across home and laboratory sessions.

Maternal Interview

The maternal interview was most commonly conducted over the telephone as soon as possible after completing the home visit. In some situations, the interview was completed after the family had made their laboratory visit, especially if the latter preceded the home visit. The interview lasted from 30 to 45 min, with the longer time required for the 14-month interview. The longer interview began with a brief history of the mother's pregnancy with the twins. This included questions about life stresses and illnesses she may have experienced in pregnancy as well as the delivery route (e.g., caesarian or vaginal). The 14-month interview also included questions regarding the twins' health and sleeping patterns during the first 14 months of life. Questions that were asked consistently from 14–36 months included (1) the twins' prosocial orientation to each other, (2) their usual responses to distress in others (especially the cotwin and mother), and (3) common emotional reactions to frustration when playing alone or with their cotwin. At 36 months, mothers were also asked to name five adjectives to describe each child and why they chose those words.

Questionnaires

Several questionnaires were given to the mother for both parents to complete before the examiners completed the home visit. In addition, two questionnaires were completed by mother during her visit to the laboratory. At 14 and 36 months, both parents were asked to complete (1) the Eysenck Personality Inventory (Eysenck & Eysenck, 1969), (2) the Family Environment Scale (Moos & Moos, 1981), (3) the Dyadic Adjustment Scale (Spanier, 1976), and (4) two copies of the Toddler Temperament Survey (Carey & McDevitt, 1978), one for each twin. Parents were requested not to discuss their responses to these questionnaires until both of them had completed their ratings. At 20 and 24 months, parents were asked to complete the Toddler Temperament Survey for each twin. In addition, at every age the father was asked to complete the Colorado Childhood Temperament Inventory (Rowe & Plomin, 1978) for each twin. During each twin's laboratory visit, the mother was asked to complete the Colorado Childhood Temperament Inventory as well as the Discrete Emotions Scale (Izard, 1972) for that twin.

Laboratory Visit Procedures

Mothers (and sometimes fathers) brought their twins to our laboratory at the Institute for Behavioral Genetics on the University of Colorado at Boulder campus. These visits generally occurred within the 2 weeks after the home visit. However, it was common for families to reschedule laboratory visits due to child illness or inclement weather, and the length of time between visits was sometimes as long as 6 or 7 weeks. Home visits always occurred before laboratory visits at 14 months. At 20 months, it occasionally was more convenient for the family to visit our lab before the home visit occurred. At 24 and 36 months, we attempted to bring two families with either male or female twins to the lab on the same day so that a peer play procedure could be conducted during a 25-min overlap period. This scheduling requirement sometimes necessitated that a family complete a laboratory assessment before the home assessment.

Procedures in the lab were administered in three different rooms: reception room, cognitive room, and inhibition room, and at 14 and 20 months each child experienced the procedures in the order mentioned. At 24 and 36 months, a fourth room, the peer play room, was added and at 36 months a duplicate inhibition/risk room was added to expedite the assessment of two families of twins. Half of the children at 24 and 36 months experienced the peer play room procedures as their initial assessment, while the other half experienced the peer play room as their final assessment.

Reception Room: 14–24 Months

The reception room was approximately 11 × 15 feet in size. A modular sofa faced a smaller modular chair in the room, a stationary video camera was mounted on the wall between the chair and sofa, and a 2 × 3 × 1.5 ft cabinet contained a VCR and other supplies adjacent to it. A basket of colorful toys and a child-size table and two chairs were in the center of the room. Two female examiners were present in the room; one examiner had met the family at their car in the adjacent parking lot and assisted them into the building while the other prepared the room and started the video recording.

Lab visits at 14 and 20 months began with an unstructured period that allowed rating of children's shyness. During this 5-min warm-up period the children were videotaped as they ventured forth from their mothers to explore their new environment. One examiner discussed the various lab procedures and gave each mother a consent form to read and sign. Neither examiner engaged the children in any manner during these first 5 min unless approached by one of the children, at which time a friendly greeting occurred. About 15 min after arrival, three electrodes were placed on the chest of one of the twins and that child and mother were escorted to the cognitive testing room. The other twin remained in the reception room with the second examiner and played freely with toys, a situation designed to be soothing during mother's absence.

At 24 months, all the children experienced a similar greeting and 15-min warm-up period in the reception room. Half of the children then followed the standard procedural order of preparation for the cognitive room procedures while the other half were brought to the adjacent peer play room.

The reception room was also used later in the laboratory visit for administrative tasks and other procedures. After each twin experienced the cognitive room procedures she or he returned to the reception room and during the subsequent (second) separation period was presented with a toy removal episode identical to the procedure administered in the home visit. At 20 and 24 months of age each twin was also presented with a mirror self-recognition task (based on Lewis and Brooks-Gunn, 1979) and a role play episode (based on Watson and Fischer, 1977; 1980) before the toy removal episode.

Self-Recognition. A lightweight wardrobe mirror, approximately 15 × 30 inches, was brought into the room and placed facing the video camera, supported by the examiner. The examiner tapped on the mirror and beckoned the child, saying "(Child's name), come look." After 30 s in front of the mirror, the child was then led away from the mirror and a dot of rouge was surreptitiously placed on his or her nose by the examiner, who claimed to need to wipe crumbs from the child's face. The forehead, chin, and cheeks were lightly brushed with a dry washcloth before the cloth wiped the nose and left a spot of rouge. The examiner then invited the child to come to the mirror for an additional 30 s, after which the child's nose was wiped again and he or she was encouraged to play freely.

Role Play. In this episode the child was seated facing the video camera and adjacent to the examiner who presented a stuffed bunny, saying, "Bunny is sleeping." The examiner then pretended to wake the toy, and this began a sequence in which the child was asked to imitate the examiner's actions. The examiner made the bunny hop, fed it a carrot, and had the bunny feed itself a carrot. Each time, following the demonstration the examiner said to the child, "Now, you do it." Next, several additional props were placed on the table and the examiner then pretended that the bunny had hurt itself. The examiner said to the child, "Oh, look, the bunny has hurt her hand. What would mommy do to make it better?" Following the child's response to this probe, the examiner hopped and then invited the child to "Be a bunny too."

Cognitive Room: 14–24 Months

In the cognitive room the infant was seated in a high chair 27 inches in front of a free-standing wall which had two cut-outs or openings that were 9 inches apart, each of which contained a rear-projection screen to present slide stimuli. A third cut-out was made for a video camera, which taped each session. The mother was seated in a chair next to the infant. Procedures were presented in the following order, consistent across and within twin pairs: (1) baseline heart rate, six trials, (2) word comprehension, nine trials, (3) sorting task, two trials, (4) word comprehension, six trials, (7) memory for locations, (8) post-test heart rate, 6 trials, (9) weight measurement.

Baseline Heart Rate. The three electrodes placed on the child in the reception room were connected to a heart rate preamplifier (Model PA-2M, Biofeedback Systems, Boulder, CO), and six pairs of slides were presented to the child for

15 s each to obtain a baseline heart rate. The preamplifier was connected directly to a Zenith model 158–43 personal computer that timed and recorded the length of the interbeat intervals.

Word Comprehension. This task was administered through the presentation of paired slides, one of which is named by the examiner and preferentially looked at by the child (Reznick, 1990). Each pair of slides was initially presented for 8 s, followed by a 1-s gap when the slides were not projected. The same slides reappeared for 8 s more, and the child was prompted by the examiner to look at one slide: "Do you see the (name of picture)? Where is the (name of picture)? Look at the (name of picture)." The examiner used a hand-held toggle switch which sent a signal to the computer to record information about the length and orientation (left, right, center) of the child's visual fixation during both 8-sec intervals. Comprehension was inferred if fixation to the named slide increased after the prompt. Nine trials were initially presented; an additional six trials followed the sorting task for a total of 15 trials.

Sorting Task. This task was used to provide a break from the focused attention of the word comprehension task and to assess the child's capacity to sort two classes of objects (Reznick and Snedley,1987). The child was offered a tray of objects of two types, people versus animals, and on a second trial, cars versus trucks. The child's manipulation of the objects was recorded for later coding of sequential touching of objects in the same category. The child was allowed to freely explore each tray of toys for 2 min.

Memory for Locations. These procedures were adapted from Kagan (1981). A toy was hidden under a cup in a variant of the "shell game" using two, four, or six cups at 14, 20, and 24 months and up to eight cups at 36 months. Cups were placed on a 24 × 5.75-inch board in inverted positions 1.5 inches apart and centered in front of the child. Each trial began with the child watching the examiner place the toy under a cup, according to a location specified in a standard protocol. A second board, 8 × 24 inches, was used as a barrier between the child and the cups to delay responding for 1, 5, 10, or 15 s. As the examiner removed the board, the child was asked, "Where's the doggie? Find the doggie."

Post-Test Heart Rate. Heart rate was gathered for a second set of six trials using a novel set of paired slides.

Weight Measurement. Before leaving the cognitive room, each child was weighed by being seated briefly on an infant-weight digital scale. The read-out of the child's weight was said aloud and recorded on both the videotape and a record form. Mother and child then returned to the reception room for that child to complete procedures there (i.e., self-recognition and role play).

Inhibition Room: 14–24 Months

After the completion of the cognitive and reception room procedures, each twin was taken separately with the mother to the inhibition room. The inhibition procedures were adapted from earlier work by Kagan et al. (1988). The twin that remained in the reception room had a free play time with the examiner. The inhibition room was arranged in a similar manner at 14, 20, and 24 months. The room contained a two-piece modular sofa, identical to that in the reception room, which faced away from a one-way observation mirror. A metal storage cabinet, containing one of the stimuli, was situated to the left of the sofa. A remotely controlled camera was mounted on the wall facing the sofa. It was used to film the child's behavior in addition to the camera behind the one-way mirror. A small audio speaker was mounted above the door and one was also located inside the metal cabinet. As the mother and child entered the room, they found 10 toys or objects arranged on the floor in the center of the room. Many of the items were duplicates wherein one object was flawed and one was intact (e.g., a doll in a pretty white dress and a doll with a ripped white dress; a clean, white washcloth and a white washcloth with a large blue stain).

Several types of observation were made while mother and child remained in the inhibition room: (1) behavioral inhibition (free play, stranger approach, monster/robot), (2) baby cry/child empathy probe, (3) mother distress/child empathy probe, (4) examiner distress/child empathy probe, and (5) response to prohibition. These types of observation occurred in a fixed order, as described below.

The mother was instructed to enter the inhibition room and sit at the far end of the couch beneath the filming window. She was given a clipboard with two questionnaires concerning that child's temperament and was asked to complete these while the child played freely for 12 min. The initial 12 min of free play provided the opportunity to observe the child's willingness to explore in an unfamiliar room, as indexed, for example, by latency to leave mother and latency to approach the toys on the floor. After approximately 3 min, the first of three empathy stimuli occurred. The sound of a baby crying was broadcast into the room from a speaker for approximately 30 s. The mother was instructed to respond to the child's bids for attention or reassurance but otherwise to continue filling out the questionnaires. About 5 min later, at a time when the child was away from mother's side, the second empathy probe, a mother distress/child empathy probe, was administered. The mother was signaled to act as if her finger had been injured by the clipboard in her lap. The same injury/recovery sequence was followed as in the home visit.

After 12 min, the second behavioral inhibition probe was administered (at 14 and 20 mos only), the stranger approach. A female stranger holding a high interest toy entered the room. She maintained a downcast gaze as she sat on the floor several feet from the child. At 1-min intervals she increased her activity with the toy until 4 min had elapsed. If the child had not approached the stranger after 3 min, the stranger raised her gaze and verbally invited the child to play. The stranger left after engaging in play with the child for 2 min.

Following this the third and final empathy probe began, the examiner distress/child empathy probe. The examiner returned to the room and feigned an injury to her foot as she repositioned a chair. After completing the standard injury/recovery sequence, the examiner opened the metal cabinet, which housed either a furry monster (at 14 months) or a tin can robot at 20 months. At 24 months, a remote-control robot entered the room through a door rather than from the metal cabinet. The monster/tin-can robot presentation constituted the third inhibition episode. The child was invited to approach and explore the object. If the child approached, or after 2 min of encouragement, a tape-recorded voice was projected through a speaker in the cabinet, inviting the child to "Come and play with me." At 24 months the remote-controlled robot included a feature that permitted a tester in the filming room to directly say this same text through the robot. The final procedure involved a prohibition that was similar to that administered during the home visit. In this situation, the examiner bounced in an enticing manner a small, colorful rubber ball taken from a small basket. The basket was placed in front of the child who was firmly told, "Now, don't touch!" After a brief period of recovery from the prohibition during which there was an opportunity to play with the balls freely, mother and child returned to the reception room, and the co-twin then accompanied mother to the Inhibition Room.

At the conclusion of each visit to the laboratory, the examiners made three ratings: (1) a rating of each twin's shyness during the entry to the reception room, which ranged from not shy to very shy, (2) a rating of each twin's response to the placement of electrodes, which ranged from a score of 1 (easily put on) to 4 (refusal), and (3) an overall mood rating ranging from a high score of 7 (marked and pervasive positive) to a low score of 1 (marked and pervasive distress).

Peer Play Room: 24 and 36 Months

The 24-month peer play involved two mothers and their same-sex twins. The room contained two couches arranged at a right angle where the mothers sat to complete questionnaires, a crawl-through-tunnel, a child-size play table, a plastic shopping cart, and a child-sized wooden sink and refrigerator. The latter was filled with plastic dinnerware, utensils, and food items. Mothers were instructed not to enter into the children's play but to intercede as they normally would should any problematic behavior occur. Peer play at this age lasted 25 mins, and 10 behaviors were coded on-line by research assistants for number or duration of occurrences. Most behaviors were coded based on the child's actions toward an unfamiliar twin or unoccupied time spent proximal to the mother.

At 36 months, laboratory procedures took place within four types of rooms: reception, cognitive, risk, and peer play rooms. Each twin was assessed in these rooms in a somewhat different order because of space and time constraints and because two families of twins of the same gender arrived for testing on the same day in order to permit observation of peer interactions.[2] The 36-month lab visit had two peer play sessions, each involving two mothers and one child each from their twins. The pairing of children for each play session was random. A larger room was used and, in addition to the 24-month materials, there was a bean bag, a set of cardboard building blocks, and a slide. Each of these sessions was 20 min mins long. The procedure was the same as at 24 months, and the coding criteria remained the same.

Reception Room: 36 Months

The reception room at 36 months was used to greet the mothers and children and as a waiting room for mothers because this was the first age at which the children were tested simultaneously, each with their own examiner and without the mother present. At various times during the lab session, this room was also used as a break and snack room. In addition, the following procedures were administered in this room: card sort, Peabody Picture Vocabulary Test (PPVT), and story stem narratives (see description in home visit procedures).

Card Sort. The Card Sort procedure was based on work by Reznick (Zelazo & Reznick, 1991; Zelazo & Pinon et al., 1995). During the card sort procedure, the child sat next to an examiner at a small table on which were two boxes constructed to facilitate sorting. A training phase included an interactive demonstration that used 10 picture cards from two familiar categories, men and birds. The examiner labeled and sorted two pictures from each of the demonstration categories into the two boxes. The examiner then labeled the remaining cards one at a time, each time asking the child to indicate in which box the name of the depicted object belonged. During this training phase, correct sorts were praised, and incorrect sorts were corrected. If the child did not comprehend the task (i.e., failed to sort the remaining six cards correctly), a second demonstration was given with pictures from two other categories, women and flowers. If the child still made errors, the first demonstration was repeated. For children who were unable to accurately complete one of the three possible training trials, the procedure ended after the third training trial.

The four remaining sets of cards used three boxes to complete the sort. The examiner identified the category associated with each of the three boxes by placing two representative exemplars from each category face-up in the boxes, labeling the categories, and drawing attention to the criteria for category inclusion. For each of the remaining cards, the examiner labeled the object depicted and asked the child to "Point to the box where the card belongs." Incorrect sorts were not corrected, and the child was usually given praise for all sorts. All four sets of cards were administered if the child was able to complete one of the three training trials.

Peabody Picture Vocabulary Test. The Peabody Picture Vocabulary Test Revised—Form L (Dunn & Dunn, 1981) was administered to the children to obtain a widely used, standardized measure of their receptive vocabulary. It was chosen for its ability to be quickly administered and rapidly scored.

Cognitive Room: 36 Months

Four procedures were conducted in this room: (1) baseline heart rate, (2) memory for locations task, (3) sentence imitation, (4) post-test heart rate, and (5) prohibition. The child and examiner walked into the cognitive room where a stuffed Mickey Mouse toy was sitting on a table. The examiner then placed three electrodes on the child's chest and seated the child in a chair across the table from her. At this time sitting and standing heart rate measurements were done to obtain an estimate of a baseline heart rate for each child. Then, as the child was seated again, the Mickey Mouse toy was placed under the table and the memory for locations task was begun. A sentence imitation task followed memory for locations, in which the examiner asked the child to imitate three sets of sentences ranging in length from two to seven words. Next, a post-test sitting and standing heart rate was done, followed by the Prohibition procedure. For the Prohibition, the table was cleared and the examiner brought out an acrylic paperweight which produced falling snow when shaken. The examiner said, "Look (child's name)," as she shook the toy out of the child's reach. Then, as the examiner lowered the toy to the table within the child's reach, she said, "Now (child's name), don't touch." The standard prohibition timing was used, releasing the prohibition in 30 s whether or not the toy was touched and then filming for an additional 30 s.

Inhibition/Risk Room: 36 Months

Three procedures were conducted in this room: (1) behavioral inhibition to perform risks, (2) examiner distress/child empathy probe, and (3) mother distress/child empathy probe. The child and examiner entered the inhibition/risk room without the mother, and the child was told that he or she could play with anything while the examiner sat on the couch and did her paperwork. The examiner allowed the child 4 min to explore. At the end of the 4 min, she feigned injuring herself by pretending to snap her finger in her clipboard. Upon recovery, the examiner proceeded to model each of the five "risks" in the room, inviting the child to imitate her. The risks were (1) a teddy bear mobile hanging from the ceiling that swayed and clinked when pushed back and forth, (2) a mask of a dog that the examiner put over her face, (3) a crib-size mattress that the examiner stood in front of (the mattress being lengthwise on the floor) and fell backward onto, (4) a 4.5 × 2-foot box covered in black contact paper with a hole in one side big enough for an arm to fit through, and (5) a balance beam that was on an incline such that one end rested on the floor and the other end was raised 6 inches off the floor. After modeling each risk for the child, the examiner walked around the room one more time, stopping at each risk in the same order as before and inviting the child to perform each risk.

The examiner then opened the door and brought in three toys, a truck, a book, and a puppet, placing them on the mattress as she named them. The child was given 1 min to explore these toys with the examiner present. The examiner then left the room to get the child's mother. When the mother returned to the room (alone), she sat on the couch and chatted with the child about what he/she had done in that room. After 4 min had elapsed, the camera person signaled for the mother to cross the room and feign injuring her knee. Filming continued for 1 min and then the examiner entered the room to conclude the inhibition/risk procedures.

CONCLUSION

This chapter has described the recruitment, sample characteristics, zygosity determination, and assessments at 14, 20, 24, and 36 months on which the analyses and results presented in this volume are based. However, this unique longitudinal twin study is a continuing enterprise and almost all of the Mac-Arthur families are continuing to actively participate in the study. In the future, we will be able to see whether the groundwork laid during the study of infancy and early childhood will also contribute to understanding later development in childhood. For now, we can turn to a consideration of what our study has revealed about the transition from infancy to early childhood.

NOTES

1. A review of family surnames suggested that an additional 4% of families may be of Hispanic origin but chose not to self-identify as Hispanic.

2. Unlike the laboratory visit at 24 months, where the second family arrived approximately 2 hr later than the first family, at 36 months the second family arrived about 1 hr later. Children were generally able to complete the entire battery of procedures within 2.5 hr; to complete both families in a single visit generally required 4.5 hr and the participation of four examiners and a camera person.

REFERENCES

Bayley, N. (1976). *Manual for the Bayley Scales of Infant Development*. New York: Psychological Corporation.

Buchsbaum, H., & Emde, R. N. (1990). Play narratives in 36-month-old children: Early moral development and family relationships. *The Psychoanalytic Study of the Child 40*, 129–155.

Carey, W. B., & McDevitt, S. C. (1978). Revision of the infant temperament questionnaire. *Pediatrics 61,* 188–194.

Dunn, L. M., & Dunn, L. M. (1981). *Peabody Picture Vocabulary Test-revised.* Circle Pines, MN: American Guidance Service.

Eysenck, H. J., & Eysenck, S. B. G. (1969). *Personality: Structure and measurement.* London: Routledge and K. Paul.

Hedrick, D. L., Prather, E. M., & Tobin, A. R. (1975). *Sequenced Inventory of Communicative Development.* Seattle: University of Washington Press.

Izard, C. (1972). *Patterns of emotion: A new analysis of anxiety and depression.* New York: Academic Press.

Kagan, J. (1981). *The second year.* Cambridge, MA: Harvard University Press.

Kagan, J., Reznick, J. L., Snidman, N., Gibbons, J., & Johnson, M. O. (1988). Childhood derivatives of inhibition and lack of inhibition to the unfamiliar. *Child Development 59,* 1580–1589.

Lewis, M., & Brooks-Gunn, J. (1979). *Social cognition and the acquisition of self.* New York: Plenum Press.

Moos, R. H., & Moos, B. S. (1981). *Family Environmental Scale manual.* Palo Alto, CA: Consulting Psychologists Press.

Nichols, R. C., & Bilbro, W. C. (1966). The diagnosis of twin zygosity. *Acta Geneticae Medicae et Gemellologiae 16;* 265–275.

Oppenheim, D., Nir, A., Warren, S., & Emde, R. N. (1997). Emotion regulation in mother-child narrative co-construction: Associations with children's narratives and adaptation. *Developmental Psychology 33;* 284–294.

Reznick, J. S. (1990). Visual preference as a test of infant work comprehension. *Applied Psycholinguistics 11;* 145–165.

Reznick, J. S., & Snedley, B. (1987, April). *Individual differences in categorization.* Paper presented at the biennial meeting of the Society for Research in Child Development, Baltimore, MD.

Rowe, D. C., & Plomin, R. (1977). Temperament in early childhood. *Journal of Personality Assessment, 41;* 150–156.

Spanier, G. (1976). Measuring dyadic adjustment: New scales for assessing the quality of marriage and similar dyads. *Journal of Marriage and the Family 38;* 15–38.

Terman, L. M., & Merrill, M. A. (1973). *Stanford-Binet Intelligence Scale: 1972 norms edition.* Boston: Houghton-Mifflin.

Watson, M. W., & Fischer, K. W. (1977). A developmental sequence of agent use in late infancy. *Child Development 48;* 828–836.

Watson, M. W., & Fischer, K. W. (1980). Development of social roles in elicited and spontaneous behavior during the preschool years. *Developmental Psychology, 16;* 483–494.

Zahn-Waxler, C., Radke-Yarrow, M., & King, R. A. (1979). Child-rearing and children's prosocial initiations toward victims in distress. *Child Development, 50,* 319–330.

Zelazo, P. D., & Reznick, J. S. (1991). Age-related asynchrony of knowledge and action. *Child Development, 62,* 719–735.

Zelazo, P. D., Reznick, J. S., & Pinon, D. E. (1995). Response control and the execution of verbal rules. *Developmental Psychology, 31,* 508–517.

Part II

Temperament

Section Editor
Jerome Kagan

4

The Structure of Temperament

Jerome Kagan

Explanations of the adult variation in mood, talents, and personality traits have cycled between an emphasis on external events, whether air, water, diet, or social encounters, on the one hand, and endogenous influences, on the other. When endogenous influences take the form of inherited physiological factors, they become synonymous with the idea of human temperaments. The concept of temperament, which was popular in the West from ancient Greece to the end of the nineteenth century, was ignored for the first half dozen decades of this century because of the political necessity of emphasizing the role of environment and minimizing biological differences among young children.

Temperamental concepts have returned to scientific discourse for a variety of reasons. One reason is the discovery that closely related strains of animals raised under identical laboratory conditions behave differently in response to the same intrusions (Scott & Fuller, 1974). Second, the rise of neuroscience has provided some theoretical scaffolding for possible explanations of individual differences that could be the result of inherited physiologies (Blanchard & Blanchard, 1988). The intellectual courage of Alexander Thomas and Stella Chess (1977), who introduced the idea of infant temperament in the late 1950s, was a third factor of influence. Although Solomon Diamond (1957) had suggested earlier that fearful, aggressive, impulsive, and apathetic behaviors in children might be influenced, in part, by temperament and experience, most investigators were not ready for this message. The revival of inquiry into children's temperaments had to wait for the important Thomas and Chess monograph published about 20 years later (Thomas & Chess, 1977).

Although there is no consensual definition of temperament, most researchers would probably agree that temperaments are inherited coherences of physiological and psychological processes that emerge early in development, although not necessarily at birth. Some profiles may require maturation of the brain before they can appear. All scientists agree that environmental conditions can modulate the initial temperamental profile; hence, behaviors with a temperamental origin are not immutable.

The first cohort of investigators in the area of temperament, including Bates (1994), Buss and Plomin (1984), Carey and McDevit (1978), Goldsmith and Campos (1990), Rothbart (1988), Strelau (1991), and Thomas and Chess (1977), agree that a temperamental category refers to a quality that varies among individuals, is moderately stable over time and situation, is under some genetic influence, and, finally, usually appears in the early years of life. A major disagreement centers around the temperamental qualities that are nominated as primary. This disagreement is due in part to the fact that investigators study individuals at different developmental stages, as well as the fact that some scientists rely on questionnaires as the primary source of evidence, while others rely primarily on behavioral observations.

At the moment, most contemporary conclusions on infant and adult temperaments are based on replies to questionnaires. However, the words used in these questionnaires have to be comprehensible to individuals with very different educational attainments, and, therefore, most questionnaires use concepts that originate in everyday conversation. Further, the questions usually do not violate ethical and moral sensitivities (e.g., questionnaires do not inquire about preferences for varied types of erotic play). For these reasons, all questionnaires seriously limit the temperamental phenomena that are studied.

A second problem is that the scientists who apply factor analysis to the replies expect a small number of factors. However, if only one item on a questionnaire measures a significant, but rare, characteristic and the answers to that question are not correlated with the answers to others, it will not emerge as a factor. Further, the use of factor analysis requires the controversial premise that the variation among individuals in a particular factor is the result of different combinations of the same set of fundamental processes. For example, a person who reported that he liked people, parties, and action would be classified as an extravert. Factor analysis assumes that the underlying processes that produce a mild extravert are the same as those that produce an extreme extravert. This assumption is controversial. For example, the cause of dwarfism is different from the usual causes of below-average height. Similarly, the causes of a psychiatric delusion that one is under surveillance by the KGB are not the same as those that lead some African-Americans to believe that their employer does not like them.

TEMPERAMENT IN INFANTS

Rothbart (1988), an influential student of temperament in infants, has suggested that ease of arousal to stimulation, called "reactivity," and the ability to modulate that arousal, called "self-regulation," are two fundamental temperaments of the first stage of development. The operational definitions of reactivity emphasize the display of motor activity, irritability, vocalization, smiling, and even physiological responses provoked by a particular stimulus. Self-regulation refers to the processes that modulate reactivity, like attention, approach, withdrawal, attack, and the ability to soothe oneself. Rothbart and her colleagues rely heavily on maternal answers to questionnaires to study temperament. Unfortunately, the correlation between parents' ratings of a particular quality in an infant or young child and direct behavioral observations are modest at best and, under some conditions, minimal (Seifer et al., 1994). Thus, it is possible that conclusions based on these two different sources of evidence do not have the same theoretical meaning.

There are several reasons for this noncorrelation between questionnaire and behavioral data. First, most parents hold an abstract, symbolic construction of their child that mutes inconsistent observations. Each parent's verbal description of his or her child competes with a nonverbal representation constructed from a history of affective experiences. Further, the psychological categories the parents describe demand a consistency to which the experientially based schemata are indifferent. That is, the mind of the adult does not like the categorical awkwardness of an infant who both cries and smiles frequently. Even though such infants exist, most parents will amplify one of these characteristics and minimize the other, and the infant will be described as frequently fretful or frequently happy, but not both.

A second problem is that parents are not equally discerning in their observations, or equally consistent in their interpretation of questions. Many mothers, for example, distort their perception of their child's behavior so that it is in accord with their understanding of the ideal child. This evaluative frame colors a parent's answers to all questions. The accuracy of a parent's descriptions also varies with parental background and personality. For example, mothers who never attended college more often report that their infants are less adaptive and less sociable than college-educated parents (Spiker et al., 1992). Depressed Irish mothers with their first infant were more likely to describe their 6 week old as irritable compared with experienced mothers, or mothers who were free from depression (Green, 1991).

Third, parents use different references in their evaluations and, as we shall see in the chapters that follow, are vulnerable to contrast affects. The young mother who has not had extensive experience with children will have a less accurate base for judging her first child than a mother with three children. This phenomenon occurs in mothers of fraternal twins, who usually rate the two siblings as much less similar than observers because the mother exaggerates the differences between them.

Fourth, psychologists can only ask parents to rate psychological qualities that they understand in words that are part of a common vocabulary. As a result, scientists must restrict their set of categories to a small number of easily understood ideas, like activity, smiling, fear, crying, and duration of attention to events. Psychologists cannot ask about subtle qualities or about temperaments that have a physiological referent.

Therefore, many unique influences on parental descriptions of children's behaviors are absent when behavior is coded by disinterested observers in standardized contexts. Perhaps that is why the agreement between parents and observers is low; the correlations between the two sources of evidence are rarely > .4 and are usually lower. One team of investigators concluded, "Our results question the assumption that these checklists measure stable traits or characteristics of young children" (Spiker et al., 1992), p. 1493). An exhaustive review of the degree of agreement among parents, teachers, and peers with respect to the occurrence of children's behavioral or emotional problems in more than 269 samples revealed poor agreement on whether a child was excessively fearful, aggressive, or impulsive. The average correlation between two different informants was < .3 (Achenbach, 1985).

A final issue in the study of temperament that remains a source of debate is the decision to conceive of temperaments as continuous behavioral dimensions or as qualitative categories. Thomas and Chess (1977) viewed the concepts "easy," "difficult," and "slow to warm-up" as temperamental categories, even though they created the variation within each of their nine dimensions as continuous. The existing data support both points of view. Although most measured variables form continuous distributions, discrete types often emerge when the investigator combines several variables to create a profile (Kagan, 1994). As Magnusson (1988) has argued, types of people, not continuous variables, are the entities that are stable over time.

WHAT QUALITIES TO STUDY

The selection of temperamental categories for the MacArthur Longitudinal Twin Study was based on reasonable criteria. The investigators chose behaviors that displayed considerable variation, were observable, easily quantifiable by laboratory staff and understood by parents, and had a theoretically reasonable link to physiology. The application of these criteria motivated the investigators to concentrate on five qualities. Behavioral inhibition (and its complement, uninhibited behavior) refers to the tendency to react with initial restraint and avoidance (or spontaneous approach) to unfamiliar events and people. Previous work had established that inhibited and uninhibited children preserved their respective qualities over time. A closely related variable is initial shyness when encountering a stranger. Shy behavior with a stranger is a component of behavioral inhibition. However, shyness refers to a very specific reaction to an unfamiliar person and is indifferent to the display of restraint and avoidance to a variety of unfamiliar events.

Variation in activity level, which is usually regarded as independent of inhibition and shyness, was chosen for study because the investigators believed it has a temperamental origin. The differences in frequency and vigor of limb activity across the first years are so dramatic, it is not surprising that observational studies have not found much preservation of variation in activity level (see Dunn & Kendrick, 1981; Feiring & Lewis, 1980; Matheny, 1983). Assessment of activity in monozygotic and dizygotic twins at 7 months and 3 years, using actometers, reported minimal preservation of activity, but modest heritability (Saudino & Eaton, 1995). However, activity level was more stable from 3 to 6 years (Saudino & Eaton, 1997). One elegant study assessed activity in 112 healthy, middle-class newborns using a pressure transducer mattress that discriminated between activity during crying and during episodes when the child was not distressed. Fifty infants were assessed again when they were between 4 and 8 years of age, using an ambulatory microcomputer as well as a parental questionnaire. There were only modest correlations between vigor of activity during the newborn period and vigor of activity during the day in the older children ($r = .29$; Korner et al., 1985). The independence of day and night activity levels suggests that the concept of general activity level may not be useful. The MacArthur Longitudinal Twin Study investigators assessed activity level by having observers rate the child during laboratory testing.

An extraverted orientation, the tendency to be highly affective and sociable, was a fourth temperamental variable rated by examiners who watched the test procedures. The fifth temperamental variable, also rated by observers, was task orientation, which referred to a persistent and conscientious approach to the tasks administered. Parental ratings of theoretically related qualities were also quantified. The four chapters that follow present in detail the bases for the following brief summary of major conclusions.

Kagan and Saudino report low correlations, on the order or .1–.03, between parental ratings of the temperamental variables described above and behavioral observations of similar characteristics. This finding is in accord with the evidence summarized earlier. The parental ratings showed high stability over time, while the behavioral observations were much less stable. But this fact is possible because the correlations between the two sources of evidence were low. More important, only a few children (< 10%) were consistently inhibited across all ages and, in accord with other research, more girls than boys were consistently inhibited.

The investigation by Manke, Saudino, and Grant, which relied on analyses of extreme groups, asked whether the heritabilities of children who were extreme in their behavior were significantly different from the heritabilities found for the entire sample. The criterion for membership in the extreme groups was a score of 1 SD above or below the mean of the distribution. The analyses of the extremes for shyness, activity, extraversion, and behavioral inhibition produced estimates of group heritability that were similar in magnitude to the estimates of group heritability across the entire sample. The authors suggest that in terms of etiology (i.e., the relative importance of genes and environments), extreme groups may not differ from the rest of the population.

The investigation by Saudino and Cherny, which examined parent ratings of temperamental qualities, reported high stability over age for all temperamental variables. This fact contrasts with the low stability of the behavioral observations. Perhaps a more important result was the negative correlation for parents' ratings of their dizygotic, fraternal twins. This fact implies contrast effects in the ratings of these twins and, therefore, a distortion of their actual behavior. These data support the earlier criticism of parental ratings.

The final chapter of this part by Saudino and Cherny contains an elegant analysis of the sources of continuity and change, using observer ratings of activity, extraversion, task orientation, and shyness, as well as the behavioral observations that formed the basis for behavioral inhibition. As noted earlier, behavioral inhibition, activity level, and extraversion were heritable at one or more ages, and genetic factors contributed to the stability of these traits. A Cholesky analysis suggested a genetic basis for continuity from 14 to 20 months and from 20 to 24 months, but not from 14 to 24 or from 14 to 36 months. Inhibited 14-month-old children changed their phenotype between 1 and 3 years of age, in accord with data on independent samples (Kagan, 1994).

The heritabilites of these traits did not change a great deal from 14 to 36 months, even though children changed their phenotype across this period of time. The changes in the behavior were due primarily to nonshared environmental influences.

The details reported in the next four chapters represent a unique contribution to our understanding of temperamental factors in the first years of life.

REFERENCES

Achenbach, T. M., (1985). *Assessment and taxonomy of child and adolescent psychopathology*. Newbury Park, CA: Sage.

Bates, J. E. (1994). Parents as scientific observers of their children's development. In S. L. Friedman & H. C. Haywood (Eds.), *Developmental follow-up: Concepts, domains, and methods* (pp. 197–216). New York: Academic Press.

Blanchard, D. C., & Blanchard, R. J. (1988). Ethoexperimental approaches to the biology of emotion. *Annual Review of Psychology, 39*, 43–68.

Buss, D.& Plomin, R. (1984). *Temperament: Early developing personality traits*. Hillsdale, NJ: Erlbaum.

Carey, W. B., & McDevitt, S. C. (1978). Stability and change in individual temperament diagnoses from infancy to early childhood. *Journal of the American Academy of Child Psychiatry, 17*, 331–337.

Diamond, S. (1957). *Personality and temperament*. New York: Harper.

Dunn, J., & Kendrick, C. (1981). Studying temperament and parent-child interaction. *Annual Progress in Child Psychiatry and Child Development,* 415–430.

Feiring, C., & Lewis, M. (1980). Sex differences in stability in vigor activity and persistence in the first three years of life. *Journal of Genetic Psychology, 136,* 65–75.

Goldsmith, H. H., & Campos, J. J. (1986). Fundamental issues in the study of early temperament: The Denver twin temperament study. In M. Lamb and A. Brown (Eds.), *Advances in Developmental Psychology* (pp. 232–283). Hillsdale, NJ: Erlbaum.

Goldsmith, H. H., & Campos, J. J. (1990). The structure of temperamental fear and pleasure in infants. *Child Development, 61,* 1944–1964.

Green, J. M. (1991). Mothers' perception of their 6 week old babies. *Irish Journal of Psychology, 12,* 133–144.

Kagan, J. (1994) *Galen's prophecy.* New York: Basic Books.

Korner, A. F., Zeanah, C. H., Linden, J., Berkowitz, R. I., Kraemer, H. C., & Argas, W. S. (1985). The relation between neonatal and later activity and temperament. *Child Development, 56,* 38–42.

Magnusson, D. (1988). *Individual development from an interactional perspective.* Hillsdale, NJ: Erlbaum.

Matheny, A. (1983). A longitudinal twin study of stability of components from Bayley's infant behavior record. *Child Development, 54,* 356–360.

Rothbart, M. K. (1988). Temperament and the development of inhibited approach. *Child Development, 59,* 1241–1250.

Saudino, K. J., & Eaton, W. O. (1995). Continuity and change in objectively assessed temperament. *British Journal of Developmental Psychology, 13,* 81–95.

Saudino, K. J., & Eaton, W. O. (1997, April). *Continuity and change in motor activity from infancy to early childhood.* Paper presented at the 62nd meeting of the Society for Research in Child Development, Washington, DC.

Scott, A. P., & Fuller, J. L. (1974) *Dog behavior: The genetic basis.* Chicago: University of Chicago Press. (Originally published 1965)

Seifer, R., Sameroff, A. J., Barrett, L. C., & Krafchuck, E. (1994). Infant temperament measured by multiple observations and mothers report. *Child Development, 65,* 1478–1490.

Spiker, D., Klebanov, P. K., & Brooks-Gunn, J. (1992, May). Environmental and biological correlates of infant temperament. Paper presented at the eighth biennial meeting of the International Society for Infant Studies, Miami, FL.

Strelau, J. (1991). Renaissance in research on temperament. In J. Strelau & A. P. Angleitner (Eds.), *Exploration in temperament* (pp. 337–349). New York: Plenum Press.

Thomas, A., & Chess, S. (1977). *Temperament and development.* New York: Brunner Mazel.

5

Extremes Analyses of Observed Temperament Dimensions

Beth Manke
Kimberly J. Saudino
Julia D. Grant

The bulk of genetic research on temperament suggests that individual differences in shyness, behavioral inhibition, affect-extroversion, activity, and task orientation are genetically influenced (see Goldsmith, 1983, Plomin, 1987 for reviews). Estimates of heritability for these temperamental dimensions fall in the moderate range (.20–40). Even though genetic influences on individual differences in temperament are well documented, behavioral genetic work in the area of temperament is far from complete. In fact, several important areas have yet to be explored. Little, if any, research on temperament has investigated the etiology of high and low extreme group membership. That is, we have yet to examine why, as a group, individuals who score high or low on measures of temperament are, on average, so different from the mean of the population.

There is some controversy surrounding the idea that the etiology of extreme-group membership (e.g., high and low temperament groups) and individual differences throughout the population are different. As a result, there can be no presupposition that the etiologies are the same or different; instead, there must be a concern to test empirically for similarities and dissimilarities. Extreme-group membership may represent the extreme of the normal distribution of genetic and environmental influences on individual differences in temperament. In this situation, the mechanisms responsible for extreme behavior would be considered quantitatively, not qualitatively, different from those that cause normal variability. Conversely, the etiology of extreme temperament groups may differ from the causes of individual differences. For example, although individual differences in activity level appear to be influenced chiefly by genetic factors and nonshared environmental influences (shared environmental influences are minimal), extreme scores on measures of activity level may be influenced instead by shared environmental factors. For example, it is possible that children with low activity levels may differ from the normal population average as a result of shared familial factors such as crowded or restrictive home environments.

It is important to note that an interest in the etiology of extreme group membership focuses on the average difference between children with extreme scores on observational temperament measures and the rest of the population. We are not asking about genetic and environmental contributions to individual differences among children in the low and high extreme groups. That is, the question as to why one child who is classified as having a low score on some temperament measure has a slightly lower score than another child classified in the same extreme group is less interesting as compared to the question of why, as a group, children rated as having lower or higher temperament scores differ from the rest of the population.

The purpose of the present chapter is to examine the etiology of high and low extreme temperamental groups defined on the basis of scores on observational temperament measures of shyness, activity, affect-extraversion, task orientation, and behavioral inhibition at 14, 20, and 24 months of age. By studying extreme groups in the context of the rest of the population, it will be possible to assess the extent to which the genetic and environmental influences on membership in high and low temperamental extreme groups are similar in magnitude to genetic and environmental influences on individual differences in temperament. In addition, the inclusion of both high and low extreme groups in the present study allows us to compare results for group membership at opposite ends of the distribution. Furthermore, by examining temperamental extreme groups at 14, 20, and 24 months of age, we are able to assess whether the pattern of results is consistent across age.

WHY EXAMINE EXTREME GROUPS OF TEMPERAMENT?

The value of examining temperament from an extremes approach derives from the potential contribution to the study of psychopathology and psychosocial disorders. Eysenck (1994) has proposed that some forms of psychopathology may be best characterized as extreme or pathological manifestations of normal personality. That is, personality disorders may represent the quantitative extreme of the normal continuum of variability in personality. The mechanisms responsible for forms of psychopathology such as affective disorders and neurotic disorders may not differ qualitatively from those that cause normal variability in personality.

Although the extremes approach has yet to be used to examine infant temperament, a few recent investigations of adult personality have explored the degree to which the abnormal is part of the normal continuum of variability.

For example, in a study of middle-age adult twins, extremes analyses of neuroticism suggested that high neuroticism or emotionality may be etiologically the high end of the distribution of individual differences (Plomin, 1991). Other areas of psychology have also begun to use the extremes analysis approach. For instance, available data indicate that some disorders such as mild mental retardation may be part of the normal continuum of variability, whereas other disorders such as severe mental retardation, reading disability, and high scores on depression are etiologically distinct (see Plomin, 1991 for a review).

Conceptual associations between normal variation in some temperamental dimensions and corresponding psychopathology are easy to draw. For example, it is reasonable to propose that a child who scores extremely high on a measure of shyness may be classified as pathologically withdrawn, whereas extremely high scores on a measure of activity may reflect diagnosable hyperactivity. For some dimensions of temperament, however, it may be difficult to derive an analog in the studies of psychopathology. Certain dimensions may have little direct bearing on behaviors that are the focus of psychosocial disorders. For example, high levels of task orientation may have no obvious disorder equivalent.

Furthermore, for some dimensions of temperament, it may be easier to draw conceptual links between normal variation and the abnormal for one end of the distribution (high versus low), but not the other. For instance, while conceptualizing severe shyness as pathological withdrawal may make sense, it is more difficult to draw a conceptual link between slight shyness and any corresponding disorder. Nevertheless, to be consistent in our presentation of results, the etiology of both high and low group membership is assessed for all five temperamental dimensions: shyness, affect-extroversion, activity, task orientation, and behavioral inhibition.

HOW DO WE INVESTIGATE THE ETIOLOGY OF EXTREME GROUP MEMBERSHIP?

A relatively new behavioral genetic approach makes it possible to examine the etiology of the mean of selected groups (such as high or low temperament groups) and its relation to the etiology of individual differences throughout the population. This approach, developed by DeFries and Fulker (1985, 1988), and called DF analysis (Plomin & Rende, 1991) was used in the present study to examine the extent to which high and low temperament are etiologically the extreme of the normal distribution of temperament. DF analysis uses a multiple regression model that leads to an estimate of group heritability (h_g^2), the proportion of the mean difference on a quantitative measure between a selected group and the unselected population that is due to genetic factors. In essence, group heritability estimates the extent to which extreme group membership is heritable.

The key issue in examining extreme group membership is whether the magnitude of group heritability for high and low temperament groups is similar to

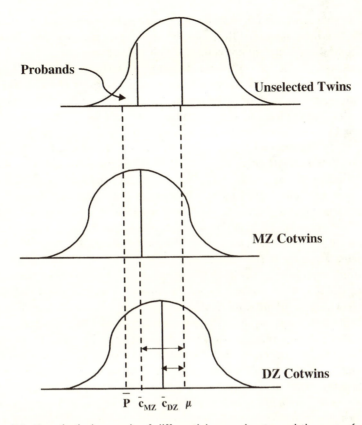

Figure 5.1. Hypothetical example of differential regression toward the mean for MZ and DZ cotwins of probands selected for low temperament scores (adapted from De-Fries & Fulker, 1985). Proband and cotwin means are represented by P and C, respectively. Under a genetic hypothesis, the mean of DZ cotwins will be closer to the mean of the unselected population (μ) than will the mean of MZ cotwins.

traditional estimates of heritability of individual differences in the population (individual heritability). If group heritability for low or high temperament groups differs from individual heritability, this suggests that extreme scores on the temperament measure in question are etiologically different from individual differences in the population. In contrast, if group heritability is similar to individual heritability, it suggests that extreme scores on the temperament measure are merely the high or low end of the normal distribution of genetic and environmental influences on individual differences in temperament.

The conceptual framework for this approach is illustrated in figure 5.1. In short, the DF approach examines differential regression to the mean, for example, for identical (monozygotic, MZ) and fraternal (dizygotic, DZ) cotwins. Twins who have been classified as extreme on a measure of temperament, by definition, have temperament scores that fall toward the extreme of the population. The mean for these twins is designated as P, the proband mean. Temperament scores of cotwins of high or low temperament probands are expected to regress toward the mean of the unselected population (μ). That is, we do not expect the scores of cotwins to be as extreme; instead, we expect cotwins' scores to more closely resemble the mean of the unselected population. However, to the extent that high or low temperament is due to genetic factors, the regression will be less for identical cotwins (C_{MZ}) than for fraternal cotwins (C_{DZ}).

Put simply, the amount of regression toward the mean is a function of the degree of similarity between the proband and the cotwin in terms of temperament. The greater the similarity, the less a cotwin's score will regress toward the mean. If extreme temperament is genetically influenced, similarity for MZ twins will be greater than that for DZ twins. Hence, DZ twins will regress farther toward the mean of the unselected population. Referring to figure 5.1, h_g^2 would be 0 if MZ and DZ cotwins regress to the population mean to the same extent. In contrast, h_g^2 would be 1.0 if MZ cotwins do not regress to the population mean and if DZ cotwins regress halfway back to the mean.

The use of the DF approach also permits an estimate of familial resemblance known as group familiality, the familial contribution to the average difference on a quantitative measure between a selected group, such as an extreme group, and the rest of the population. In short, group familiality refers to the extent to which the mean difference between cotwins of selected probands and the unselected population approaches the mean difference between the probands and the population. An index of group familiality can be calculated from the mean scores for probands, cotwins, and the population (Plomin, 1991). Genetic influences are implied when estimates of group familiality for MZ twins (MZ r_{group}) exceed those for DZ twins (DZ r_{group}).

The DF approach also allows one to estimate the extent to which shared environmental influences (c_g^2) contribute to extreme group membership. In twin analyses, shared environmental influences are estimated as the extent to which twin resemblance cannot be explained by heredity. In the case of group parameters, c_g^2 is the difference between the MZ group familiality estimates and h_g^2. The remainder of the difference between probands and the population is ascribed to nonshared environment and measurement error.

In summary, examination of group familiality estimates and the estimates derived from the DF analyses (i.e., group heritability and group shared environmental influences) afford us the opportunity to evaluate whether and to what extent the difference between the probands and the unselected population is due to genetic and environmental differences.

MEASURES

Observational temperament data gathered at 14, 20, and 24 months of age as part of the MacArthur Longitudinal Twin Study were used in the present analyses. The Infant Behavior Record (IBR; Bayley, 1969) was used to assess infant activity, task orientation, and affect-extroversion. Additional temperament measures included a home assessment of infant shyness and an assessment of infant behavioral inhibition. These measures, procedures, and sample characteristics are discussed in more detail in chapter 7.

ANALYSES

Temperament scores were standardized across all subjects to a mean of 0 and a standard deviation of 1.0. A selection criterion of ± 1.0 SD from the mean of the full sample was then used to identify low and high temperament groups. Subjects whose scores were ≥ 1.0 were identified as probands for the high-extreme groups; subjects whose scores were ≤ − 1.0 were classified as probands for the low-extreme groups. In those cases where both members of a twin pair met the criterion for an extreme group, the data were double entered such that each twin served as a proband and a cotwin. The standard errors were then adjusted for the true sample size (see Stevenson, 1992). High and low groups were identified separately for each of the five temperament dimensions and at each 14, 20, and 24 months of age.

Group familiality estimates for high and low groups were calculated for MZ and DZ twins as part of a simple data transformation used in the basic DF model (see DeFries and Fulker, 1988, for an explanation). When there are no significant mean differences between MZ and DZ twins in the unselected population, standardized scores can be transformed by dividing them by the mean proband score for each twin group. For each twin group, group familiality is then the mean of the cotwins' transformed scores. In only 1 of the 15 cases (shyness at 20 months of age) did t-tests reveal significant mean differences for MZ versus DZ twins. In this case, the transformation was altered to adjust for group mean differences (see Saudino et al., 1994, for details).

When the scores have been transformed, the DF analysis fits a multiple regression model to the data that takes into account mean differences between the probands of each twin group, yields a direct estimate of group heritability, and provides the standard error for this estimate. The basic DF regression model is represented by the following equation:

$$C = \beta_1 P + \beta_2 R + A,$$

where C (cotwin's temperament scores) is predicted by P (proband's temperament score) and R (coefficient of genetic relatedness, which is 1.0 for MZ twins and .50 for DZ twins). Under this model, the regression coefficient, β_2, estimates h_g^2. The group shared environmental parameter c_g^2, is then estimated as the difference between the familiality estimate for identical twins and group heritability.

Results of the DF regression analyses for each temperament dimension will be discussed separately. As part of the discussion of each temperament dimension, the results for high and low group membership across 14, 20, and 24 months are presented. In addition, a comparison of results for high and low group membership is made. Finally, estimates of group heritability and group shared environmental influences are compared to estimates of genetic and shared environmental influences on individual differences in the population (estimates derived from the full MacArthur Longitudinal Twin Study sample and presented in chapter 7).

SHYNESS

Proband means and the means of their cotwins for both high- and low-extreme groups across 14, 20, and 24 months of age are presented in table 5.1. There were no significant differences between MZ and DZ proband means for either high- or low-extreme groups across all three ages. As described earlier, the basis of the DF analysis is the comparison of MZ and DZ cotwin means. Di-

Table 5.1. Proband and cotwin means for MZ and DZ high and low shyness twin groups

Group/age	Proband mean	Cotwin mean	Cotwin distance from group mean[a]	n
High extreme				
14 Months				
MZ	1.65	1.03	1.01	58
DZ	1.70	0.36	0.39	47
20 Months				
MZ	1.52	0.85	0.76	74
DZ	1.59	0.82	0.93	39
24 Months				
MZ	1.48	0.99	0.94	66
DZ	1.49	0.89	0.95	46
Low extreme				
14 Months				
MZ	−1.42	−0.82	0.84	64
DZ	−1.44	−0.55	0.52	60
20 Months				
MZ	−1.48	−1.17	1.26	44
DZ	−1.46	−0.88	0.77	58
24 Months				
MZ	−1.43	−1.00	1.05	59
DZ	−1.44	−0.96	0.90	63

[a] Group mean = control group mean for that twin group.

Table 5.2. Group familiality (r_{group}), group heritability (h_g^2), and group shared environment (c_g^2) for high and low shyness

Group/age	r_{group}	h_g^2	c_g^2
High extreme			
14 Months			
MZ	.63	.83 ± .26*	0
DZ	.21		
20 Months			
MZ	.53	−.04 ± .26	.57 ± .21*
DZ	.54		
24 Months			
MZ	.67	.15 ± .22	.52 ± .18*
DZ	.60		
Low extreme			
14 Months			
MZ	.58	.40 ± .23**	.18 ± .21**
DZ	.38		
20 Months			
MZ	.80	.46 ± .22*	.34 ± .18*
DZ	.57		
24 Months			
MZ	.70	.08 ± .22	.62 ± .17*
DZ	.66		

*$p < .05$; **$p < .10$, one tailed.

zygotic cotwins will regress farther back to the mean of the unselected population (group mean) to the extent that extreme group membership is genetically influenced.

At 14 months of age, the DZ cotwins in the high-extreme group regressed farther back to the group mean than do the MZ cotwins. This pattern suggests that a high level of shyness is influenced genetically. This pattern, however, is not evident for high levels of shyness at 20 and 24 months of age; MZ and DZ cotwins regressed back toward the group mean to a similar extent. That is, at 20 and 24 months of age, both MZ and DZ cotwins regressed just over halfway back to the population mean. This suggests that shared environmental influences, instead of genetic influences, are in part responsible for the difference between the probands and the population mean at 20 and 24 months. The interpretations of the cotwin means are confirmed by the estimates of group familiality and the results of the DF multiple regression analyses (see table 5.2). At 14 months, MZ group familiality for high levels of shyness exceeded DZ group familiality (estimates are .63 and .21 for MZ and DZ twins, respectively). As expected from these group familiality estimates, group heritability was significant (.83), whereas the estimate of group shared environment was

0. In contrast, at 20 and 24 months, MZ and DZ group familiality estimates for high levels of shyness were moderate and remarkably similar in magnitude. The DF analyses indicated that group heritability at both 20 and 24 months is nonsignificant, whereas shared environmental influences are significant.

For low levels of shyness, the pattern of results was similar across 14 and 20 months of age: DZ cotwin means regressed farther back to their group means than do MZ cotwin means, group familiality estimates for MZ twins exceeded those for DZ twins, and the DF regression analyses suggested both significant group heritability and group shared environmental influences. At 24 months of age, results differed slightly in that the DZ cotwin mean did not regress farther back to the group mean as compared to the MZ cotwin mean, although both MZ and DZ cotwin means still regressed less than halfway back to their respective group means. This pattern is suggestive of significant shared environmental influences, but not genetic influences. The presence of significant group shared environmental influences (.62) and nonsignificant group heritability (.08) for low shyness at 24 months was confirmed by the DF analyses.

Comparison of results for high and low shyness groups across 14, 20, and 24 months of age revealed a relatively consistent pattern of significant group shared environmental influences. With the exception of high levels of shyness at 14 months, shared environmental influences appear to be responsible, in part, for the difference between high and low probands and the rest of the population. Results concerning group heritability were less consistent. Group heritability was significant only at 14 months of age for both high and low shyness and at 20 months for low shyness. In short, group shared and non-shared environmental influences appear to be more important for extreme shyness than are group genetic influences.

How do these results compare to what we know about genetic and shared environmental influences on individual differences in shyness? The results concerning genetic and environmental influences on individual differences in shyness derived from the full sample (presented in chapter 7) reveal that the etiology of group membership appears similar to that for individual differences in the population. Significant environmental influences (both shared and non-shared) are present and significant. Although somewhat less consistent, genetic influences are also present. These results suggest that the mechanisms responsible for extreme group membership (both low levels and high levels of shyness) are quantitatively, not qualitatively, different from those that cause normal variability throughout the population.

ACTIVITY

Proband and cotwin means for both high and low IBR activity across 14, 20, and 24 months of age are presented in table 5.3. Although mean differences between MZ and DZ probands were detected for high activity at 24 months of age, a correction was made as part of the DF analysis to adjust for these mean

Table 5.3. Proband and cotwin means for MZ and DZ high and low Infant Behavior Record activity twin groups

Group/age	Proband mean	Cotwin mean	Cotwin distance from group mean[a]	n
High extreme				
14 Months				
MZ	1.97	0.81	0.77	38
DZ	2.05	0.16	0.20	29
20 Months				
MZ	2.06	0.76	0.74	37
DZ	1.86	0.05	0.07	32
24 Months				
MZ	2.08	0.64	0.64	37
DZ	1.80	0.18	0.18	32
Low extreme				
14 Months				
MZ	−1.90	−0.85	0.89	38
DZ	−1.71	−0.21	0.17	38
20 Months				
MZ	−1.86	−0.25	0.27	25
DZ	−2.10	−0.09	0.07	16
24 Months				
MZ	−1.92	−0.29	0.29	28
DZ	−1.83	0.01	0.01	15

[a] Group mean = control group mean for that twin group (from table 5.1).

differences. For both high and low activity across all three ages, DZ cotwins regressed farther back toward the mean of the population than did MZ cotwins, suggesting genetic influence. It should also be noted that in a few cases DZ cotwins regressed almost completely back to the mean of the population, suggesting little, if any, shared environmental influence on either low or high extreme group membership. The tendency of DZ cotwin means to regress almost completely back to the population mean also suggests that nonadditive, instead of additive, genetic influences or rater contrasts effects may have been responsible for the difference between the probands and the unselected population. Contrast effects arise when a rater compares two siblings, thereby magnifying existing differences between the two. For genetically influenced traits, MZ twins, who are more behaviorally alike, would be less prone to rater contrasts. Dizygotic twins, who share only 50% of their segregating genes, are more prone to rater contrasts and are thus more likely to be rated differently. This tendency to contrast DZ cotwins and rate them as less similar results in DZ correlations too low as compared to MZ correlations.

Table 5.4. Group familiality (r_{group}), group heritability (h_g^2), and group shared environment (c_g^2) for high and low Infant Behavior Record activity

Group/age	r_{group}	h_g^2	c_g^2
High extreme			
14 Months			
MZ	.41	.67 ± .30*	0
DZ	.08		
20 Months			
MZ	.37	.69 ± .26*	0
DZ	.02		
24 Months			
MZ	.31	.41 ± .29	0
DZ	.10		
Low extreme			
14 Months			
MZ	.45	.65 ± .36**	0
DZ	.12		
20 Months			
MZ	.13	.18 ± .20	0
DZ	.04		
24 Months			
MZ	.15	.32 ± .33	0
DZ	−.00		

*p < .05; **p < .10, one tailed.

Inspection of group familiality, group heritability, and group shared environmental estimates presented in table 5.4 confirm our interpretations of the proband and cotwin means. In all cases, nonadditive genetic variance is suggested by DZ familiality estimates that are less than one-half the corresponding moderate MZ familiality estimates. Accordingly, estimates of group heritability are in many cases significant, whereas all estimates of group shared environment are 0.

Lack of significance for the group heritability estimates for high activity at 24 months and low activity at both 20 and 24 months may have been due to small sample sizes. Small samples reduce the power to detect significant group heritability. Although the estimates of group heritability are not statistically significant in three of the six cases, the pattern of results across ages and across both high and low activity levels suggest that as much as two thirds of the mean difference between probands and the population is due to genetic factors. The remainder of the difference can be ascribed to nonshared environment and measurement error; in all cases shared environmental influences are unimportant.

Upon initial comparison, the group heritability estimates appear to be somewhat larger in magnitude than the traditional individual heritability estimates
found for the full sample (see chapter 7). However, once the standard errors
around the estimates of group heritability have been taken into account, it is
clear that these estimates of group heritability are similar in magnitude to individual heritability estimates. This suggests that, as was the case for shyness,
in terms of genetic influence, high and low activity level may merely be the
extremes of the normal continuum. In other words, the mechanisms responsible for high and low activity levels do not appear to differ from those that
cause normal variability in activity level.

AFFECT-EXTRAVERSION

Proband and cotwin means for high and low IBR affect-extraversion groups
across 14, 20, and 24 months are depicted in table 5.5. In all but one case (i.e.,

Table 5.5. Proband and cotwin means for MZ and DZ high and low Infant
Behavioral Record affect-extraversion twin groups

Group/age	Proband mean	Cotwin mean	Cotwin distance from group mean[a]	n
High extreme				
14 Months				
MZ	1.25	0.39	0.41	44
DZ	1.25	0.02	0	40
20 Months				
MZ	1.30	0.50	0.51	57
DZ	1.25	0.09	0.08	45
24 Months				
MZ	1.31	0.58	0.56	38
DZ	1.29	0.08	0.11	37
Low extreme				
14 Months				
MZ	−1.82	−0.65	0.63	63
DZ	−1.76	−0.26	0.28	53
20 Months				
MZ	−1.81	−0.52	0.51	49
DZ	−2.18	−0.41	0.42	33
24 Months				
MZ	−1.83	−0.90	0.92	51
DZ	−1.98	−0.51	0.48	42

[a] Group mean = control group mean for that twin group (from table 5.1).

Table 5.6. Group familiality (r_{group}), group heritability (h_g^2), and group shared environment (c_g^2) for high and low Infant Behavioral Record affect-extraversion

Group/age	r_{group}	h_g^2	c_g^2
High extreme			
14 Months			
MZ	.31	$.60 \pm .36$**	0
DZ	.01		
20 Months			
MZ	.39	$.62 \pm .31$*	0
DZ	.08		
24 Months			
MZ	.44	$.77 \pm .35$*	0
DZ	.06		
Low extreme			
14 Months			
MZ	.36	$.43 \pm .29$	0
DZ	.14		
20 Months			
MZ	.29	$.20 \pm .30$	$.09 \pm .26$
DZ	.19		
24 Months			
MZ	.49	$.46 \pm .31$*	$.03 \pm .28$
DZ	.26		

*$p < .05$; **$p < .10$, one tailed.

low affect-extraversion at 20 months of age), there were no significant mean differences between MZ and DZ probands for either high- or low-extreme groups across age. Once again, the DF analyses take into account these mean differences between MZ and DZ probands. The pattern of cotwin means for high extreme group membership was consistent across 14, 20, and 24 months of age; DZ cotwin means regress farther back to the population mean than do MZ cotwin means, suggesting genetic influence.

Group familiality estimates (table 5.6) also presented a pattern indicative of group heritability for high-extreme group membership. The MZ group familiality estimates were .31, .39, and .44 for 14, 20, and 24 months, respectively. The DZ group familiality estimates were only slightly greater than 0. As was the case for activity level, nonadditive genetic influences or rater contrast effects are implicated in that DZ familiality estimates were less than half the corresponding MZ familialities. Accordingly, the DF regression analyses revealed significant and substantial estimates of group heritability; genetic influences appear to account for more than 60% of the difference between the probands and the population mean. Shared environmental influences appear to be unimportant for high affect-extraversion, as evidenced by zero estimates of c_g^2.

Results for low affect-extraversion were somewhat less consistent. Although DZ cotwins regressed farther back to the population mean than did MZ cotwins at all three ages, suggesting the presence of genetic influence, the estimates of group heritability for low affect-extraversion at 14 and 20 months were not significant (see table 5.6). It should be noted, however, that when one takes into account the standard errors around these estimates of group heritability, they are similar in magnitude to the estimates obtained for low affect-extraversion at 24 months of age and similar to the estimates obtained for high-extreme-group membership across all three ages. In essence, although less consistent, and somewhat lower in magnitude, the results for low-group membership are similar to those obtained for high-group membership. That is, genetic and nonshared environmental influences, not shared environmental influences, are chiefly responsible for the mean difference between low affect-extraversion probands and the population.

Despite the generally elevated estimates of group heritability for high affect-extraversion, both high and low affect-extraversion appear to be etiologically similar to individual differences in affect-extraversion. Both extreme group membership and individual differences appear to be influenced more by genetic factors than by shared environmental contributions, suggesting, once again, that extreme-group membership is merely the tail end of the normal distribution of genetic and environmental factors.

TASK ORIENTATION

Proband and cotwin means for both high and low IBR task orientation across 14, 20, and 24 months of age are presented in table 5.7. There were no significant differences between MZ and DZ proband means for either high- or low-extreme groups across all three ages. For high-extreme-group membership at 14 months of age, both MZ and DZ cotwins regressed 90% of the way back to the population mean, suggesting little group heritability or shared environmental influences. It would appear that nonshared environmental influences and measurement error are chiefly responsible for the difference between the probands and the unselected population at 14 months of age. In contrast, at 20 months of age, MZ cotwins regressed less than halfway back to the population mean, whereas DZ cotwins regressed almost completely back to the population mean, indicating substantial group heritability and no shared environmental influences. The pattern of cotwin means at 24 months of age is similar to that at 20 months in that the DZ cotwins regressed farther back to the population mean than do MZ cotwins. The results at 24 months, however, differ from those at 20 months in that MZ cotwins regressed more than halfway back

Table 5.7. Proband and cotwin means for MZ and DZ high and low Infant Behavior Record task orientation twin groups

Group/age	Proband mean	Cotwin mean	Cotwin distance from group mean[a]	n
High extreme				
14 Months				
MZ	1.44	0.15	0.13	67
DZ	1.42	0.16	0.18	59
20 Months				
MZ	1.54	0.81	0.78	53
DZ	1.41	0.07	0.10	36
24 Months				
MZ	1.36	0.28	0.26	52
DZ	1.42	0.01	0.02	56
Low extreme				
14 Months				
MZ	−1.46	−0.13	0.15	76
DZ	−1.52	−0.25	0.23	65
20 Months				
MZ	−1.45	−0.41	0.44	66
DZ	−1.58	−0.05	0.02	56
24 Months				
MZ	−1.57	−0.33	0.35	43
DZ	−1.50	−0.05	0.03	56

[a] Group mean = control group mean for that twin group (from table 5.1).

to the population mean, indicating that although genetic influences may still be operating, these influences may be less substantial at 24 months as compared to those at 20 months of age.

Table 5.8 contains the group familiality, group heritability, and group shared environmental estimates for both high and low task orientation. As expected from the pattern of cotwin means, MZ and DZ group familiality estimates for high task orientation at 14 months are low and remarkably similar in magnitude. The DF analyses indicate that both group heritability and shared environmental influences are negligible and nonsignificant. In contrast, at 20 and 24 months of age, MZ familiality estimates are moderate and exceed DZ familiality estimates. The DF analyses reveal, however, that only the group heritability estimate at 20 months (.96) is substantial and statistically significant. Estimates of shared environmental contributions to high task orientation at 20 and 24 months are nonsignificant, indicating that the environmental influences responsible for the mean difference between low task orientation probands and the unselected population are nonshared.

Table 5.8. Group familiality (r_{group}), group heritability (h_g^2), and group shared environment (c_g^2) for high and low Infant Behavior Record task orientation

Group/age	r_{group}	h_g^2	c_g^2
High extreme			
14 Months			
MZ	.10	.01 ± .28	.09 ± .24
DZ	.11		
20 Months			
MZ	.53	.96 ± .34*	0
DZ	.05		
24 Months			
MZ	.20	.40 ± .29	0
DZ	.01		
Low extreme			
14 Months			
MZ	.09	−.15 ± .24	.09 ± .23
DZ	.17		
20 Months			
MZ	.28	.50 ± .24*	0
DZ	.03		
24 Months			
MZ	.21	.35 ± .26	0
DZ	.04		

*$p < .05$, one tailed.

The pattern of results for low-extreme task orientation across the three ages is remarkably similar to the pattern found for high extreme group membership. At 14 months of age (see table 5.7), DZ cotwins do not regress farther back to the population mean as compared to MZ cotwins, suggesting little group heritability. In fact, although both cotwin types regress more than 80% of the way back toward the mean of the unselected population, MZ cotwins actually regress farther back than do DZ cotwins. The group familiality estimates and DF analyses confirm the presence of little genetic or shared environmental contributions to low task orientation at 14 months (see table 5.8). That is, MZ and DZ group familiality estimates are similar and low in magnitude (estimates are .09 and .17 for MZ and DZ twins, respectively), and the group heritability and shared environmental estimates are negligible and nonsignificant.

The results for low task orientation at 20 and 24 months are similar to each other and to the results found for high task orientation at the same ages. In all cases the DZ cotwin mean regresses farther back to the mean of the unselected population than does the MZ cotwin mean, and the MZ group familiality estimate exceeds the DZ group familiality estimate, suggesting genetic influence.

The DF analyses confirm these interpretations by revealing moderate group heritability estimates for low task orientation at the later two ages (group heritability estimates are .50 and .35 at 20 and 24 months of age, respectively), although the estimate at 24 months of age was not statistically significant. Once again, group shared environmental influences are estimated to be zero at 20 and 24 months of age.

As noted previously, the results for high- and low-extreme task orientation are quite similar. It appears that high and low group membership at 14 months is influenced primarily by nonshared environmental contributions. In contrast, at 20 and 24 months of age, genetic contributions emerge in importance, whereas shared environmental influences remain negligible.

Is the etiology of high and low task orientation similar to that for individual differences in task orientation? Comparison of the above results to the results for individual differences (reported for the full sample in chapter 7) suggests, once again, similar etiologies. Like extreme-group membership, individual differences at 14 months of age appear to be influenced primarily by nonshared environmental influences, with genetic factors increasing in importance at 20 and 24 months of age. The similarity in findings suggests that those factors responsible for the difference between the probands and the unselected population mean are quantitatively, not qualitatively, different from the factors influencing normal variability in task orientation.

BEHAVIORAL INHIBITION

Proband means and the means of their cotwins for both high and low extreme groups across 14, 20, and 24 months of age are presented in table 5.9. There were no significant differences between MZ and DZ proband means for either high or low extreme groups across all three ages. Inspection of cotwin means in table 5.9 suggests that with the exception of low behavioral inhibition at 14 months, the pattern of results is similar across high and low group membership and across age, that is, genetic influences appear to be important for high and low levels of behavioral inhibition, as evidenced by DZ cotwin means that regress farther back to the population mean as compared to MZ cotwin means. In contrast, group shared environmental influences seem to be important for low-extreme group membership at 14 months of age, as demonstrated by DZ and MZ cotwins means that regress back toward the population mean to the same extent.

Impressions concerning both high- and low-group membership gained from the cotwin means are confirmed by the group familiality estimates and the estimates of group heritability and group shared environmental influences (see table 5.10). In all cases, except for low behavioral inhibition at 14 months of age, MZ group familiality estimates exceeded estimates of DZ group familiality. In addition, estimates of group heritability are substantial and significant (ranging from .48 to .87). As expected, group shared environmental influences are

Table 5.9. Proband and cotwin means for MZ and DZ high and low behavioral inhibition twin groups

Group/age	Proband mean	Cotwin mean	Cotwin distance from group mean[a]	n
High extreme				
14 Months				
MZ	1.79	1.03	1.02	49
DZ	1.93	0.44	0.45	38
20 Months				
MZ	1.79	0.82	0.75	50
DZ	1.80	0.19	0.28	34
24 Months				
MZ	1.86	1.04	1.03	37
DZ	1.50	0.18	0.19	42
Low extreme				
14 Months				
MZ	−1.30	−0.64	0.65	56
DZ	−1.34	−0.64	0.63	35
20 Months				
MZ	−1.26	−0.69	0.76	31
DZ	−1.29	−0.34	0.25	44
24 Months				
MZ	−1.31	−0.62	0.63	48
DZ	−1.34	−0.31	0.30	36

[a] Group mean = control group mean for that twin group (from table 5.1).

estimated to be zero. In contrast, results for low-group membership at 14 months of age reveal moderate and similar MZ and DZ group familiality estimates and corresponding negligible group heritability and moderate group shared environmental influences. Although the estimate of group shared environmental influences at 14 months of age is not significant, the fact that it is larger than the estimate of group heritability suggests that environmental influences, both shared and nonshared, may be more important than genetic influences for low group membership at this age.

The results for both high- and low-extreme-group membership parallel the data concerning individual differences in behavioral inhibition (presented in chapter 7). Genetic and nonshared environmental influences are chiefly responsible for both individual differences throughout the population as well as for extreme-group membership. In other words, the mechanisms responsible for high and low behavioral inhibition do not appear to differ from those that cause normal variability in behavioral inhibition. The only exception seems to be low group membership at 14 months of age. Genetic and nonshared envi-

Table 5.10. Group familiality (r_{group}), group heritability (h_g^2), and group shared environment (c_g^2) for high and low behavioral inhibition

Group/age	r_{group}	h_g^2	c_g^2
High extreme			
14 Months			
MZ	.57	.69 ± .27*	0
DZ	.23		
20 Months			
MZ	.34	.58 ± .29*	0
DZ	.15		
24 Months			
MZ	.56	.87 ± .28*	0
DZ	.12		
Low extreme			
14 Months			
MZ	.49	.03 ± .29	.46 ± .24
DZ	.48		
20 Months			
MZ	.57	.71 ± .35*	0
DZ	.21		
24 Months			
MZ	.47	.48 ± .30**	0
DZ	.23		

*p < .05; **p < .10, one tailed.

ronmental influences are responsible for individual differences in behavioral inhibition at 14 months of age, whereas low behavioral inhibition at 14 months appears to be influenced more by shared environmental influences than genetics. Perhaps in this case, extreme-low-group membership does not reflect the extreme of the normal distribution.

SUMMARY AND DISCUSSION

For the most part, results of the DF analyses suggest that the etiology of extreme-group membership (both high and low) across several dimensions of temperament are not qualitatively different from those factors that influence individual differences throughout the population. With few exceptions, estimates of h_g^2 and c_g^2 are similar in magnitude to traditional individual heritability and shared environmental estimates. We conclude that the mechanisms responsible for high and low extreme temperament scores do not appear to differ from those that cause normal variability. Furthermore, the results presented in this chapter suggest that the pattern in which the etiology of extreme-group membership parallels that of individual differences is relatively consistent across 14, 20, and 24 months of age.

It must be emphasized, however, that to detect significant differences between group familiality and individual familiality requires a much larger sample size than the one in this study. For example, to identify differences between group and individual familialities on the order of .15 would require samples of approximately 160 pairs of probands and their cotwins for each extreme group and 350 pairs of unselected twins.

One notable exception to our conclusion pertains to low behavioral inhibition at 14 months of age. In this case, the etiology of extreme-group membership appears to differ qualitatively rather than quantitatively from the factors that affect individual differences throughout the population, that is, shared and nonshared environmental influences are important for low extreme group membership, whereas individual differences in behavioral inhibition are due primarily to genetic and nonshared environmental factors. Although further research is needed to replicate these findings, they appear in line with recent evidence suggesting the presence of qualitatively distinct categories of behaviorally inhibited children (Kagan, 1998).

It is important to reiterate that our interest in the etiology of extreme-group membership focuses on the average difference between children with extreme scores on observational temperament measures and the rest of the population. Our investigation did not focus on the genetic and environmental contributions to individual differences among children in low- and high-extreme groups. This is a different type of analysis, one that focuses on why one child who is classified as having an extreme score on some temperament measure has a slightly lower (or higher) score than another child classified in the same extreme group.

When discussing the results derived from the MacArthur Longitudinal Twin Study sample, we must recognize that results for other samples could differ. Ours is a normative sample, and even its most extreme individuals may not show clinically relevant levels of temperament. As a result, the DF results reported here might differ from those obtained from a sample of clinically diagnosed children. Furthermore, the DF results might have varied had we used a different selection criterion for extreme group membership. Our selection criterion of ± 1 SD was chosen primarily for pragmatic reasons (i.e., to allow for large enough sample sizes to conduct the analyses of interest). Clearly, a more extreme group within the same population (e.g., children whose scores were ≥ 2 SD) might have produced different results.

Future research on extreme temperament ought to focus on which specific shared environmental, genetic, and nonshared environmental factors are responsible for extreme group membership. Identifying anonymous components of variance is only the first step toward understanding the etiology of extreme temperament scores. This is particularly important in regard to our investigation of nonshared environmental contributions. Typical investigations of environmental and genetic factors are unable to disentangle substantively interesting sources of nonshared environment from measurement error. Given our findings that in several cases nonshared environmental factors are chiefly responsible for extreme group membership, it is imperative that we be able to identify and incorporate specific measures of nonshared environment into our models.

Future research on extreme temperament might also include not only the application of the DF approach to more extreme samples, but also multivariate extensions of these analyses. For example, one might analyze the heterogeneity and comorbidity of several different temperament disorders. Or one might examine the extent to which etiology of extreme temperament defined on the basis of parental or teacher reports differs from the etiology of extreme-group membership defined on the basis of observational scores. We conclude that the comparison between group and individual heritability in the present and future applications of the DF approach are valuable in identifying potential etiological links between temperament disorders and the normal continuum of variability.

REFERENCES

Bayley, N. (1969). *Manual for the Bayley Scales of Infant Development.* New York: The Psychological Corporation.

DeFries, J. C., & Fulker, D. W. (1985). Multiple regression analysis of twin data. *Behavior Genetics, 15,* 467–473.

DeFries, J. C., & Fulker, D. W. (1988). Etiology of deviant scores versus individual differences. *Acta Geneticae Medicae et Gemellologiae, 37,* 205–216.

Eysenck, H. J. (1994). Normal-abnormality and the three-factor model of personality. In S. Strack & M. Lorr (Eds.), *Differentiating normal and abnormal personality* (pp. 3–25). New York: Springer.

Goldsmith, H. H. (1983). Genetic influences on personality from infancy to adulthood. *Child Development, 55,* 1005–1019.

Kagan, J. (1998). Biology and the child. In W. Damon and N. Eisenberg (Eds.), *Handbook of child psychology* (5th ed., pp. 177–235). New York: Wiley.

Plomin, R. (1987). Developmental behavioral genetics and infancy. In J. Osofsky (Ed.), *Handbook of infant development* (2nd ed., pp. 363–417). New York: Wiley.

Plomin, R. (1991). Genetic risk and psychosocial disorders: Links between the normal and abnormal. In M. Rutter & P. Casear (Eds.), *Biological risk factors for psychosocial disorders.* (pp. 101–138). Cambridge: Cambridge University Press.

Plomin, R., & Rende, R. (1991). Human behavioral genetics. *Annual Review of Psychology, 42,* 161–190.

Saudino, K. J., Plomin, R., Pedersen, N. L., & McClearn, G. E. (1994). The etiology of high and low cognitive ability during the second half of the life span. *Intelligence, 19,* 359–371.

Stevenson, J. (1992). Evidence for a genetic etiology in hyperactivity in children. *Behavior Genetics, 12,* 535–542.

6

Parental Ratings of Temperament in Twins

Kimberly J. Saudino
Stacey S. Cherny

Parental ratings offer a number of advantages to the researcher interested in studying temperament in infants or young children. Because parents know their children well and spend much time with them, they are able to provide valuable information about their child's typical behaviors across many situations. As compared to behavioral observations, which typically involve brief samples of behavior, parental ratings are based on more extensive behavioral sampling and therefore avoid potential problems associated with situational influences. In addition, parental rating scales are inexpensive, easy to administer, and display good psychometric properties in terms of reliability and internal consistency (Goldsmith et al., 1991; Hubert et al., 1982; Slabach et al., 1991). Therefore, it is not surprising that parental rating measures are the most common method of assessing temperament in infancy and childhood.

Parental rating scales are not, however, without their weaknesses. Parental ratings require judgments based on the parents' knowledge of their own and other children's behavior, yet parents may not have sufficient experience to accurately rate their child's behaviors in relation to other children. Additionally, parental ratings of temperament are not independent of the child–caregiver interaction (Rothbart, 1981). Therefore, how the child's temperament affects the parent may influence how the parent rates the child's behavior. Similarly, there is evidence to suggest that parental characteristics such as Socioeconomic status, race, personality, and mental health can bias perceptions of temperament (e.g., Bates, et al., 1979; Daniels, et al., 1984; Vaughn et al., 1987). Expectations of parents may also influence parental ratings of temperament (e.g., Mebert, 1989, 1991; Wolk et al., 1992; Zeanah et al., 1985).

Finally, there is mixed evidence regarding convergent validity. Although there is substantial agreement between parental ratings of temperament across different measures (Bates et al., 1979; Goldsmith et al., 1991; Worobey, 1986), there is minimal convergence between parental and observer ratings on the same scale (e.g., Bates & Bayles, 1984; Goldsmith et al., 1991). Moreover, parental ratings of temperament are only weakly correlated with observational or mechanical measures of temperament (e.g., Hubert et al., 1982; Rothbart, 1986; Saudino & Eaton, 1991, 1995; Seifer et al., 1994).

Parental ratings pose an additional problem in twin studies. The strongest evidence for genetic influences on individual differences in temperament comes from research using the twin design; however, parental rating scales produce unusual results. With such measures, intraclass correlations for identical (monozygotic; MZ) twins are typically high, whereas intraclass correlations for fraternal (dizygotic; DZ) twins are much lower than one-half the MZ correlation, as would be predicted from an additive genetic model (e.g., Buss et al., 1973; Emde et al., 1992; Neale & Stevenson, 1989; Plomin & Rowe, 1977; Plomin et al., 1993; Stevenson & Fielding, 1985; Torgersen, 1981, 1985; Torgersen & Kringlen, 1978). Indeed, DZ intraclass correlations for parent-rated temperament are often near zero or negative, which implies that DZ twin pairs are perceived as no more similar, and in some instances, less alike, than two randomly paired children. The problem of too low DZ correlations suggests that parent ratings of temperament may be prone to contrast effects (Buss & Plomin, 1984). Parents of DZ twins may judge one twin's behavior in the context of the other's, thereby magnifying their behavioral differences and lowering DZ within-pair similarity. If this is the case, twin studies will exaggerate the difference between MZ and DZ correlations and will overestimate the magnitude of genetic influence.

The present chapter examines parental ratings of temperament on the Colorado Childhood Temperament Inventory (CCTI; Rowe & Plomin, 1977). The inclusion of both parental ratings and observed behavioral measures of temperament in the MacArthur Longitudinal Twin Study provides a unique opportunity to examine issues of convergent validity and possible parental rating biases. The results from longitudinal behavior genetic analyses examining sources of continuity and change in parent-rated temperament are compared and contrasted to those from similar analyses of observer-rated temperament presented in chapter 7. In addition, genetic and environmental sources of agreement between mother and father ratings are explored.

COLORADO CHILDHOOD TEMPERAMENT INVENTORY

At 14, 20, 24, and 36 months, both parents rated the temperament of their twins on the CCTI (Rowe & Plomin, 1977), which had been modified to include separate and distinct scales of shyness and sociability (Buss & Plomin, 1984).

The revised measure contains 30 general statements describing the temperament dimensions of emotionality, activity, sociability, shyness, persistence, and soothability (e.g., "child cries easily" or "child is very energetic"). Parents were asked to rate each statement on a 5-point Likert scale ranging from 1 = strongly disagree; not at all like the child to 5 = strongly agree; a lot like the child. Because aggregating across raters increases the reliability of a measure by reducing the error variance associated with a single rater (Epstein, 1983), mid-parent scores were created by averaging across mothers and fathers.

PARENT–OBSERVER AGREEMENT

We examined convergent validity by correlating mid-parent ratings for five CCTI dimensions with observational measures that assess conceptually similar dimensions of temperament. Admittedly, parental and observer ratings assess children in different contexts, but given that temperament is presumed to show some consistency across situations (Goldsmith et al., 1987), it is reasonable to expect that the two types of measures would be related. For example, parental ratings of emotionality on the CCTI evaluate the child's tendency to become upset easily and intensely (Buss & Plomin, 1984). As a measure of the child's general level of distress, fear, and anger, CCTI emotionality should be related to the affect-extraversion factor on the Bayley Infant Behavior Record (IBR; Bayley, 1969), which includes observations of the child's emotional tone, fearfulness, and tension during administration of the Bayley mental scales (see chapter 7 for a description of observational measures of temperament). However, CCTI emotionality assesses the negative pole of emotionality, whereas the IBR affect-extraversion factor is scored in the direction of positive emotionality, therefore, the two measures should correlate negatively. In addition, because the IBR affect-extraversion factor also contains observations of the child's degree of social orientation and cooperativeness, we also expected it to be related to parental ratings of sociability, a preference to be with other people. Activity on the CCTI refers to the tempo, energy, and vigor with which the child behaves (Buss & Plomin, 1984) and thus should be related to the IBR activity factor, which assesses the child's levels of energy and body movement. Similarly, parental ratings of shyness on the CCTI assess the child's tendency to be inhibited or withdrawn around strangers and should be related to behavioral observations of the child's initial reaction to the entrance of two female examiners to the home. Finally, both the CCTI attention-persistence scale and IBR task orientation assess the child's tendency to attend and persist when working on tasks and therefore should be positively correlated.

Correlations between parental ratings and measures of observed temperament are presented in table 6.1. At all ages, the correlations were low, but in the expected direction. Consistent with previous research, there is little convergence between observational and parental ratings. Only activity and shy-

Table 6.1. Correlations between parental ratings on the Colorado Childhood
Temperament Inventory and observed temperament measures

Measure	14 Months	20 Months	24 Months	36 Months
Emotionality	−.16*	−.10*	−.07	—[a]
Activity	.20*	.23*	.09*	.14*
Sociability	.06	.06	.10*	—[a]
Shyness	.36*	.35*	.39*	.41*
Persistence	.11*	.07	.18*	.07

[a] No Infant Behavioral Record affect-extraversion factor at 36 months. Based on full sample at all ages. $N = 674$–685 at 14 months, 585–597 at 20 months, 591–604 at 24 months, and 409–520 at 36 months.
*$p < .05$.

ness displayed significant correlations between observational and parental ratings at each age; however, even for these dimensions, the effect sizes are
modest.

AGE-TO-AGE STABILITY OF PARENTAL RATINGS

Most studies of infant and child temperament find moderate stability across
age (Hubert et al., 1982; McDevitt, 1986; Slabach et al., 1991); however, few
studies have used measures other than parental ratings. When temperament is
assessed by the same rater on each occasion, as is the case for parental ratings,
it is difficult to disentangle the stability of the rater's perceptions or rating
behaviors from the stability of the child's temperament behaviors. A striking
example comes from the work of Zeanah and colleagues. Parental ratings of
temperament taken before birth, when parents have no knowledge of their
unborn infant's temperament, have been found to correlate significantly with
parental ratings of temperament during infancy (e.g., Mebert, 1989; Wolk et
al., 1992; Zeanah et al., 1985). Although it is possible that parents' prenatal
ratings of their infants reflect their own temperament and thus their child's
temperament via genetic transmission, the stability between pre- and postnatal
ratings of temperament has generally been interpreted as indicating that, very
early on, parents form enduring perceptions of their child's behavior.

MacArthur Longitudinal Twin Study age-to-age correlations for parental ratings on the CCTI are presented in table 6.2. A number of features are worth
noting. First, across all age intervals, there is substantial stability in parent-
rated temperament. Second, all dimensions show similar stabilities across age.
Third, in contrast to previous research which suggests increases in stability
across age (McDevitt, 1986; Plomin et al., 1988), stability correlations across
adjoining intervals (i.e., 14–20 months, 20–24 months, and 24–36 months) are
reasonably similar.

Table 6.2. Phenotypic stability correlations for parent-rated temperament

Measure	14–20 Months	14–24 Months	14–36 Months	20–24 Months	20–36 Months	24–36 Months
Emotionality	.53	.47	.39	.59	.49	.54
Activity	.68	.65	.56	.70	.60	.68
Sociability	.52	.47	.39	.58	.39	.54
Shyness	.61	.53	.41	.65	.52	.59
Persistence	.54	.45	.40	.62	.56	.57

Mid-parent ratings of temperament on the CCTI. N = 402–404 individuals with complete data across all ages. All correlations are significant at $p < .01$.

These stability correlations are particularly interesting when compared to the age-to-age correlations for behavioral observations of similar temperament dimensions presented in chapter 7. Consistent with previous research (Plomin & DeFries 1985; Rothbart, 1986; Wilson & Matheny, 1986), observational measures of temperament displayed substantially lower correlations than parental ratings; ranging from .09 to .39 across the same age intervals. The lower stabilities for observational measures do not appear to be due to lower measurement reliability (i.e., arising as a result of brief behavioral sampling or specific situational factors), the reliabilities for observed temperament (ranging from .64 to .88) were only slightly lower than those for parental ratings (ranging from .73 to .89). Moreover, even when corrected for attenuation due to unreliability, age-to-age stabilities for observed temperament remained lower than those for parental ratings (ranging from .11 to .45). These results suggest that the stability of temperament on the CCTI may reflect parents expectations in addition to reflecting actual child behavior.

TWIN INTRACLASS CORRELATIONS

Twin intraclass correlations for CCTI dimensions (see table 6.3) provide additional evidence suggesting that parental ratings of temperament may be biased by parents expectations. Across age and dimension, mid-parent ratings on the CCTI produced a pattern of moderate MZ intraclass correlations and near zero or negative DZ twin intraclass correlations. The finding of significant negative intraclass correlations is particularly interesting because, for random pairings of unrelated children, we would not expect correlations below zero. Thus, parents of identical twins perceive their twin children as being somewhat similar in temperamental disposition, whereas parents of fraternal twins not only view their twins as not similar with regard to temperament, DZ co-twins are actually perceived as having opposing behavioral tendencies. For

Table 6.3. Twin intraclass correlations for mid-parent ratings of temperament on the Colorado Childhood Temperament Inventory

	14 Months		20 Months		24 Months		36 Months	
Measure	MZ	DZ	MZ	DZ	MZ	DZ	MZ	DZ
Emotionality	.23*	−.18	.47*	−.09	.28*	−.12	.30*	−.18
Activity	.53*	−.26*	.51*	−.23*	.55*	−.22*	.56*	−.32*
Sociability	.38*	.02	.48*	.07	.42*	.01	.38*	−.17
Shyness	.35*	−.28*	.44*	−.15	.49*	.06	.47*	−.11
Persistence	.47*	−.09	.50*	−.29*	.22*	−.23*	.32*	−.39*

N = 101 MZ, 93 DZ pairs for emotionality; and 101 MZ, 95 DZ pairs for activity, sociability, shyness, and persistence.
*$p < .05$.

example, one twin is considered to be active, whereas the other is considered to be inactive.

These twin intraclass correlations for parental ratings are in sharp contrast to the intraclass correlations for observed temperament dimensions reported in chapter 7. When temperament was assessed by observational ratings, DZ correlations were positive, and the pattern of MZ–DZ correlations was generally consistent with genetic expectations. Thus, it appears that parental ratings, but not observational ratings, are subject to contrast effects that magnify existing behavioral differences of DZ twins. It is noteworthy that activity is the dimension that consistently displays significant negative DZ intraclass correlations in our MacArthur Study sample because previous research has suggested that activity is the temperament dimension most prone to contrast effects (Buss & Plomin, 1984).

GENETIC AND ENVIRONMENTAL CONTRIBUTIONS TO AGE-TO-AGE STABILITY

Could parental contrast effects be contributing to the high age-to-age stability on the CCTI in our sample? As indicated in chapter 7, for most observational measures of temperament, stability was almost entirely mediated by genetic factors. Observed shyness in the home was an exception in that both genetic and shared environmental factors contributed to stability. Longitudinal behavior genetic analyses of the CCTI dimensions present a different pattern of results (see chapter 7 for a description of the longitudinal model). Figure 6.1 shows the genetic and environmental contributions to the phenotypic age-to-age correlations. For all dimensions, phenotypic stability is mediated by significant genetic and nonshared environmental influences. Nonshared environmental influences are those environmental factors that are unique to each individual that make family members different from each other. Significant

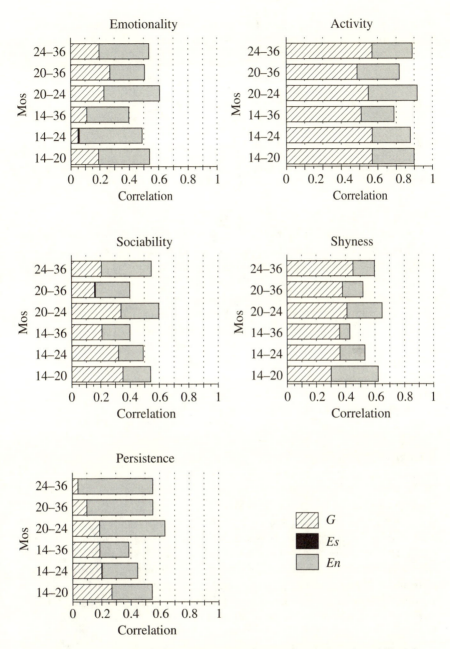

Figure 6.1. Genetic and environmental contributions to the phenotypic stability of parent-rated temperament. Each bar represents the age-to-age stability correlation that is decomposed into its genetic component (*G*), shared environmental component (*Es*), and nonshared environmental component (*En*).

Table 6.4. Mother–father agreement for Colorado Childhood Temperament Inventory temperament ratings

Measure	14 Months	20 Months	24 Months	36 Months
Emotionality	.41	.38	.41	.45
Activity	.48	.52	.47	.33
Sociability	.25	.42	.35	.49
Shyness	.53	.51	.50	.55
Persistence	.35	.28	.41	.41

N = 512–516 individuals at 14 months; 400–404 at 20 months; 352–360 at 24 months; and 248–250 at 36 months. All correlations are significant at $p < .01$.

nonshared contributions to stability for parental ratings but not for observational measures of the same temperament dimensions suggest the presence of stable parental rating biases such as contrast effects (Plomin et al., 1993). For example, the tendency for parents to label one twin as active and the other as inactive might persist across age intervals from 14 to 36 months. Thus, contrast effects do appear to be contributing to the high phenotypic stability of parent-rated temperament; the stability of parental ratings reflects not only stable genetic influences on temperament but also stable parental perceptions about differences in their twins' behavior. Moreover, because contrast effects result in increased differences between MZ and DZ twins, thereby leading to overestimates of genetic influence, their presence throws some doubt on the finding of apparent stable genetic influences.

MOTHER–FATHER AGREEMENT

So far, we have used mid-parent ratings of temperament on the CCTI, however, mothers and fathers may have different opportunities to observe their children, and this may result in differences in their reports of temperament. Indeed, as indicated by the correlations presented in table 6.4, mother–father agreement for CCTI temperament dimensions is moderate. Although previous research has found variable interparent agreement across dimensions within a single instrument (Slabach et al., 1991), in our sample, mother–father correlations were not dramatically different across dimensions. In addition, mother–father agreement was generally similar at all ages.

ARE CONTRAST EFFECTS SHARED ACROSS PARENTS?

The moderate correlations in table 6.4 make it reasonable to ask whether contrast effects are unique to a particular rater or shared across parents. There is some evidence suggesting that mothers might be more discriminating raters than fathers (Martin & Halverson, 1991). If this is the case, it raises the question of whether mothers are more prone to contrast effects than fathers.

Previous research suggests that contrast effects are shared across parents. In one study (Plomin, 1974, cited in Buss & Plomin, 1984), twin correlations for mid-parent ratings of temperament were compared with cross-rating twin correlations (i.e., mother's rating of twin A was correlated with father's rating of twin B). Both types of correlations yielded the same pattern of results: negative DZ twin correlations and MZ–DZ differences that were larger than would be predicted from a genetic hypothesis. These results suggest that the contrast bias is shared by both parents.

Data from the MacArthur Longitudinal Twin Study also suggest that contrast effects might be shared across parents. Table 6.5 presents twin intraclass correlations for maternal and paternal ratings of temperament on the CCTI and maternal–paternal cross-rating twin correlations. A pattern of moderate MZ correlations and near zero or negative DZ correlations was apparent for both

Table 6.5. Twin intraclass correlations and cross-correlations for maternal and paternal ratings of temperament on the Colorado Childhood Temperament Inventory

Measure	14 Months		20 Months		24 Months		36 Months	
	MZ	DZ	MZ	DZ	MZ	DZ	MZ	DZ
Mother								
Emotionality	.47*	.02	.52*	.08	.34*	.04	.20	−.22
Activity	.36*	−.19*	.33*	−.21*	.49*	−.14	.44*	−.11
Sociability	.44*	.13	.33*	.07	.26*	−.17	.52*	−.19
Shyness	.45*	−.23*	.32*	−.03	.35*	−.14	.44*	−.10
Persistence	.36*	.05	.32*	−.03	.16	−.03	.16	−.30*
Father								
Emotionality	.57*	.07	.40*	−.14	.39*	−.15	.38*	−.03
Activity	.51*	−.22*	.49*	−.26*	.45*	−.09	.66*	−.10
Sociability	.45*	−.03	.56*	−.16	.51*	−.19	.42*	−.20
Shyness	.32*	−.16	.42*	−.39*	.52*	−.13	.53*	−.25*
Persistence	.42*	.06	.60*	−.06	.31*	−.08	.46*	−.09
Mother–father cross–correlations								
Emotionality	.14	−.14	.28*	−.14	.07	−.07	.07	−.14
Activity	.15	−.23*	.24*	−.32*	.26*	−.32*	.22	−.18
Sociability	.10	−.11	.18	−.12	.01	−.11	.18	−.25*
Shyness	.29*	−.21*	.16	−.23*	.23*	−.02	.32*	−.13
Persistence	.15	.01	.16	−.16	.07	−.11	.20	−.22*

N = 139–140 MZ, 117–119 DZ pairs at 14 months; 109–110 MZ, 90–93 DZ pairs at 20 months; 92–95 MZ, 83–85 DZ pairs at 24 months; and 58 MZ, 66–67 DZ pairs at 36 months.
*p < .05.

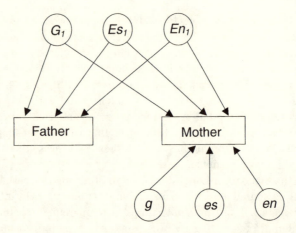

Figure 6.2. Bivariate Cholesky model of genetic and environmental associations between paternal and maternal ratings of temperament. G_1, Es_1, and En_1 represent the genetic, shared environmental, and nonshared environmental factors, respectively, that are common to both paternal and maternal ratings; g, es, and en are the genetic, shared environmental, and nonshared environmental factors that are unique to maternal ratings of temperament.

maternal and paternal ratings, therefore, contrast effects appear to be operating for both parents. Interestingly, mothers do not appear to be more prone to contrast effects than fathers. If anything, it is the other way around: the average difference between MZ and DZ correlations was .44 for mothers and .60 for fathers. Twin cross-rating correlations address the question of shared contrast effects. As expected given the moderate correlation between maternal and paternal ratings, the twin cross-rating correlations are lower than the intraclass correlations. Nonetheless, the pattern is similar. MZ cross-rating correlations, although moderately low, were positive, whereas DZ correlations were generally negative. Moreover, approximately one-half of the negative DZ correlations were significant. The finding of "too-low" DZ cross-rating correlations suggests that contrast effects are shared across parents.

To more directly test the question of whether contrast effects are shared across parents, bivariate model-fitting procedures were used to explore genetic and environmental sources of covariance between maternal and paternal ratings separately at each age. The model, a bivariate Cholesky model, decomposes the phenotypic variance of paternal and maternal ratings for a particular temperament dimension into genetic and environmental variance that is common to both parents and that is unique to maternal ratings. The model is illustrated as a path diagram in figure 6.2. The phenotypic variances of father-rated temperament and mother-rated temperament are represented by rectangles. The circles represent latent genetic and environmental factors. G_1, Es_1, and En_1 are the common genetic (additive) factor, the common shared environmental factor, and the common nonshared environmental factor, respectively; g, es, and en are the genetic, shared environmental, and nonshared environmental residual components of variance that are unique to maternal ratings once the common variance is removed.

Using this model, the proportion of variance due to genetic effects (h^2), shared environment (c^2) and nonshared environment (e^2), can be estimated. In addition, and more central to the question of shared parental contrast bias, the phenotypic correlation between paternal and maternal ratings of temperament can be decomposed into genetic, shared environmental, and nonshared environmental components. If contrast effects are shared across parents, we would expect to find significant nonshared environmental mediation of the correlation between maternal and paternal ratings of temperament.

Table 6.6 presents the variance components for maternal and paternal ratings of temperament on the CCTI. In general, both maternal and paternal ratings suggest that individual differences in temperament during infancy and early childhood are genetically influenced. There were, however, some differences between mothers and fathers. With the exception of shyness at 20 months, paternal ratings of temperament displayed significant genetic influences at each age. Maternal ratings were more variable; nevertheless, a pattern of genetic influences on temperament at 14, 20, and 24 months of age did emerge.

Genetic and environmental contributions to the phenotypic correlation between paternal and maternal ratings are also presented in table 6.6. Overall, genetic factors do not substantially contribute to the phenotypic correlation between paternal and maternal ratings of temperament. In fact, only shyness and persistence demonstrated significant genetic contributions at more than one age. However, even for these two dimensions, no more than 50% of the phenotypic correlation between paternal and maternal ratings could be explained by genetic influences. Perhaps more interesting is the lack of significant genetic mediation of the correlation between paternal and maternal ratings for temperament dimensions in which both ratings of temperament displayed significant genetic influence (i.e., emotionality at 14 and 24 months; activity at 14, 24, and 36 months; sociability at 14 and 24 months). This suggests that for these dimensions, the genetic factors that influence maternal ratings are not the same as those that influence paternal ratings. It is likely that mothers and fathers are rating different aspects of the same temperament dimensions.

Shared environmental influences cannot contribute to mother–father agreement because, as indicated in table 6.6, neither maternal nor paternal ratings of temperament displayed significant shared environmental variance. What mediates the correlation between maternal and paternal ratings of temperament? At all ages, there are significant nonshared environmental contributions to the phenotypic correlation between paternal and maternal ratings of each temperament dimension. Mother–father agreement on the CCTI is chiefly due to those factors that are unique to each child and that make cotwins different from each other. Given the pattern of negative DZ correlations and cross-

Table 6.6. Variance components (h^2, c^2, e^2) for maternal and paternal ratings of temperament on the Colorado Childhood Temperament Inventory and genetic and environmental contributions to the correlation between maternal and paternal CCTI ratings (G, Es, En)

Measure/Age	Maternal Ratings			Paternal ratings			Mother–father agreement		
	h^2	c^2	e^2	h^2	c^2	e^2	G	Es	En
Emotionality									
14 Months	.37*	.04	.59*	.28*	.07	.65*	.15	−.05	.30*
20 Months	.47*	.07	.47*	.31*	.03	.65*	.33*	−.05	.11*
24 Months	.31*	.00	.69*	.27*	.00	.73*	.06	.00	.34*
36 Months	.11	.00	.89*	.28*	.00	.72*	.01	.00	.45*
Activity									
14 Months	.22*	.00	.78*	.35*	.00	.65*	.09	.00	.40*
20 Months	.14	.02	.84*	.28*	.01	.71*	.07	−.01	.45*
24 Months	.34*	.03	.63*	.30*	.04	.66*	.10	−.03	.38*
36 Months	.35*	.00	.65*	.56*	.00	.44*	.14	.00	.20*
Sociability									
14 Months	.42*	.00	.58*	.41*	.00	.59*	.05	.00	.20*
20 Months	.24*	.09	.66*	.48*	.00	.52*	.23*	−.02	.22*
24 Months	.36*	.00	.64*	.38*	.00	.62*	.08	.00	.28*
36 Months	.04	.00	.96*	.24*	.00	.76*	−.01	.00	.51*
Shyness									
14 Months	.34*	.00	.66*	.25*	.00	.75*	.22*	.00	.32*
20 Months	.16*	.08	.76*	.19	.00	.81*	.09	.00	.43*
24 Months	.25*	.00	.75*	.45*	.00	.55*	.19*	.00	.30*
36 Months	.33*	.00	.67*	.37*	.00	.63*	.21*	.00	.33*
Persistence									
14 Months	.33*	.00	.67*	.41*	.00	.59*	.15*	.00	.19*
20 Months	.30*	.00	.70*	.59*	.00	.41*	.15*	.00	.14*
24 Months	.12	.00	.88*	.24*	.00	.76*	.00	.00	.40*
36 Months	.00	.00	1.00*	.36*	.00	.64*	.05	.00	.37*

*p < .05.

correlations, contrast effects are a likely nonshared environmental candidate mediating the correlation between ratings. In other words, the present results are consistent with the notion that the tendency to contrast one twin with the other and magnify existing behavioral differences is shared across parents.

DISCUSSION AND SUMMARY

Parental rating scales have been the major method of assessing infant and child temperament because of their convenience, extensive behavioral sampling, ecological validity, and sound psychometric properties. Recently, however, the usefulness of this temperament measure has been called into question as more and more research suggests that parental ratings include a subjective as well as an objective component (Seifer et al. 1994). That is, parental ratings may reflect parental characteristics and expectations in addition to actual child behavior. The present chapter examined parental expectation biases within the context of the twin design.

When temperament was assessed via parental ratings, MZ cotwin resemblance was moderate, but DZ twins were often perceived as having opposing temperaments. More important, this pattern of results did not emerge when temperament was assessed through observational methods. For these observed measures, DZ cotwins were approximately half as similar as MZ cotwins. Taken together, these results suggest that parental ratings are prone to contrast effects that magnify existing behavioral differences of twins. As indicated earlier, contrast effects arise when one twin is rated in the context of the other (i.e., the parent has knowledge of both twins and compares and contrasts them). Contrast effects would not be expected for observational ratings where an observer is familiar only with the one twin whose behavior is being rated.

Presumably, the greater the actual behavioral difference between twins, the greater the tendency for contrast effects to occur. Thus, for genetically influenced traits, such as temperament, MZ twins, who are more behaviorally alike, would be less prone to rater contrasts (Buss & Plomin, 1984). Because contrast effects operate more strongly for DZ twins, their correlations will be too low as compared to MZ correlations, and twin studies will overestimate the magnitude of genetic influence. The potential for contrast biases occurs whenever parental ratings of twins are used, and therefore such measures, by themselves, may be inadequate for assessing genetic hypotheses.

Little is known about how contrast effects operate. Our analysis of the MacArthur Longitudinal Twin Study data suggests that parental contrast effects persist across age and contribute significantly to the phenotypic stability of parent-rated temperament. Thus, as has been proposed by other researchers (e.g., Saudino & Eaton, 1995; Seifer & Sameroff, 1986), the stability of temperament reflects the stability of parents perceptions and expectations in addition to the actual behaviors of the child. It is particularly interesting that this bias is not limited to one parent. In fact, the agreement between maternal and paternal ratings is due, to a great extent, to shared contrast effects. Perhaps labeling twins as "*X*" and "not *X*" based on subtle behavioral differences provides parents with a heuristic for understanding and interacting with their different children. Similarly, contrast effects may reflect that parent's value and seek to promote the development of individuality for each of their children.

Does the finding of contrast effects mean that parental ratings have no utility? Absolutely not. Parents are a rich source of information about their children's typical behaviors across many situations—information that may be lost when relying solely on observational or mechanical measures of temperament. For example, observational measures are typically conducted by strangers during a brief testing session and therefore may not provide a full picture of the child's behavior. Similarly, mechanical measures, such as motion recorders, fail to provide information about the nature and quality of temperamental behaviors. Although biases might be operating, we believe that, at their core, parental ratings provide some accurate information about their child's temperament. Clearly, discarding parental ratings because of possible contrast effects is akin to throwing the baby out with the bath water. We suggest, however, that our finding of contrast effects highlights the need to know more about the factors that affect the parent-rating process. Carey (1986) suggests that sibling constellation variables such as sex, age, birth order, or spacing might affect the contrast process. Only by studying more than one child per family will we be able to learn more about these issues.

REFERENCES

Bates, J. E., & Bayles, K. (1984). Objective and subjective components in mothers' perceptions of their children from age 6 months to 3 years. *Merrill-Palmer Quarterly, 30,* 111–130.

Bates, J. E., Freeland, C. A., & Lounsbury, M. L. (1979). Measure of infant difficultness. *Child Development, 50,* 794–803.

Bayley, N. (1969). *Manual for the Bayley Scales of Infant Development.* New York: Psychological Corporation.

Buss, A. H., & Plomin, R. (1984). *Temperament: Early developing personality traits.* Hillsdale, NJ: Erlbaum.

Buss, A. H., Plomin, R., & Willerman, L. (1973). The inheritance of temperaments. *Journal of Personality, 41,* 513–524.

Carey, G. (1986). Sibling imitation and contrast effects. *Behavior Genetics, 16,* 319–341.

Daniels, D., Plomin, R., & Greenhalgh, J. (1984). Correlates of difficult temperament in infancy. *Child Development, 55,* 1184–1194.

Emde, R. N., Plomin, R., Robinson, J., Reznick, J. S., Campos, J., Corley, R., DeFries, J. C., Fulker, D. W., Kagan, J., & Zahn-Waxler, C. (1992). Temperament, emotion, and cognition at 14 months: The MacArthur Longitudinal Twin Study. *Child Development, 63,* 1437–1435.

Epstein, S. (1983). Aggregation and beyond: Some basic issues on the prediction of behavior. *Journal of Personality, 51,* 360–392.

Goldsmith, H. H., Rieser-Danner, L. A., & Briggs, S. (1991). Evaluating convergent and discriminant validity of temperament questionnaires for preschoolers, toddlers, and infants. *Developmental Psychology, 27,* 566–579.

Hubert, N. C., Wachs, T. D., Peters-Martin, P., & Gandour, M. J. (1982). The study of early temperament: Measurement and conceptual issues. *Child Development, 53,* 571–600.

McDevitt, S. C. (1986). Continuity and discontinuity of temperament in infancy and early childhood: A psychometric perspective. In R. Plomin & J. Dunn (Eds.), *The study of temperament: Changes, continuities and challenges.* (pp. 27–38). Hillsdale, NJ: Erlbaum.

Martin & Halverson (1991). Mother-father agreement in temperament ratings. In J. Strelau & A. Angleitner (Eds.), *Explorations in temperament. International perspectives on theory and measurement* (pp. 235–248). New York: Plenum.

Mebert, C. J. (1989). Stability and change in parents' perceptions of infant temperament: Early pregnancy to 13.5 months postpartum. *Infant Behavior and Development, 12,* 237–244.

Mebert, C. J. (1991). Dimensions of subjectivity in parents' ratings of infant temperament. *Child Development, 62,* 352–361.

Neale, M. C., & Stevenson, J. (1989). Rater bias in the EASI temperament scales: A twin study. *Journal of Personality and Social Psychology, 56,* 446–455.

Plomin, R., & DeFries, J. C. (1985). *Origins of individual differences in infancy: The Colorado Adoption Project.* Toronto: Academic Press.

Plomin, R., DeFries, J. C., & Fulker, D. W. (1988). *Nature and nurture during infancy and early childhood.* New York: Cambridge University Press.

Plomin, R., Emde, R., Braungart, J. M., Campos, J., Corley, R., Fulker, D. W., Kagan, J., Reznick, S., Robinson, J., Zahn-Waxler, C., & DeFries, J. C. (1993). Genetic change and continuity from 14 to 20 months: The MacArthur Longitudinal Twin Study. *Child Development, 64,* 1354–1376.

Plomin, R., & Rowe, D. C. (1977). A twin study of temperament in young children. *Journal of Psychology, 97,* 107–113.

Rothbart, M. K. (1981). Measurement of temperament in infancy. *Child Development, 52,* 569–578.

Rothbart, M. K. (1986). Longitudinal observation of infant temperament. *Developmental Psychology, 22,* 356–365.

Rowe, D. C. & Plomin, R. (1977). Temperament in early childhood. *Journal of Personality Assessment, 41,* 150–156.

Saudino, K. J., & Eaton, W. O. (1991). Infant temperament and genetics: An objective twin study of motor activity level. *Child Development, 62,* 1167–1174.

Saudino, K. J., & Eaton, W. O. (1995). Continuity and change in objectively assessed temperament: A longitudinal twin study of activity level. *British Journal of Developmental Psychology, 13,* 81–95.

Seifer, R., & Sameroff, A. J. (1986). The concept, measurement, and interpretation of temperament in young children: A survey of research issues. *Advances in Developmental and Behavioral Pediatrics, 7,* 1–43.

Seifer, R., Sameroff, A. J., Barrett, L. C., & Krafchuck, E. (1994). Infant temperament measured by multiple observations and mother reports. *Child Development, 65,* 1487–1490.

Slabach, E. H., Morrow, J., & Wachs, T. D. (1991). Questionnaire measurement of infant and child temperament. In J. Strelau & A. Angleitner (Eds.), *Explorations in temperament. International perspectives on theory and measurement* (pp. 205–234). New York: Plenum.

Stevenson, J., & Fielding, J. (1985). Ratings of temperament in families of young twins. *British Journal of Developmental Psychology, 3,* 143–152.

Torgersen, A. M. (1981). Genetic factors in temperamental individuality. A longitudinal study of same-sexed twins from two months to six years of age. *Journal of the American Academy of Child Psychiatry, 20,* 702–711.

Torgersen, A. M. (1985). Temperamental differences in infants and 6-year-old children: A follow-up study of twins. In J. Strelau, F. H. Farley, & A. Gale (Eds.), *The biological basis of personality and behavior: Theories, measurement techniques, and development* (Vol. 1, pp. 227–239). Washington: DC: Hemisphere.

Torgersen, A. M., & Kringlen, E. (1978). Genetic aspects of temperamental differences in infants. A study of same-sexed twins. *Journal of the American Academy of Child Psychiatry, 17,* 433–444.

Vaughn, B. E., Bradley, C. F., Joffe, L. S., Seifer, F., & Barglow, P. (1987). Maternal characteristics measured prenatally predict ratings of temperamental difficulty on the Cary Infant Temperament Questionnaire. *Developmental Psychology, 23,* 152–161.

Wilson, R. S., & Matheny, A. P., Jr. (1986). Behavior genetics research in infant temperament: The Louisville twin study. In R. Plomin & J. Dunn (Eds.), *The study of temperament: Changes, continuities and challenges* (pp. 81–97). Hillsdale, NJ: Erlbaum.

Wolk, S., Zeanah, C. H., Garcia Coll, C. T., & Carr, S. (1992). Factors affecting parents' perceptions of temperament in early infancy. *American Journal of Orthopsychiatry, 62,* 71–82.

Worobey, J. (1986). Convergence among assessments of temperament in the first month. *Child Development, 57,* 47–55.

Zeanah, C. H., Keener, M. A., Stewart, L., & Anders, T. F. (1985). Prenatal perception of infant personality: A preliminary investigation. *Journal of the American Academy of Child Psychiatry, 24,* 204–210.

7

Sources of Continuity and Change in Observed Temperament

Kimberly J. Saudino
Stacey S. Cherny

Longitudinal studies of infant temperament consistently yield moderate cross-age correlations (see reviews by Hubert et al., 1982; Salbach et al., 1991). This finding has two important implications for developmentalists. First, there is some consistency across age for those behaviors that we define as temperamental. Second, from an individual differences perspective, there is substantial change in temperament. That is, although age-to-age correlations have typically been interpreted in terms of stability, correlations that are less than the reliable variance of the measure reflect genuine developmental change (Clarke & Clarke, 1984). The present chapter uses developmental behavioral genetic strategies to explore genetic and environmental sources of temperamental continuity and change across the transition from infancy to early childhood.

Although there is abundant evidence indicating that individual differences in temperament are genetically influenced (see Goldsmith, 1983; Plomin, 1987 for reviews), few studies have explored developmental changes in the etiology of individual differences in temperament or the etiology of developmental changes in these individual differences. Genes are dynamic in nature, changing in the quantity and quality of their effects as the organism changes developmentally (Plomin, 1986). Thus, genetic factors can be a source of change as well as a source of continuity in behavioral development.

The issue of genetic effects and developmental change can be approached in two ways. The first method explores changes in the relative contribution of genetic and environmental influences to variance at each age. That is, does the proportion of individual differences that can be attributed to genetic influences vary across age? Current research suggests that the heritability of temperamental dimensions tends to remain constant, or possibly increases with age (e.g., Cyphers et al., 1990; Matheny, 1980; McCartney et al., 1990; Plomin et al., 1993; Stevenson & Fielding, 1985).

The second approach explores genetic influences on continuity and change. This issue can be addressed by assessing genetic contributions to phenotypic continuity across age. Genetic influences on phenotypic continuity imply that there is some overlap between the genetic factors that affect a trait across age (Plomin, 1986). Age-to-age genetic change is indicated by the extent to which genetic factors that affect a trait at one age differ from genetic factors that affect the same trait at another age independent of changes in heritability across age. This approach is particularly interesting to developmentalists because it addresses the etiological processes by which developmental change takes place (Plomin et al., 1993), but it requires longitudinal research, and such behavioral genetic research is relatively rare (Goldsmith, 1984). Examination of age-to-age change profiles in the Louisville Twin Study suggests that during infancy, genetic influences may contribute to change as well as to continuity in temperament (Matheny, 1983, 1989; Wilson & Matheny, 1986); however, the analysis of change profiles does not reveal the extent to which genetic effects on a trait at one age are similar to genetic effects on the same trait at another age (Plomin & Nesselroade, 1990).

In the present chapter, we examine changes in heritability across age and explore genetic and environmental sources of continuity and change in observed temperament during early childhood in the MacArthur Longitudinal Twin Study (Plomin et al., 1990). Although parental rating scales are the most commonly used measures of infant temperament, in twin studies, parental ratings of temperament frequently yield odd patterns of results such as negative correlations between fraternal twins, which suggest the possible presence of rater biases (Plomin et al., 1991). Therefore, we focused on observed measures of temperament. Although behavioral observations typically involve a limited sample of the child's behavior, they offer a number of distinct advantages compared to parental ratings. First, because testers are not personally involved with the child that they are rating, they may be more objective in their ratings. Second, testers have experience with a range of children and therefore have a broader basis from which to make comparisons for rating child behaviors. Third, observing children in the same standard situation facilitates comparisons among children. Finally, assessing the infant's reactions to mildly stressful situations may enrich observations of behaviors related to temperament (Schmitz et al., 1996).

OBSERVATIONAL MEASURES OF TEMPERAMENT

The Infant Behavior Record

At 14, 20, and 24 months of age, examiners used the Infant Behavior Record (IBR) to rate each child's behavior during administration of the Bayley Scales of Infant Development (Bayley, 1969). Items were aggregated on three scales, activity, task orientation, and affect-extraversion, as suggested by Matheny (1980). The activity factor includes the infant's general level of body motion and degree of energy exhibited during the test situation. Task orientation includes attention span, persistence, goal directedness, and responsiveness to test materials. The third factor, affect-extraversion, relates to emotionality and sociability; items loading on this factor include social responsiveness, emotional tone, and cooperativeness. At 36 months of age, twins were rated on a modified version of the IBR following administration of the Stanford-Binet Intelligence Scale (Terman & Merrill, 1973). Although factor analysis of the modified IBR items yielded no clear affect-extraversion factor, factors related to activity and task orientation did emerge. It should be noted, however, that the factor structures of activity and task orientation at 36 months were not identical to their corresponding factors at earlier ages.

Shyness

An objective measure of shyness at 14, 20, 24, and 36 months of age was obtained based on the children's initial reactions to the examiners' arrival in the home (laboratory assessments of shyness were also collected at 14 and 20 months and are discussed in chapter 20). We videotaped and scored each child's behavior for the occurrence of discrete behaviors during each minute of the 5-min shyness episode. Behaviors included approach to the examiner, approach to a proffered toy, proximity to mother, clinging to mother, self-soothing, vocalization, and crying. In addition, global ratings of shyness and hesitation were completed by raters. We then used an unrotated principal component score as a composite measure of shyness at each age (see Plomin et al., 1990 for details).

Behavioral Inhibition

Inhibited and uninhibited behavior to the unfamiliar was assessed by behavioral reactions to unfamiliar events in a playroom at 14, 20, 24, and 36 months of age (see Robinson et al., 1992 for a description of the procedures). The behavioral inhibition paradigm is based on previous research by Kagan and colleagues (e.g., Garcia-Coll et al., 1984; Kagan et al., 1984, 1988). At 14 and 20 months of age, we derived a behavioral inhibition aggregate by averaging Z-score values for seven behaviors observed from videotape: the latency to leave the mother upon entering the playroom; the latency to approach toys; the latency to approach a stranger; the latency to approach an unfamiliar object; and the time spent proximal to the mother during free play, stranger, and unfamiliar object episodes. At 24 months, the behavioral inhibition paradigm did not include the stranger episodes, and thus these two behavioral measures were not included in the aggregate index at this age. At 36 months, the be-

Table 7.1. Phenotypic stability correlations for observed temperament

Measure	14–20 Months	14–24 Months	14–36 Months	20–24 Months	20–36 Months	24–36 Months
Behavioral inhibition	.28	.10	.00	.23	.25	.13
Shyness in home	.24	.26	.24	.38	.36	.39
Activity	.21	.27	.14	.25	.24	.25
Task orientation	.18	.22	.09	.28	.23	.22
Affect-extraversion	.24	.34	—	.27	—	—

All nonzero correlations are significant at p < .05. All longitudinal analyses are based on a subsample of children for whom there are complete longitudinal data. N = 416 for behavioral inhibition; 530 for shyness; 448 for activity and task orientation; and 626 for affect-extraversion.

havioral inhibition aggregate includes the number of examiner-presented toys that the child touches and the time spent proximal to mother.

PHENOTYPIC CONTINUITY AND CHANGE

Age-to-age correlations are presented in table 7.1. With the exception of the 14–36-month correlation for behavioral inhibition, all correlations were significant. The pattern of low to moderate correlations across each time interval indicates some continuity in the face of substantial change in the rank order of individuals from infancy to early childhood.

TWIN CORRELATIONS

Intraclass Correlations

Genetic influences are implied when cotwin similarity covaries with the degree of genetic relatedness. Thus, if heredity affects a trait, the twofold greater genetic similarity of monozygotic (MZ) twins is expected to make them more similar than dizygotic (DZ) twins. Intraclass correlations serve as indices of cotwin similarity. Genetic influences are implied when an MZ intraclass correlation for a trait is greater than the DZ correlation.

Figure 7.1 presents the intraclass correlations at each age for each observed temperament dimension. As seen by the difference between identical (MZ) and fraternal (DZ) twin correlations, behavioral inhibition, activity, and affect-extraversion demonstrate substantial heritabilities across infancy and early childhood. For these dimensions, there are no consistent developmental trends (i.e., no evidence of increasing or decreasing heritability). Intraclass correlations for shyness and task orientation are more variable across age. At 14 and 36 months of age, there is evidence of genetic influences on shyness; however,

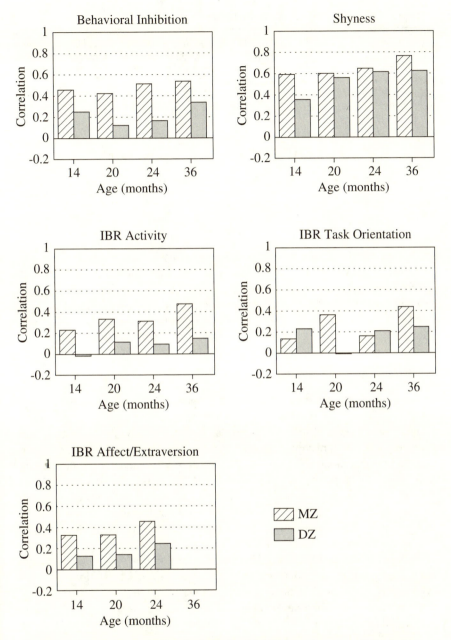

Figure 7.1. Intraclass correlations for behavioral inhibition, shyness, activity, task orientation, and affect-extraversion.

Table 7.2. Twin cross-age intraclass correlations

Measure	14–20 Months		14–24 Months		14–36 Months		20–24 Months		20–36 Months		24–36 Months	
	MZ	DZ	MZ	DZ	MZ	DZ	MZ	DZ	MZ	DZ	MZ	DZ
Behavioral inhibition	.18*	.15	.07	.08	−.05	.07	.26*	.02	.11	.12	.16	.11
Shyness in home	.31*	.07	.26*	.21*	.22*	.15	.46*	.28*	.39*	.25*	.39*	.29*
Activity	.29*	.17	.28*	−.06	.08	.05	.27*	.11	.18*	.05	.21*	.13
Task orientation	.24*	.10	.12	.16	.13	.02	.37*	.16	.20*	.10	.22*	.05
Affect-extraversion	.37*	.25*	.20*	.09	—	—	.43*	.29*	—	—	—	—

$N = 111$ pairs MZ, 97 pairs DZ for behavioral inhibition. $N = 145$ pairs MZ, 120 pairs DZ for shyness. $N = 118$ pairs MZ, 106 pairs DZ for activity and task orientation. $N = 167$ pairs MZ, 146 pairs DZ for affect-extraversion.
*p < .05.

the substantial intraclass correlations for both MZ and DZ twins at 20 and 24 months suggest that shared environmental factors are contributing to individual differences in shyness at these ages. The lack of genetic influence on task orientation at 14 and 24 months is somewhat surprising because previous twin and adoption studies have found significant genetic influence on this dimension at 12 and 24 months of age (Braungart et al., 1992; Matheny, 1980).

Cross-age Intraclass Correlations

Cross-age intraclass correlations are the essence of a genetic analysis of continuity. This statistic is actually a cross-twin, cross-age intraclass correlation where a twin's score at one age is correlated with its cotwin's score at another age using a double-entry procedure. A genetic contribution to phenotypic continuity is suggested to the extent that cross-age intraclass correlations are greater for MZ twins than for DZ twins. Table 7.2 presents cross-age intraclass correlations for each pair of ages. Generally, MZ correlations are greater than DZ correlations, suggesting that genetic factors contribute to age-to-age stability in temperament across infancy and early childhood. An exception to this is behavioral inhibition, where for four of the six intervals, neither MZ nor DZ correlations were significantly different from zero. In this case, genetic change is implied. That is, as indicated by the simple intraclass correlations in figure 7.1, behavioral inhibition is genetically influenced at all ages, but the cross-age intraclass correlations indicate that there is little genetic continuity. Thus, genetic factors contribute to change, not to continuity.

LONGITUDINAL MODELS

Model-fitting procedures provide an elegant analysis of continuity and change because they simultaneously analyze genetic contributions to covariance be-

tween ages and take into account the heritabilities at all ages and the phenotypic stabilities between ages (Boomsma et al., 1989; Hewitt et al., 1988; Loehlin et al., 1989; Phillips & Fulker, 1989). The longitudinal Cholesky model used
in this chapter decomposes the phenotypic variance of a trait at 14, 20, 24,
and 36 months into genetic (additive) and environmental variance that is common to all four ages, common to 20, 24, and 36 months only, common to 24,
and 36 months, and unique to 36 months (Cherny et al., 1994).

The longitudinal model is illustrated as a path diagram in figure 7.2. For
clarity of presentation, genetic, shared environment, and nonshared environment parameters are presented separately. The phenotypic variances at each
age are represented by rectangles. The circles represent latent genetic and environmental variables. G_1, Es_1, and En_1 are the common genetic factor, the
common shared environmental factor, and the common nonshared environmental factor, respectively, that is shared at all four ages. G_2, Es_2, and En_2 are
the respective common factors that are shared at 20, 24, and 36 months of age
independent of what is shared at all four ages; G_3, Es_3, and En_3 are the common
factors that are shared at 24 and 36 months of age; independent of the first
two factors; and g_4, es_4, and en_4 are factors that are unique to 36 months of
age. Genetic continuity is suggested when two or more path coefficients within
a common G factor are significant. For example, genetic continuity across all
ages is implied when all the paths in the G_1 factor are significant. However, if
only the h_{12} and h_{13} paths were significant, it would suggest genetic continuity
from 14 to 20 months, but not from 14 to 24 or 36 months. Under this model
genetic contributions to change are conceptualized as new genetic effects at
one age that are independent of genetic effects at an earlier age.

The model fitting results for the full longitudinal model are presented in
figures 7.3–7.7. Asterisks indicate significant paths (i.e., parameters that were
retained in the best-fitting reduced model). The model-fitting results for behavioral inhibition (figure 7.3) are consistent with our interpretation of the
cross-age intraclass correlations. There is genetic continuity from 14 to 20
months and from 20 to 24 and 36 months, but not from 14 to 24 or 36 months
of age. In addition to genetic continuity, there is also evidence of genetic
change as indicated by significant new genetic effects emerging at 20, 24, and
36 months of age. Genetic change after 20 months must, however, be interpreted cautiously because the measures of behavioral inhibition differ at 24
and 36 months of age. That is, the new genetic variance at 24 and 36 months
may be a function of measurement differences and not developmental change.
Nonshared environmental effects contribute to change but not to continuity,
whereas shared environmental factors do not appear to contribute significantly
to either continuity or change in behavioral inhibition.

Also in agreement with our correlational analyses, genetic factors contribute
to continuity in shyness. As can be seen in figure 7.4, all paths from the G_1
factor are significant, indicating that the same genetic factors are influencing
individual differences in shyness at all ages. No significant new genetic effects
emerge after 14 months. In contrast to behavioral Inhibition, shared environ-

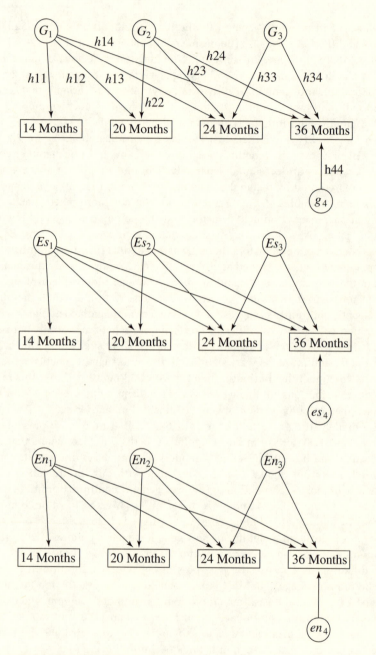

Figure 7.2. Longitudinal path model of genetic and environmental sources of change and continuity.

Behavioral Inhibition

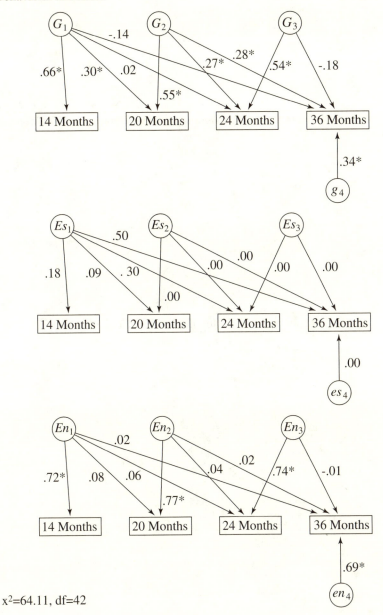

x²=64.11, df=42

Figure 7.3. Longitudinal model fitting results for behavioral inhibition.

Shyness

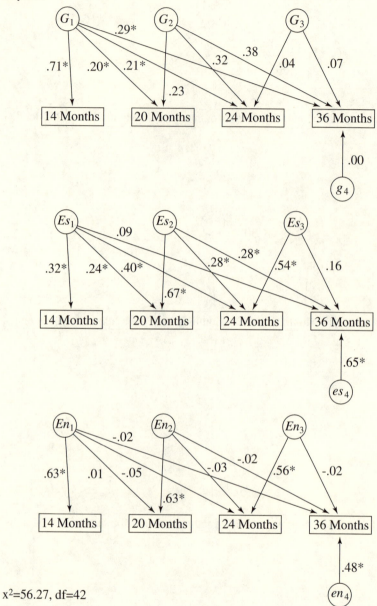

x²=56.27, df=42

Figure 7.4. Longitudinal model fitting results for shyness.

mental influences also contribute to the phenotypic continuity between 14, 20, and 24 months, and between 20, 24, and 36 months. In addition, new shared environmental effects emerge at each age, indicating that shared environmental factors also contribute to phenotypic change in shyness. Once again, non-shared environmental factors contribute to change and not continuity.

The pattern of results for the IBR activity and task orientation, the two IBR measures for which we were able to derive 36-month temperament factors, were remarkably similar (figures 7.5 and 7.6). The significant parameter estimates for the paths from the common genetic factor, G_1, indicate that there are genetic influences on activity and task orientation that persist across age. That is, there is significant genetic continuity from 14 to 20, 24, and 36 months of age. There was also evidence of significant genetic change at 36 months of age. For these two dimensions, the G_2 factor is defined by a single, significant path to the 36-month phenotype; all other paths could be dropped from the model without a decrement in fit. Thus, for these two dimensions, no new significant genetic effects emerge until 36 months of age. It is noteworthy that genetic change in these dimensions coincides with changes in the measures. This once again raises the question of whether the observed genetic change reflects true developmental change or differences in the factor structures at 36 months. Consistent with previous findings with the IBR, shared environmental influences are nonsignificant at all ages (Braungart et al., 1992; Emde et al., 1992); however, there are substantial nonshared environmental influences unique to each age.

Affect-Extraversion (figure 7.7) also demonstrates significant genetic continuity across all periods of assessment. Although not apparent in the less powerful analysis of change profile correlations, there is evidence of genetic change for affect-extraversion. The significant G_2 paths indicate that new genetic effects emerge at 20 months and continue to influence individual differences in affect-extraversion at 24 months of age. As was the case with the other IBR temperaments, shared environmental influences are nonsignificant, and non-shared environmental influences contribute to change and not to continuity across age.

Variance Components

The path coefficients in figures 7.4–7.7 are standardized partial regressions that indicate the relative influence of the latent variables on the phenotype. The percent variance explained by a path is the square of the path coefficient. The proportion of variance at each age due to genetic influences is then the sum of all squared G path coefficients leading to the phenotype. For example, in figure 7.2, the heritability for behavioral inhibition at 14 months is estimated as the square of the genetic path coefficient at 14 months ($.66^2 = .44$). Similarly, heritability at 36 months is .25 ($-.14^2 + .28^2 + -.18^2 + .34^2$). The proportions of variance due to shared and nonshared environmental influences at each age

Activity

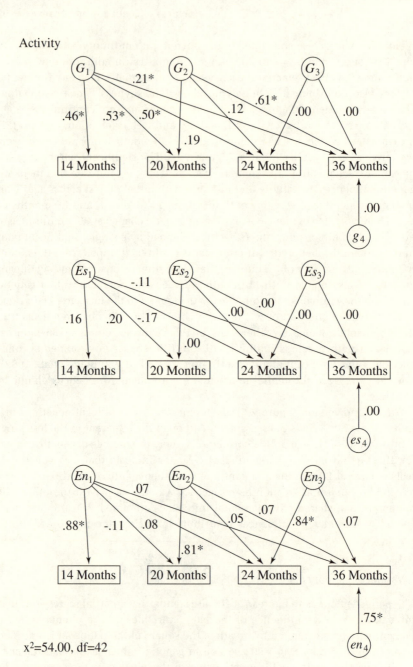

x²=54.00, df=42

Figure 7.5. Longitudinal model fitting results for activity.

Task Orientation

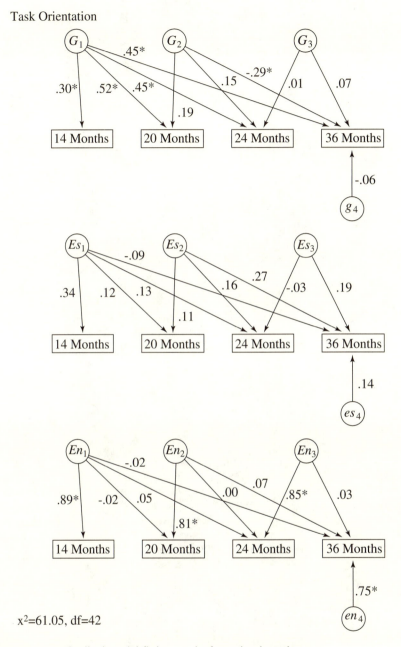

Figure 7.6. Longitudinal model fitting results for task orientation.

$x^2 = 61.05$, df=42

Affect/Extraversion

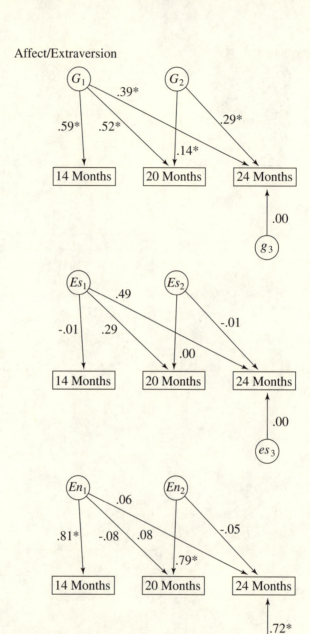

$x^2=27.92$, df=24

Figure 7.7. Longitudinal model fitting results for affect-extraversion.

Table 7.3. Variance components from full longitudinal model

Measure	14 Months			20 Months			24 Months			36 Months		
	h^2	c^2	e^2	h^2	c^2	e^2	h^2	c^2	e^2	h^2	c^2	e^2
Behavioral inhibition	.44*	.03	.53*	.39*	.01	.60*	.36*	.09	.55*	.25*	.25	.49*
Shyness	.50*	.10*	.40*	.09*	.51*	.40*	.15*	.53*	.32*	.24*	.53*	.23*
Activity	.21*	.03	.77*	.30*	.04	.67*	.26*	.03	.71*	.41*	.01	.57*
Task orientation	.09*	.12	.79*	.31*	.03	.67*	.23*	.04	.73*	.29*	.14	.57*
Affect-extraversion	.35*	.00	.56*	.29*	.08	.63*	.24*	.24	.52*	—	—	—

h^2 = heritability, c^2 = shared environmental variance, e^2 = nonshared environmental variance. *p < .05.

is the sum of all squared *Es* and *En* path coefficients, respectively, leading to the phenotype.

Variance components (i.e., heritability estimates and estimates of shared and nonshared environmental variance) for each dimension at 14, 20, 24, and 36 months of age are presented in table 7.3. Consistent with our interpretation of the twin intraclass correlations, behavioral inhibition, activity, and affect-extraversion demonstrate significant heritability at each age. Between one quarter and one-third of the total variance in these temperament dimensions can be attributed to genetic factors. Nonshared environment, which includes measurement error, accounts for most of the remaining variance. Also consistent with our interpretation of the intraclass correlations, heritability estimates for shyness are low at 20 and 24 months, and after 14 months of age, shared environmental factors have a significant influence on individual differences in shyness. Although not apparent from the intraclass correlations, task orientation is significantly heritable at all ages. This finding demonstrates the power of the model-fitting analysis, which analyzes MZ and DZ covariances between ages in addition to MZ and DZ variances at each age. When this information is taken into consideration, genetic influences on task orientation at 14 and 24 months of age are unveiled. Overall, for all variables there was no consistent pattern of increasing or decreasing heritability, and it would appear that for these dimensions heritability does not change dramatically during the transition from infancy to early childhood.

Genetic Contributions to Phenotypic Continuity

Path coefficients can also be used to estimate the extent to which the same genetic and environmental influences contribute to the variability in temperament across age—that is, genetic and environmental contributions to phenotypic continuity (table 7.4). The genetic mediation of a cross-age phenotypic correlation is obtained by multiplying the path coefficients for both time points within each common *G* factor and summing across the common *G* factors that are shared across time points. Using behavioral inhibition (figure 7.3) as an

Table 7.4. Genetic and environmental contributions to the phenotypic correlations (r) and genetic correlations between ages (r_G)

Measure/age	r	G	Es	En	r_G
Behavioral inhibition					
14–20 Months	.28*	.20*	.02	.06	.49*
14–24 Months	.10*	.01	.06	.05	.03
14–36 Months	.00	−.10	.09	.01	−.30
20–24 Months	.23*	.16*	.03	.03	.43*
20–36 Months	.25*	.11*	.04	.09	.35*
24–36 Months	.13*	−.02	.15	.00	−.07
Shyness					
14–20 Months	.24*	.14*	.08*	.01	.71*
14–24 Months	.26*	.15*	.13*	−.03	.55*
14–36 Months	.24*	.21*	.03	−.01	.61*
20–24 Months	.38*	.12*	.28*	−.02	1.00*
20–36 Months	.36*	.15*	.21*	−.01	1.00*
24–36 Months	.39*	.19*	.20*	−.01	1.00*
Activity					
14–20 Months	.21*	.24*	.03	−.10	.95*
14–24 Months	.27*	.23*	−.03	.07	.98*
14–36 Months	.14*	.09*	−.02	.06	.31*
20–24 Months	.25*	.28*	−.03	.03	1.00*
20–36 Months	.24*	.19*	−.02	.05	.54*
24–36 Months	.25*	.18*	.02	.06	.55*
Task orientation					
14–20 Months	.18*	.16*	.04	−.02	.96*
14–24 Months	.22*	.14*	.05	.05	.97*
14–36 Months	.09*	.13*	−.03	−.01	.80*
20–24 Months	.28*	.26*	.03	.00	.97*
20–36 Months	.23*	.18*	.02	.06	.52*
24–36 Months	.22*	.16*	.03	.03	.62*
Affect-extraversion					
14–20 Months	.24*	.31*	.00	−.06	.98*
14–24 Months	.34*	.23*	.00	.05	.80*
20–24 Months	.27*	.24*	.14	−.04	.91*

*$p < .05$.

example, the genetic mediation of the phenotypic correlation between 14 and 20 months is estimated by $.66 \times .30 = 20$. Because the phenotypic correlation between 20 and 24 months can be genetically related via the G_2 factor as well as the G_1 factor, the genetic mediation between the two ages is estimated by summing the two genetic contributions [i.e., $(.30 \times .02) + (.55 \times .27) = .15$]. Similarly, the phenotypic correlation between 24 and 36 months can be genetically related via, G_1, G_2, and G_3, thus the genetic contribution to the phenotypic correlation is $(.02 \times -.14) + (.27 \times .28) + (.54 \times -.18) = -.02$.

As can be seen in table 7.4, the age-to-age phenotypic correlations for behavioral inhibition at 14–20 months, 20–24 months, and 20–36 months are mediated by genetic factors. At all other age intervals, the cross-age phenotypic correlations are low, and hence our model does not have sufficient power to distinguish between genetic, shared, or nonshared environmental factors contributions. Although genetic factors significantly contribute to the phenotypic correlations in shyness across all ages, shared environmental factors also contribute to the phenotypic continuity of this temperament dimension. In contrast, the phenotypic continuities of each IBR temperament are almost entirely mediated by genetic factors.

Age-to-Age Genetic Correlations

Age-to-age genetic correlations (r_G), indicating the extent to which genetic effects at one age correlate with genetic effects at another, independent of the heritability of the trait at each age, are also presented in table 7.4. This statistic is calculated as the genetic contribution to phenotypic stability divided by the product of the square roots of the heritabilities at each age (Plomin, 1986). The genetic factors that influence a trait at two ages can co-vary perfectly even though the genetic factors at each age contribute only slightly to the phenotypic variance. Thus, r_G can be 1.0 even though the genetic contribution to phenotypic continuity is only modest if the heritability at each age is only modest and the same genetic effects operate at the two ages.

The modest, and generally nonsignificant, age-to-age genetic correlations for behavioral inhibition suggest that there is substantial genetic change in this dimension across age. Thus, although behavioral inhibition is genetically influenced at each age, there is only modest overlap between the genetic effects that govern the trait at any two ages. In contrast, the high age-to-age genetic correlations for shyness, activity, task orientation, and affect-extraversion suggest that there is considerable concordance between the genetic factors that affect each temperament across age. However, the differential pattern of correlations within each dimension warrants comment. For shyness, the genetic correlations between 14 months and subsequent ages are moderate, yet for later intervals, the genetic correlations are unity. This suggests that new, albeit nonsignificant, genetic effects emerge at 20 months and persist at 24 and 36 months. Similarly, for activity and task orientation, the genetic correlations for intervals involving 36 months are lower than other intervals. This pattern is consistent with our earlier finding of new genetic variance at 36 months.

SUMMARY AND DISCUSSION

Longitudinal twin studies such as the MacArthur Longitudinal Twin Study provide a rare opportunity to examine the genetics of temperament from a developmental perspective. Developmental behavioral genetics approaches the question of genetic change in two ways. The first asks about changes in heritability across age and may serve to identify points of causal transition. The second explores the role of genetic influences on age-to-age continuity and change and addresses the etiological processes by which developmental change takes place. These questions have seldom been examined in the literature on temperament.

In our longitudinal sample, tester-rated temperament demonstrated significant heritability at 14, 20, 24, and 36 months of age. These results are consistent with previous research that found genetic influences on individual differences in temperament. More interesting, however, is the finding of no consistent pattern of change in the magnitude of genetic influence across age and the finding that, overall, heritability does not appear to change dramatically during the transition from infancy to early childhood.

Previous twin studies of temperament in early infancy have found some evidence of age-related increases in genetic influence; however, small sample sizes and reliance on parental ratings, which frequently show twin contrast effects, make it difficult to interpret the findings (e.g., Buss, et al., 1973; Stevenson & Fielding, 1985; Torgersen, 1981). Contrast effects, the tendency for a rater to contrast DZ twins, thereby magnifying existing behavioral differences, are a common problem with parent-rating measures of temperament. Studies of parent-rated temperament that suggest age-related increases in heritability also show a trend toward increasing contrast effects (e.g., Buss et al., 1973; Stevenson & Fielding, 1985; Torgersen, 1981). If this bias increases with age, the DZ cotwins would become increasingly dissimilar, and heritability, which is based on the magnitude of the difference between MZ and DZ cotwin resemblance, would rise. The use of observer ratings of temperament in our large longitudinal sample circumvents the problem of contrast effects by providing an objective assessment of temperament on standard measures, especially when, as in the present study, different testers rate each twin. Thus, our results suggest that previous findings of developmental increases in the heritability of early temperament may be due to measurement artifact.

Phenotypically, change predominates over continuity in tester-rated temperament during the transition from infancy to early childhood. Stability correlations for each dimension were low, indicating a marked reordering of individual differences across age. Despite this, for behavioral inhibition, activity, task orientation, and affect-extraversion, genetic influences appear to be mediating whatever continuity exists. Shyness is an exception in that both genetic and shared environmental factors contribute to the phenotypic continuity of this dimension.

If genes contribute to continuity, what brings about change? The answer to this question varies according to temperament dimension. For behavioral inhibition, there was evidence that across 14 and 20 months, developmental change from an individual differences perspective is due to genetic and non-shared environmental influences (which includes measurement error). New genetic effects at 24 and 36 months may also be due to developmental change in behavioral inhibition; however, as indicated earlier, measurement differences across age may be confounding developmental differences. For shyness, new shared and nonshared environmental influences emerge at each age, whereas the same genetic factors operate across age. Thus, for shyness, developmental change is a result of shared and nonshared environmental influences. For activity and task orientation, change is due primarily to nonshared environmental factors. Both dimensions displayed near perfect genetic overlap between 14, 20, and 24 months of age, and although new genetic effects on these traits emerge at 36 months, as is the case for behavioral inhibition, it is difficult to tell whether this new genetic variance is due to developmental change or due to changes in the construct as a result of measurement differences. Similarly, although new genetic effects emerged at 20 months for affect-extraversion, the high genetic correlations between ages implies that there is relatively little genetic change. Again, change appears to be largely due to nonshared environmental influences.

The possibility of measurement differences resulting in new genetic effects for behavioral inhibition, activity, and task orientation warrants further comment. As indicated earlier, there were subtle changes in the measurement of these dimensions across age. Although the measures are presumed to reflect the same temperamental dimension at each age, there were differences in item content and/or factor structure. Changes in the construct across age may result in the emergence of new genetic effects from one measurement period to the next simply because different behaviors are being assessed. Because for behavioral inhibition, activity, and task orientation the emergence of new genetic effects coincide with changes in the measures for these dimensions, it is difficult to distinguish between changes due to development and changes due to measurement.

Our infant temperament results are consistent with longitudinal twin studies of adult personality that find nonshared environmental influences on personality change and genetic contributions to the stability of personality (e.g., McGue et al., 1993). This is somewhat surprising because, in contrast to adulthood, infancy is a period of rapid change. A possible reason for the finding of environmental change and genetic continuity may lie in the measures of temperament that, like those of adult personality, focus on stability. That is, with these measures, there may not be enough reliable change variance to partition into genetic and environmental components. However, the low age-to-age phenotypic correlations imply that there is much differential change in our measures of temperament. This leaves open more interesting possibilities for non-shared environmental influences on change in temperament. For example, individual differences in changes in temperament may be due to differences within the family environment, such as differential treatment, experiences, or accidents. Identifying nonshared environmental factors that influence developmental change is an important goal for future temperament research.

REFERENCES

Bayley, N. (1969). *Manual for the Bayley Scales of Infant Development*. New York: Psychological Corporation.

Boomsma, D. I., Martin, N. G., & Molennar, P. C. M. (1989). Factor and simplex models for repeated measures: Application to two psychomotor measures of alcohol sensitivity in twins. *Behavior Genetics, 19*, 79–96.

Braungart, J. M., Plomin, R., DeFries, J. C., & Fulker, D. W. (1992). Genetic influence on tester-rated infant temperament as assessed by Bayley's Infant Behavior Record: Nonadoptive and adoptive siblings and twins. *Developmental Psychology, 28*, 40–47.

Buss, A. H., Plomin, R., & Willerman, L. (1973). The inheritance of temperaments. *Journal of Personality, 41*, 513–524.

Cherny, S. S., Fulker, D. W., Emde, R. N., Robinson, J., Corley, R. P., Reznick, J. S., Plomin, R., & DeFries, J. C. (1994). A developmental-genetic analyis of conitnity and change in the Bayley Mental Development Index from 14 to 24 months: The MacArthur Longitudinal Twin Study. *Psychological Science, 5*, 354–360.

Clarke, A. D. B., & Clarke, A. M. (1984). Constancy and change in the growth of human characteristics. *Journal of Child Psychology and Psychiatry, 25*, 191–210.

Cyphers, L. H., Phillips, K., Fulker, D. W., & Mrazek, D. A. (1990). Twin temperament during the transition from infancy to early childhood. *Journal of the American Academy of Child and Adolescent Psychiatry, 29*, 392–397.

Emde, R. N., Plomin, R., Robinson, J., Reznick, J. S., Campos, J., Corley, R., DeFries, J. C., Fulker, D. W., Kagan, J., & Zahn-Waxler, C. (1992). Temperament, emotion, and cognition at 14 months: The MacArthur Longitudinal Twin Study. *Child Development, 63*, 1437–1455.

Garcia-Coll, C., Kagan, J., & Reznick, J. S. (1984). Behavioral inhibition in young children. *Child Development, 55*, 1005–1019.

Goldsmith, H. H. (1983). Genetic influences on personality from infancy to adulthood. *Child Development, 54*, 331–355.

Goldsmith, H. H. (1984). Continuity of personality: A genetic perspective. In R. N. Emde & R. J. Harmon (Eds.), *Continuities and discontinuities in development* (pp. 403–413). New York: Plenum.

Hewitt, J. K., Eaves, L. J., Neale, M. C., & Meyer, J. M. (1988). Resolving causes of developmental continuity or "tracking." I. Longitudinal twin studies during growth. *Behavior Genetics, 18*, 133–151.

Hubert, N. C., Wachs, T. D., Peters-Martin, P., & Gandour, M. J. (1982). The study of early temperament: Measurement and conceptual issues. *Child Development, 53,* 571–600.

Jöreskog, K. G., & Sörbom, D. (1993). *LISREL 8 User's Guide.* Chicago: Scientific Software.

Kagan, J., Reznick, J. S., Clarke, C., Snidman, N., & Garcia-Coll, C. (1984). Behavioral inhibition to the unfamiliar. *Child Development 55,* 2212–2225.

Kagan, J., Reznick, J. S., & Snidman, N. (1988). Biological bases of childhood shyness. *Science, 240,* 167–171.

Loehlin, J. C., Horn, J. M., & Willerman, L. (1989). Modeling IQ change: Evidence from the Texas Adoption Project. *Child Development, 60,* 993–1004.

Loehlin, J. C., Horn, J. M., & Willerman, L. (1990). Heredity, environment, and personality change: Evidence from the Texas Adoption Project. *Journal of Personality, 58,* 221–243.

Matheny, A. P., Jr. (1980). Bayley's Infant Behavior Record: Behavioral components and twin analysis. *Child Development, 51,* 1157–1167.

Matheny, A. P., Jr. (1983). A longitudinal twin study of stability of components from Bayley's Infant Behavior Record. *Child Development, 54,* 356–360.

Matheny, A. P., Jr. (1989). Children's behavioral inhibition over age and across situations: Genetic similarity for a trait during change. *Journal of Personality, 57,* 215–235.

McCartney, K., Harris, M. J., & Bernieri, F. (1990). Growing up and growing apart: A developmental meta-analysis of twin studies. *Psychological Bulletin, 107,* 226–237.

McGue, M., Bacon, S., & Lykken, D. T. (1993). Personality stability and change in early adulthood: A behavioral genetic analysis. *Developmental Psychology, 29,* 96–109.

Phillips, K., & Fulker, D. W. (1989). Quantitative genetic analysis of longitudinal trends in adoption designs with application to IQ in the Colorado Adoption Project. *Behavior Genetics, 19,* 621–658.

Plomin, R. (1986). Multivariate analysis and developmental behavioral genetics: Developmental change as well as continuity. *Behavior Genetics, 16,* 25–43.

Plomin, R. (1987). Developmental behavioral genetics and infancy. In J. Osofsky (Ed.), *Handbook of infant development* (2nd ed., pp. 363–417). New York: Wiley.

Plomin, R., Campos, J. J., Corley, R., Emde, R. N., Fulker, D. W., Kagan, J., Reznick, J. S., Robinson, J., Zahn-Waxler, D., & DeFries, J. C. (1990). Individual differences during the second year of life: The MacArthur Longitudinal Twin Study. In J. Colombo & J. Fagen (Eds.), *Individual differences in infancy: Reliability, stability, and predictability* (pp. 431–455). Hillsdale, NJ: Erlbaum.

Plomin, R., Coon, H., Carey, G., DeFries, J. C., & Fulker, D. (1991). Parent-offspring and sibling adoption analyses of parental ratings of temperament in infancy and early childhood. *Journal of Personality, 59,* 705–732.

Plomin, R., Emde, R., Braungart, J. M., Campos, J., Corley, R., Fulker, D. W., Kagan, J., Reznick, S., Robinson, J., Zahn-Waxler, C., & DeFries, J. C. (1993). Genetic change and continuity from 14 to 20 months: The MacArthur Longitudinal Twin Study. *Child Development, 64,* 1354–1376.

Plomin, R., & Nesselroade, J. R. (1990). Behavioral genetics and personality change. *Journal of Personality, 58,* 191–220.

Robinson, J. L., Kagan, J., Reznick, J. S., & Corley, R. (1992). The heritability of inhibited and uninhibited behavior: A twin study. *Developmental Psychology, 28,* 1030–1037.

Salbach, E. H., Morrow, J., & Wachs, T. D. (1991). Questionnaire measurement of infant and child temperament. In J. Strelau & A. Angleitner (Eds.), *Explorations in temperament. International perspectives on theory and measurement.* New York: Plenum.

Schmitz, S., Saudino, K. J., Plomin, R., Fulker, D. W., & DeFries, J. C. (1996). Genetic and environmental influences on temperament in middle childhood: Analyses of teacher and tester ratings. *Child Development 67,* 409–422.

Stevenson, J., & Fielding, J. (1985). Ratings of temperament in families of young twins. *British Journal of Developmental Psychology, 3,* 143–152.

Terman, L. H., & Merrill, M. A. (1973). *Stanford-Binet Intelligence Scale: 1972 norms edition.* Boston: Houghton-Mifflin.

Torgersen, A. M. (1981). Genetic factors in temperamental individuality. A longitudinal study of same-sexed twins from two months to six years of age. *Journal of the American Academy of Child Psychiatry, 20,* 702–711.

Wilson, R. S., & Matheny, A. P., Jr. (1986). Behavior genetics research in infant temperament: The Louisville twin study. In R. Plomin & J. Dunn (Eds.), *The study of temperament: Changes, continuities and challenges* (pp. 81–97). Hillsdale, NJ: Erlbaum.

8

Behavioral Inhibition and Related Temperaments

Jerome Kagan
Kimberly J. Saudino

Most psychologists agree that temperaments refer to stable moods and behavioral profiles that are seen in infants or young children, each partially controlled by the individual's genotype. It is important to appreciate that the referential meaning of temperament is not a particular collection of genes but the psychological profile that emerges when an environment acts upon a child who has inherited a particular physiology. Although it is likely that future research will discover a large number of temperamental types, two temperamental constructs that have been a popular target of inquiry are called "inhibited" and "uninhibited" to the unfamiliar.

These two temperamental categories are easily detected in the second year of life when children are exposed to unfamiliar people, procedures, and situations. The inhibited child is typically avoidant, affectively subdued, and remains proximal to a caregiver during initial encounters with an unfamiliar event. In contrast, the uninhibited child is minimally avoidant and approaches the unfamiliar person or object, often with a display of positive affect. Previous research suggests that these two categories are moderately stable over the first decade of life (Kagan et al., 1988, 1992).

This definition of inhibited and uninhibited temperaments implies that a child's response to novel people, objects, or events should be associated with measures of shyness with strangers, approach or withdrawal to unfamiliar places or objects, and timidity in dangerous situations. One purpose of this chapter is to examine the coherence of inhibited and uninhibited behaviors across different contexts. A second purpose is to examine the stability of behavioral inhibition from infancy to early childhood. Because most psychologists prefer to view the differences between inhibited and uninhibited children as quantitative rather than qualitative, we first treat inhibited and uninhibited behavior as continuous variables. However, there is some support for the view that inhibited and uninhibited children are members of qualitatively different temperamental groups (Kagan & Snidman, 1991). Therefore, we also examine the stability of behavioral inhibition from a categorical perspective.

MEASURES

Behavioral Inhibition in the Laboratory

We assessed behavioral inhibition (BI) through observations of each child's behavioral reactions to varied, unfamiliar events in a laboratory playroom at 14, 20, 24, and 36 months of age (see chapter 3 for a description of the procedures). An aggregate index of BI was derived by averaging Z-score values for total time a child remained proximal to the mother, amount of vocalization and play, and long latencies to approach unfamiliar but interesting objects or people in an unfamiliar laboratory situation. We standardized the behavioral inhibition aggregate to a mean of 0 and an SD of 1 at each age.

Shyness in the Home

A measure of shyness (SHY) in the home was based on each child's initial reactions to the entrance of two female examiners into the home at each age. We scored the videotapes of the children's behavior for a number of discrete responses, especially latency to approach and latency to talk to the stranger. A composite measure of shyness was based on an unrotated principal component score derived from the ratings of the observed behaviors.

Parental Ratings

A parental rating (CCTI Shy) index of shyness was obtained from the Colorado Childhood Temperament Inventory (CCTI; Rowe & Plomin, 1977), which had been modified to include a scale for shyness as distinct from sociability (Buss & Plomin, 1984). The CCTI shyness subscale includes five items that require the parents to make judgments about each child's behavior with strangers (e.g., child takes a long time to warm up to strangers). The Toddler Temperament Survey (TTS) (Carey & McDevitt, 1978) provided a parental rating measure of the related but more general construct of approach–withdrawal. This subscale contains 12 items that ask parents about their child's typical behavior when exposed to unfamiliar situations, people, or objects (e.g., "The child is fearful of being put down in an unfamiliar place [supermarket cart, new stroller, play-pen] with parent present"). The higher the score on the TTS approach–withdrawal scale, the more the child tends to withdraw from novelty. Both mother and father completed the CCTI and the TTS for each child, and scores were averaged across the mother and father to improve reliability.

We treat the three major variables separately, even though they may share variance, because the two observational variables (BI and SHY) were assessed in different settings and were based on unequal amounts of information. Parental report, however differs from the first two in its informational source.

RESULTS

Descriptive Statistics

Table 8.1 presents the means and standard deviations for the full MacArthur Longitudinal Twin Study sample by age and gender for BI, SHY, CCTI shy, and TTS approach–withdrawal. To evaluate mean differences, repeated-measures analyses of variance with age as a within-subjects variable and gender as a between-subjects variable were conducted for each measure. For the BI measure, there was a significant main effect for gender ($F_{1,408} = 4.10$, $p < .05$); however, this is best interpreted in light of the significant age-by-gender interaction ($F_{3,406} = 3.59$, $p < .05$). Follow-up tests indicated that boys and girls differed significantly on the BI measure only at 20 and 24 months of age. At these ages, girls were significantly more inhibited than boys. Shyness at home (SHY) also displayed a main effect for gender ($F_{1,540} = 10.32$, $p < .01$), with girls being significantly shyer than boys at home.

In contrast to the variables based on observed behaviors, neither parental ratings of shyness nor approach–withdrawal displayed significant main effects for gender. Both parental rating measures did, however, display significant main effects for age (CCTI shy, $F_{3,339} = 4.87$, $p < .01$; TTS approach–withdrawal, $F_{3,216} = 9.38$, $p < .001$). According to parents' reports, both shyness and the tendency to withdraw increase with age.

The fact that the analyses of behavior and the parental ratings of presumably related variables yielded dissimilar results deserves comment. The finding of gender differences for observed BI and SHY, but not for parental ratings of shyness and approach–withdrawal, suggests that the measures may have different meanings (i.e., may be tapping different aspects of the same construct). The differences in age effects cannot be interpreted the same way because the observed BI and shy variables were standardized within age. This is not the case for parental ratings.

Intercorrelations among Measures

The intercorrelations between BI, SHY, CCTI Shy, and TTS approach–withdrawal for girls and boys at ages 14, 20, 24, and 36 months are presented in

Table 8.1. Means and standard deviations by gender

Variable/age	Girls			Boys		
	N	Mean	SD	N	Mean	SD
Behavioral inhibition						
14 Months	302	−0.01	0.95	331	−0.09	0.89
20 Months	273	0.12	0.96	294	−0.22	0.85
24 Months	289	0.05	0.84	308	−0.20	0.85
36 Months	295	−0.10	0.67	291	−0.14	0.67
Observed shyness						
14 Months	349	0.08	0.99	377	−0.11	0.98
20 Months	314	0.12	1.01	345	−0.13	0.96
24 Months	322	0.05	1.03	335	−0.08	0.98
36 Months	325	0.13	0.98	313	−0.15	1.00
CCTI shyness						
14 Months	326	12.89	4.08	358	12.08	3.86
20 Months	291	13.13	4.32	317	12.59	4.15
24 Months	301	13.15	4.01	314	12.75	4.13
36 Months	269	13.42	4.37	270	12.66	4.55
TTS approach–withdrawal						
14 Months	298	3.31	0.93	322	3.23	0.89
20 Months	256	3.46	0.96	275	3.40	0.94
24 Months	224	3.43	0.94	251	3.38	1.01
36 Months	175	3.31	0.97	169	3.14	0.99

CCTI, Colorado Childhood Temperament Inventory; TTS, Toddler Temperament Survey.

table 8.2. Although the intercorrelations are slightly higher for girls than for boys, the differences were not significant. As expected, shyness (both observed and parent rated) and approach–withdrawal were significantly related to observed behavioral inhibition. However, the relations among the three types of variables are generally modest (between .2 and .3), suggesting that each is measuring somewhat different aspects of the child's disposition to be timid, shy, or sociable. An exception is the high correlation (.7–.8) between the two parent-reported measures. This result is consistent with previous research that finds substantial agreement between parental ratings of a temperament across different measures (Bates et al., 1979; Goldsmith et al., 1991; Worobey, 1986) but minimal correlation between parent ratings and observational measures of temperament (e.g., Rothbart, 1986; Saudino & Eaton, 1991, 1995 Seifer et al., 1994).

Table 8.2. Intercorrelations between behavioral inhibition (BI), observed shyness (SHY), Colorado Childhood Temperament Inventory (CCTI) shyness, and Toddler Temperament Survey (TTS) approach–withdrawal at each age

Age/measure	BI	SHY	CCTI shyness	TTS approach–withdrawal
14 Months				
BI	—	.22	.30	.31
SHY	.25	—	.39	.43
CCTI shyness	.27	.36	—	.77
TTS approach–withdrawal	.19	.40	.80	—
20 Months				
BI	—	.38	.27	.31
Shy	.27	—	.45	.53
CCTI shyness	.31	.26	—	.79
TTS approach–withdrawal	.31	.34	.77	—
24 Months				
BI	—	.30	.27	.36
SHY	.16	—	.41	.53
CCTI shyness	.15	.37	—	.77
TTS approach–withdrawal	.22	.42	.81	—
36 Months				
BI	—	.35	.30	.34
SHY	.33	—	.43	.51
CCTI shyness	.29	.37	—	.83
TTS approach–withdrawal	.30	.47	.78	—

Girls ($N = 352$–295) in bold above the diagonal; boys ($N = 356$–297) below the diagonal. All correlations significant at $p < .05$.

Stability Treating Behavioral Inhibition as a Continuous Variable

Table 8.3 presents the stability of each variable across each pair of ages. Several generalizations seem reasonable. First, once again, consistent with previous research (e.g., Plomin & DeFries, 1985; Rothbart, 1986; Wilson & Matheny, 1986), the stability coefficients for the parent-reported measures (i.e., CCTI Shy and TTS approach–withdrawal) were higher than those based on the behavioral observations (BI or SHY). For example, the stability correlations for the period 14–36 months were .41 for CCTI Shy and .51 for TTS approach–withdrawal. The comparable correlations for observed BI and SHY were .01 and .21.

Table 8.3. Stability correlations

Measure/age	Girls	Boys	Full sample
Behavioral inhibition			
14–20 Months	.25*	.34**	.29**
14–24 Months	.14*	.10	.12**
14–36 Months	−.04	.07	.01
20–24 Months	.20**	.19*	.21**
20–36 Months	.30**	.24**	.27**
24–36 Months	.30**	.07	.18**
Observed shyness			
14–20 Months	.37**	.13*	.26**
14–24 Months	.36**	.19**	.28**
14–36 Months	.30**	.10	.21**
20–24 Months	.51**	.34**	.43**
20–36 Months	.41**	.32**	.38**
24–36 Months	.43**	.33**	.38**
CCTI shyness			
14–20 Months	.65**	.57**	.62**
14–24 Months	.52**	.55**	.53**
14–36 Months	.41**	.41**	.41**
20–24 Months	.64**	.71**	.68**
20–36 Months	.53**	.54**	.54**
24–36 Months	.57**	.65**	.61**
TTS approach–withdrawal			
14–20 Months	.71**	.66**	.68**
14–24 Months	.64**	.65**	.65**
14–36 Months	.59**	.42**	.51**
20–24 Months	.77**	.84**	.81**
20–36 Months	.69**	.63**	.66**
24–36 Months	.77**	.74**	.75**

CCTI, Colorado Childhood Temperament Inventory; TTS, Toddler Temperament Survey.
*$p < .05$; ** $p < .01$.

Second, the stability of BI across the 14–24 month and 14–36 month intervals was low for both girls and boys. This pattern suggests that a shift occurs in the second year. Kagan and colleagues (1994) have supporting data for this conclusion. After the emergence of self-awareness in the middle of the second year, many children change their social behavior. Some children who were not inhibited at 14 months become inhibited at 20 months because of the emergence of a concern over adult evaluation rather than fear of the unfamiliar. Further, a proportion of boys and girls who were inhibited at 14 months were able, perhaps through experience and the exercise of will, to control their tendency to withdraw from strangers and from unfamiliar events.

It is also important to note that the stability correlations for observed shyness (SHY) are modest in magnitude (.2–.5). Moreover, shy behavior tends to be more stable for girls than for boys (correlations of .3–.5 for girls versus .1–.3 for boys). The stability correlations for girls were significantly greater than those for boys across the intervals 14–24 months, 14–36 months, and 2–24 months.

Stability Treating Behavioral Inhibition as a Categorical Variable

We also examined the stability of observed BI and SHY using a categorical strategy. Children with scores on BI or SHY that exceeded # 1 SD from the mean for that age were identified as extreme for that age. For example, a child with a BI score of 1.4 at 14 months would be considered to be high BI; a child with a BI score of-1.4 would be identified as behaviorally uninhibited. We evaluated whether membership in these two extreme qualitative categories was preserved across age.

Overall, there was significant but modest stability for both variables. No more than 10% of the children retained an extreme degree of BI or SHY across any pair of ages. Only 5% of the girls and 7% of the boys remained extreme on BI from 14 to 36 months; 9% of the girls and 8% of the boys maintained an extreme BI score from 24 to 36 months. Similar results were found for the two categorical groups classified as extremely high or low shy. Most children changed their level of BI or SHY across the 2 years of observation from 14 to 36 months. It is important to note that when stability was defined as the preservation of membership in the extreme category, parental reports yielded only slightly higher stability estimates than did observational data. For example, according to parental reports, only 12% of girls and 13% of boys remained extremely shy or fearful from 14 to 36 months.

A small group of children were extreme on BI at three or more ages; these children preserved their extreme BI status. Eleven children were consistently uninhibited, and 20 were consistently inhibited, using a criterion of #1 SD from the mean. We compared the consistently inhibited and uninhibited groups on a large number of variables that included cognitive abilities, parental reports of behavior, empathy, height, weight, heart rate, and smiling in the laboratory. As expected, the inhibited, compared with uninhibited, children were more shy in the home and were rated by parents as being more shy and more likely to withdraw at all ages. But, with the exception of these obvious marker variables, inhibited and uninhibited children did not differ consistently on any of the other variables. There were no differences, for example, between inhibited and uninhibited children on Bayley mental developmental scores, Stanford-Binet IQ scores, height, weight, or heart rate.

It is important to note, however, that at 20 months the consistently inhibited children smiled less often than uninhibited children in the laboratory. Further, there were interactions between gender and inhibition for empathy and emotionality at 36 months. Consistently inhibited girls were rated as more emotional but less empathic than the inhibited boys. Among uninhibited children the pattern was reversed.

SUMMARY AND CONCLUSIONS

This corpus of data on a reasonably large sample of twin siblings leads to several important conclusions. First, as suggested by Saudino and Cherny in chapter 6, parent-reported measures yield information that is different from the information revealed by behavioral observations. Research that relies on just one type of measure may not represent a full picture of the child's temperament. However, we would also caution that parental ratings may reflect parental expectations and biases as well as representations of the child's behavior.

Second, both observed behavioral inhibition and shyness are modestly related, and both show modest stability. As indicated in chapter 7, continuity and change in both measures is due to a combination of genetic and environmental influences (Plomin & DeFries, 1985; Wilson & Matheny, 1986). The child's phenotype reflects both inherent temperamental biases as well as the products of socialization. An inhibited child can learn to be sociable with a stranger but still show some fear of an unfamiliar laboratory setting. Further, any child can learn, through experience, a fear of unfamiliar events. The emergence of self in the middle of the second year makes children anxious over adult evaluations of their appearance and competence; this emotional state is different from fear of the unfamiliar. However, the reaction to anxiety over evaluation (a coy, shy style with others) is similar, in some ways, in its phenotypic appearance to behavioral inhibition. Hence, there is a change in the stability of BI at about 20 months.

Finally, it appears that behavioral inhibiton in the laboratory and shy behavior at home are different constructs (Goldsmith & Campos, 1990). The differences between behavioral inhibition in the laboratory and shyness at home can be reconciled if one considers how the two constructs are measured. Behavioral inhibition is based on more than 1 hr of observation of a child in an unfamiliar laboratory. In contrast, shyness is based on a few minutes of behavior when the child encounters two strangers in the home. The effect of parental socialization regarding how children should behave with strangers who enter their home could make siblings living in the same family more similar in their behavioral style with strangers.

REFERENCES

Bates, J. E., Freeland, C. A., & Lounsbury, M. L. (1979). Measure of infant difficultness. *Child Development, 50*, 794–803.

Buss, A. H., & Plomin, R. (1984). *Temperament*. Hillsdale, NJ: Erlbaum.

Carey, W. B., & McDevitt, S. C. (1978). Stability and change in individual temperament diagnoses from infancy to early childhood. *American Academy of Child Psychiatry, 17*, 331–337.

Goldsmith, H. H., & Campos, J. J. (1990). The structure of temperamental fear and pleasure in infants. *Child Development, 61*, 1944–1464.

Goldsmith, H. H., Rieser-Danner, L. A., & Briggs, S. (1991). Evaluating convergent and discriminant validity of temperament questionnaires for preschoolers, toddlers, and infants. *Developmental Psychology, 27*, 566–579.

Kagan, J., Reznick, J. S., & Snidman, N. (1988). Biological bases of childhood shyness. *Science, 240*, 167–171.

Kagan, J., & Snidman, N. (1991). Temperamental factors in human development. *American Psychologist, 46*, 865–862.

Kagan, J., Snidman, N., & Arcus, D. M. (1992). Initial reactions to unfamiliarity. *Current Directions in Psychological Science, 1*, 171–174.

Plomin, R., & DeFries, J. C. (1985). *Origins of individual differences in infancy: The Colorado Adoption Project*. Toronto: Academic Press.

Rothbart, M. K. (1986). Longitudinal observation of infant temperament. *Developmental Psychology, 22*, 356–365.

Rowe, D. C., & Plomin, R. (1977). Temperament in early childhood. *Journal of Personality Assessment, 41*, 150–156.

Saudino, K. J., & Eaton, W. O. (1991). Infant temperament and genetics: An objective twin study of motor activity level. *Child Development, 62*, 1167–1174.

Saudino, K. J., & Eaton, W. O. (1995). Continuity and change in objectively assessed temperament: A longitudinal twin study of activity level. *British Journal of Developmental Psychology, 13*, 81–95.

Seifer, R., Sameroff, A. J., Barrett, L. C., & Krafchuck, E. (1994). Infant temperament measured by multiple observations and mother reports. *Child Development, 65*, 1487–1490.

Wilson, R. S., & Matheny, A. P., Jr. (1986). Behavior genetics research in infant temperament: The Louisville twin study. In R. Plomin, & J. Dunn (Eds.), *The study of temperament: Changes, continuities and challenges*, (pp. 81–97). Hillsdale, NJ: Erlbaum.

Worobey, J. (1986). Convergence among assessments of temperament in the first month. *Child Development, 57*, 47–55.

Part III

Emotion

Section Editor
JoAnn L. Robinson

9

Emotional Development in the Twin Study

JoAnn L. Robinson
Robert N. Emde

The past 50 years have witnessed increasing interest in the child's early emotional development, but the focus of interest has repeatedly changed in the face of new discoveries and new opportunities for research. At mid-century, attention was directed to the importance of early emotional development when the world became aware of the devastating effects of deprivations in early caregiving. Observations of such effects pointed to serious deficits and distortions in emotional processes (Bowlby, 1951; Goldfarb, 1945; Spitz, 1945, 1946). Infant emotions could be observed directly, and filmed, conveying not only sadness and depression, but also other emotions such as distress (evidenced by crying) and pleasure (evidenced by smiling). Many studies were stimulated by these observations, which then documented normal developmental milestones in the appearance of emotional expressions such as crying-distress and smiling and the circumstances under which they occurred. This in turn led to another focus of research, namely, developmental transitions that often introduced new emotional expressions as well as the periods of adaptive organization in between such transitions.

During this time, theory changed. Most came to realize that emotions were more appropriately regarded as active, ongoing, and emotional processes than as reactive, intermittent, and disruptive states (Emde et al., 1972; Spitz, 1959). This in turn led to another change in focus that was stimulated by the classic cross-cultural studies of adult facial expressions of emotion carried out by Ekman et al. (1972) and Izard (1972). Since a set of basic emotions seemed to occur across a wide variety of cultures and settings, including those that had not been exposed to mass media or common socialization influences, many began to wonder about early facial expressions of emotion in infancy, prior to the development of language and the forms of later socialization that could occur through language. Considerable research attention was then directed to mapping such expressions in infancy (for a review, see Campos et al., 1983).

During the past decade, interest in the adaptive functions of emotions in relation to biological and environmental contexts have deepened. This has resulted from two influences. First, the above-noted cross-cultural work led to a revival of the evolutionary theories of adaptiveness for emotions that originated with Darwin (1872) and that was later given new theoretical coherence from the works of Tomkins (1962; 1963) and Lazarus (1991). Second, researchers became aware that facial expression is only one aspect of emotional expressiveness and that adaptive functions of communication are linked to adaptively organized, multifaceted motivational states (see discussions in Lewis & Haviland, 1993; Scherer & Ekman, 1984).

Currently, a focus of interest has returned to organizational questions about emotion. Because development involves increasingly organized complexity, the questions take the form of describing and understanding complex emotional systems over time. Here are a few examples of such questions: To what extent might we identify early components of emotion systems that have consistency over time? To what extent might early components of emotion be related to specific contexts? To what extent will components of emotional systems during the early years become reconfigured and connected later?

Organizational questions have resulted in a renewed interest in individual differences that are highlighted by our advances in knowledge about how genes work in relation to the environment. We can now ask questions about genetic and environmental influences across development. Moreover, we can target information about specific developmental processes that influence components of emotional systems even as they develop and become reconfigured. New opportunities for research also result from our combining knowledge that we obtained at the beginning of the past half century with knowledge we have recently gained. For example, since the time of early deprivation studies, we have known that emotional development is embedded in the context of caregiving relationships. Two generations of attachment research and probes into other aspects of social development have deepened and broadened that insight (for recent reviews, see Lewis & Haviland, 1993, Sroufe, 1995). In a similar vein, current research has given us knowledge that has directed increasing attention to the context of development because of our understanding of the way genes work, and there is also a renewed interest in the differentiation and organization of behavioral systems (e.g., see Smith and Thelen, 1993).

The three contributions of this part represent inquiries into different areas of emotional development. Empathy, anger, and cheerfulness all play an important role in the establishment of relationships with others. Adaptive functions of these emotions reflect both motivations and social communications. Moreover, the regulation of these emotions can be considered a vital adaptive task of early development and caregiving. It is not surprising, therefore, that the findings of the three chapters indicate that emotion is expressed in complex ways that are highly dependent on context.

The first chapter by Emde et al. grapples with a fundamental challenge to emotion researchers. Emotions are expressed by diverse means. Anger, for example, is conveyed through facial expression, actions (such as overt aggression), and other verbal and nonverbal means. Sources of data in this chapter are varied, with both maternal report and observational measures. The different measures do not cohere into a unitary construct of anger expression but instead suggest varied, loosely associated means of expressing anger (as suggested by weak intercorrelations among variables). Still, there is considerably strong evidence for coherence suggested by longitudinal correlations for each phenotypic measure. The chapter then reveals that important variations in genetic and environmental sources of individual differences are indicated across data-gathering methods. In some cases, strong continuity and environmental influences are indicated (such as in maternal reports of anger expression on the Differential Emotions Scale), while in other cases, changing genetic influences are indicated (as in the response to restraint observational measure). This chapter makes salient the moderating role of context of measurement and, by implication, the context of changing influences across development. The former reminds us that our operational definitions can make a critical impact on our results. The chapter illustrates that context can exert its influence, both at the phenotypic level, as well as the etiologic level, where we explore genetic and environmental sources of individual differences.

The chapter by Zahn-Waxler et al. is on empathic development and also continues the theme about the role of measurement context. Maternal report data yielded different sources of individual differences than did observational measures of the components or discrete aspects of children's empathic responses. In addition, interesting developmental patterns of change were found in the component behaviors of empathy at the phenotypic level; for some measures mean levels increased, while for others there were decreases. Even more strikingly, patterns of developmental change varied by measures at the level of genetic and environmental sources of individual differences.

The third contribution of this part by Robinson et al. suggests that traitlike positive affect can be observed in child behavior during the second year of life. Context did not operate as strongly at the phenotypic level in analyses of what is referred to as "cheerfulness" as it did in the analysis of anger-related behaviors in the first chapter of this section. Continuity across measurement contexts was seen in the expressed positive affects of children within each age. Still, genetic and environmental sources of influence differed across the low- and high-stress contexts in which the positive affect was observed, suggesting differentiation at the etiologic level. It may be that phenotypic continuity was seen in the analyses of cheerfulness because it relied on observational measurement strategies in three situations assessed in both home and laboratory. In contrast, the analyses of anger drew from several maternal reports of fairly distinct types of anger-related behavior and included an observational construct as well from two different types of situations that yielded measures in the home and the lab. Examining the phenotypic correlations within just that observational construct suggests a result similar to the Robinson et al. analyses of cheerfulness: moderate continuity does exist in response to restraint across home and laboratory situations.

The chapters of this part offer some first views of emotion in children. Taken together, they provide much to stimulate further thinking and research.

REFERENCES

Bowlby, J. (1951). *Maternal care and mental health.* Geneva: World Health Organization.

Campos, J. J., Barrett, K. C., Lamb, M. E., Goldsmith, H. H., & Stenberg, C. (1983). Socioemotional development. In M. Haith & J. J. Campos (Eds.), *Handbook of Child Psychology* (Vol. 1, pp. 783–915). New York: Wiley.

Darwin, C. (1872). *The expression of emotions in man and animals.* London: John Murray.

Ekman, P., Friesen, W., & Ellsworth, P. (1972). *Emotion in the human face.* New York: Pergamon Press.

Emde, R. N., Gaensbauer, T. J., & Harmon, R. J. (1972). Emotional expression in infancy: A biobehavioral study. *Psychological Issues, A Monograph Series 10*(37).

Goldfarb, (1945). Effects of psychological deprivation in infancy and subsequent stimulation. *American Journal of Psychiatry, 102,* 18–33.

Izard, C. (1972). *Patterns of emotion: A new analysis of anxiety and depression.* New York: Academic Press.

Lazarus, R. S. (1991). *Emotion and adaptation.* New York: Oxford University Press.

Lewis, M., & Haviland, J. M. (Eds.). (1993). *Handbook of emotions.* New York: Guilford Press.

Scherer, K. R., & Ekman, P. (Eds.). (1984). *Approaches to emotion.* Hillsdale, NJ: Erlbaum.

Smith, L. B., & Thelen, E. (Eds.). (1993). *A dynamic systems approach to development: Applications.* Cambridge, MA: MIT Press.

Spitz, R. A. (1945). Hospitalism: An inquiry into the genesis of psychiatric conditions in early childhood. *Psychoanalytic Study of the Child, 1,* 53–74.

Spitz, R. (1946). Anaclitic depression. *Psychoanalytic Study of the Child, 2,* 313–342.

Spitz, R. A. (1959). *A genetic field theory of ego formation.* New York: International Universities Press.

Sroufe, L. A. (1995). *Emotional development: The organization of emotional life in the early years.* Cambridge: Cambridge University Press.

Tomkins, S. S. (1962). *Affect, imagery, consciousness,* vol. 1. New York: Springer.

Tomkins, S. S. (1963). *Affect, imagery, consciousness,* vol. 2. New York: Springer.

10

Reactions to Restraint and Anger-Related Expressions during the Second Year

Robert N. Emde
JoAnn L. Robinson
Robin P. Corley
Diana Nikkari
Carolyn Zahn-Waxler

Anger expressions in infancy have been described in a variety of different measurement contexts, including arm restraint (Camras et al., 1992; Fox, 1989; Stenberg & Campos, 1990); a stranger's approach (Fox & Davidson, 1988); extinction of instrumental learning (Lewis et al., 1990); cookie withdrawal (Stenberg et al., 1983); frustration of needs and dressing restrictions (Klinnert et al., 1984); play (Goodenough, 1931); and socialization experiences with parents (Miller & Sperry, 1987; Radke-Yarrow & Kochanska, 1990). Controversy remains, however, as to whether anger is best regarded as a discrete, biologically based response system. Thus, on the one hand, some programmatic researchers find it useful to take a discrete emotions point of view of anger in infancy (Emde et al., 1985; Fischer & Tangney, 1995; Izard et al., 1980; Goldsmith & Campos, 1982). On the other hand, some researchers favor a constructivist view of anger. Anger might better be regarded as a response system that is socially and cognitively constructed from an arousal of general distress that occurs in particular contexts. Stein and Levine (1990), for example, point out that both anger and sadness can be elicited by similar situations. In their view, anger is expressed because an individual refuses to accept being in a state of unexpected loss or failure—a view that bears some similarities to the more general constructionist views of Averill (1979) and Mandler (1990).

The functionalist framework of Lazarus (1991) as articulated for infancy by Campos and colleagues (Barrett & Campos, 1987; Campos et al., 1994) may provide a clarification for this controversy. In the functionalist framework, emotions are considered in terms of the adaptive goals of the individual, with particular goals understood in terms of their consequences or, more broadly, in terms of their person–environment relations. Goals can vary in different situations and contexts. Thus, anger, according to this framework, may be an inborn response tendency that is mobilized under particular circumstances when a goal is frustrated. The most common context for eliciting anger in infancy has been the reaction to physical restraint, where activity goals are frustrated (Camras et al., 1992; Fox, 1989; Goldsmith & Rothbart, 1988; Stenberg, 1982; Stenberg & Campos, 1990). Another context for eliciting anger, especially during the child's second year, are situations when interpersonal goals are frustrated in the midst of conflict. The latter occurs during situations of parental prohibition (Emde et al., 1987, Klinnert et al., 1984) and during situations of sibling rivalry (Dunn & Kendrick, 1982).

The functionalist adaptive view of anger also allows for the possibility of individual differences in anger propensity. Goldsmith and Rothbart presume that individual differences in the propensity for expressing anger are an important aspect of temperament. Questions remain, however, even if we incorporate the functionalist-adaptive view. Among these questions are the following: To what extent do individual differences in anger-related responses depend upon the context of measurement? To what extent are anger-related responses characteristic of individuals across time? To what extent are individual differences in anger-related responses different from other negative emotion responses? Do anger-related responses show increases across age? Are there gender differences, such that females have less expressed anger responses as compared with males by the end of the second year (perhaps because of differential socialization)?

If anger-related responses characterize individuals, other research questions about genetic and environmental influences follow. To what extent are individual differences in anger-related responses under genetic influence? Under environmental influences? How do such influences change across development, and can we specify quantitative models for continuity and change?

The analyses presented in this chapter make use of measures obtained during the child's second year in the MacArthur Longitudinal Twin Study to answer these questions. The first set of questions posed above are not genetic, and hence our analyses are phenotypic, making use of this unique and large data set. The second set of questions are genetic, and hence our analyses involve cross-sectional and longitudinal quantitative genetic approaches. Because the second year is a time of major developmental change, we hypothesize that there will be new genetic and environmental contributions to anger-related responses at each age. Moreover, in line with other measures involving temperament, emotion, and cognition previously reported in preliminary analyses (Plomin et al., 1993), we hypothesize that genetic influence should contribute to the stability of anger-related responses.

We focus on our observations of reactions to restraint because this measurement context has been one of the most salient for eliciting anger in infancy. We also include multiple contexts from mothers' reports. These are designed to tap maternal observations of the child's interpersonal conflict, outbursts, and anger expressions. Maternal reports are especially important because they can sample child behavior over a longer time frame than can researcher-based observations.

METHOD

Sample

Analyses are based on a maximum sample of 351 twin pairs—190 Monozygotic (MZ) and 161 dizygotic (DZ) twin pairs. They represent all twins who were assessed in home and laboratory visits at 14, 20, and 24 months at the time of analysis along with corresponding maternal reports. Specific analyses may be based on a somewhat smaller sample because of noncompletion of procedures. The characteristics of the sample are described in chapter 3.

Procedures

Four types of restraint were implemented; two in the home and two in the laboratory. In the home, the first restraint conditions occurred following the entry of the two testers into the home. After the standardized entry sequence, one examiner put an identifying vest or bib on each twin. In circumstances where the child repeatedly resisted the examiner's attempt at dressing the child, the mother was asked to do so. The waist-length, colorful vests (one yellow, one red) were designed to slip over the child's arms and fastened by Velcro in the back. The second restraint condition in the home, the measurement-restraint, occurred shortly after the vest-restraint condition. Each twin, one at a time, was led into a room away from the cotwin but with the mother present for this procedure. The measurement began with mother, child, and examiner sitting on the floor. The examiner picked up the child in a comfortable way and laid the child on his or her back on the floor with the mother seated behind the child. Using a small, spring-loaded measuring tape, the examiner quickly measured the child from a point at the top of the child's head (created by the mother holding a videocassette box perpendicular to the head) to the sole of the foot. After this measurement, typically lasting up to 20s, the examiner placed her hand flat on the child's abdomen and said in a firm but pleasant tone of voice, "(Name of child), lie still." This restraint lasted up to 3 min or until the child had a strong vocal and physical protest (e.g., pushing the examiner's hand away while negatively vocalizing or evidencing clear distress). The mother was instructed not to engage the child verbally or physically during this time and was engaged in conversation by the examiner. The examiner repeated the restraint command if the child squirmed energetically without strong protest.

Laboratory restraint conditions included (1) a repetition of the vest restraint 5 min after entry into the reception area of the laboratory and (2) an electrode placement. The latter procedure involved the adhesion of three small electrodes to the child's chest in preparation for heart-rate monitoring. Mothers had been instructed, prior to arrival at the lab, to outfit the twins in two-piece outfits that would permit less intrusive placement of the electrodes. The examiner lifted the child's shirt/vest and placed the three adhesive electrodes under each nipple and in the center of the chest.

Three maternal report measures of anger-related responses were also obtained at each age. During the laboratory visit, the mother endorsed the occurrence of anger expressions on the Differential Emotions Scale (DES; Izard, 1982). A telephone interview was conducted after the home visit in which mother responded to two interview questions concerning the frequency of angry outbursts and the initiation of physical fights with the cotwin. The mother's response to the interview questions concerning the frequency of angry outbursts and the initiation of physical fights by each twin were rated by the interviewer on a seven point scale.

Measures

For the measurement-restraint procedure, we used a 5-point, global rating of protest strength to evaluate the child's reaction. This rating included consideration of the intensity of protest actions, pushing away the hand, squirming, and staccato vocalizations indicative of anger. On this scale, 1 indicated no protest occurred, 3 indicated some/moderate protest (two or three attempts to remove the examiner's hand, plus one anger vocalization), and 5 indicated forceful protest (more than three attempts to end the restraint, including loud anger vocalizations). The child's reactions to the three other restraint procedures, the two vest restraints and the electrode placement, were based on 4-point scales as follows: 1, child accepts the vest/electrodes with little or no hesitation; 2, child resists but examiner is able to complete placement; 3, child resists forcefully and mother completes placement; and 4, child resists forcefully and placement cannot be completed by mother or examiner.

Because the observational measures of reaction to restraint were conceptually related and moderately correlated with each other, they were aggregated by extracting a principal component that accounted for 40% or more of the variance at each age (40%, 54%, and 47%, respectively). The maternal report measures assessed responses in different contexts and were not aggregated.

RESULTS

Our findings are presented first with respect to phenotypic analyses and second with respect to genetic analyses.

Table 10.1. Anger-related responses: means by age

Age	Observed		Maternal report		
	Restraint protest	DES anger	Initiate fights	Outbursts
14 Months	7.33 (1.7)	2.42 (0.9)	3.51 (1.6)	2.47 (1.8)
20 Months	8.28 (2.6)	2.61 (0.9)	3.86 (1.4)	2.63 (1.7)
24 Months	7.35 (2.6)	2.57 (0.9)	4.00 (1.3)	2.42 (1.7)

N = 651–742 subjects at each age; standard deviations are in parentheses.

Phenotypic analyses

Means and standard deviations of the mean for our anger-related responses are presented in table 10.1. Mean levels of all four measures increased from 14 to 20 months. A repeated-measures analysis of variance showed significant age trends for restraint protest, DES anger and initiation of physical fights (F-51, 10, and 21, respectively). The analysis of variance did not reveal an age trend for angry outbursts (F = 2.7). The incremental differences across age were relatively small—typically less than one-half of a pooled standard deviation unit. A curvilinear trend was suggested in the mean levels of observed restraint protest and DES anger, with an increase in values from 14 months to 20 months and then a decline in values to 24 months. The specific contrast testing whether 20 months differed from 14 and 24 months within the repeated-measures analysis of variance was highly significant for these two variables (20 versus 14 months, t = 9.98, p < .001; 20 versus 24 months, t = 3.27, p <.005). Changes over these three ages for reported initiation of physical fights were also significant, but the pattern was increasing and linear (t = 6.29, p <.001). Gender differences were not substantial. Nine of the 12 comparisons for the 4 measures showed no difference. Moreover, the significant differences did not indicate any coherent shift related to gender. Boys were reported to initiate more fights at 24 months (p <.01) and more DES anger at 14 months (p < .01); girls were observed to have more restraint protest at 14 months (p < .05).

Longitudinal phenotypic correlations within measurement context are indicated in table 10.2. Stability was moderate across the three ages for all measures, with the highest levels for the maternal report of DES anger.

Surprisingly, the observed restraint protest measure did not correlate with the three maternal report measures, nor did the latter show a substantial pattern of significant correlations with each other. These findings held at 14, 20, and 24 months of age. Thus, although context-specific measures demonstrated relative stability across time, the findings indicate no support for a general anger-related propensity reflected in individual differences across contexts using the measures we employed. These results are illustrated in table 10.3.

Table 10.2. Longitudinal phenotypic correlations within measurement
context

Measurement		20 Months	24 Months
Observed restraint	14 Months	.36**	.19**
protest	20 Months		.46**
DES anger	14 Months	.55**	.53**
	20 Months		.61**
Initiate fights	14 Months	.30**	.22**
	20 Months		.32**
Outbursts	14 Months	.32**	.28**
	20 Months		.29**

$N = 528$–640.
**$p < .01$.

With respect to discriminant validity, we asked whether our anger-related responses correlated with other nonanger measures of negative emotional response. For the observational measure of restraint protest we chose two other observational measures, the behavioral inhibition principal component aggregate and an index of distress from the Infant Behavior Record (IBR). The latter was obtained by using the negative scale points from the General Emotional Tone Scale of the IBR. The measure of restraint protest did show low correlation with the two other observational measures that documented individual

Table 10.3. Interrelatedness of responses

Measurement	DES anger	Initiate fights	Outbursts
14 Months			
Restraint protest	−.05	.09	−.01
DES anger		.10**	.19**
Initiate fights			.14**
20 Months			
Restraint protest	−.05	.01	.06
DES anger		.15**	.20**
Initiate fights			.09
24 Months			
Restraint protest	.00	.04	−.01
DES anger		.24**	.11**
Initiate fights			.06

$N = 556$–734.
**$p < .01$.

Table 10.4. Correlations of restraint protest with behavioral inhibition (BI) and Infant Behavior Record (IBR) negative emotion at each age

Age	BI	IBR negative emotion
14 Months	.12**	.18**
20 Months	.25**	.19**
24 Months	.31**	.21**

N = 595–714.
**p < .01.

differences in distress. Table 10.4 indicates these correlations at each of the three ages. Our observational measure of restraint protest does seem to be associated with a negative emotional expressive feature of individual differences. We conclude that evidence for discriminant validity of our observed measure, beyond the expressed distress component, is modest or questionable.

For our maternal report measures of anger-related responses, we chose two other maternal report measures to assess discriminant validity, DES fearfulness and negative emotionality from the Colorado Childhood Temperament Inventory (CCTI). Results of these analyses are displayed in table 10.5. For initiation of physical fights and for outbursts, correlations were low or nonexistent, indicating evidence of discriminant validity. For DES anger, there were moderate

Table 10.5. Correlations of maternal-report measures with fearfulness and negative emotion

Measurement/age	DES fearfulness	CCTI negative emotion
DES anger		
14 Months	.21**	.36**
20 Months	.35**	.45**
24 Months	.32**	.39**
Initiate fights		
14 Months	.01	.05
20 Months	.00	.03
24 Months	.12**	.10*
Outbursts		
14 Months	.05	.09*
20 Months	.09*	.12**
24 Months	.08	.08

N = 593–704.
*p < .05; **p < .01.

Table 10.6. Genetic analyses: twin correlations

Measurement/age	MZ	DZ	h^2	c^2
Restraint protest				
14 Months	.45	.36	.19	.26
20 Months	.68	.51	.33	.34*
24 Months	.77	.61	.32*	.45**
DES anger				
14 Months	.66	.35	.62**	.04
20 Months	.69	.44	.51**	.19
24 Months	.74	.55	.36*	.37**
Initiate fights				
14 Months	.87	.65	.44**	.43**
20 Months	.76	.61	.31*	.45**
24 Months	.74	.52	.44**	.30*
Outbursts				
14 Months	.75	.59	.32*	.42**
20 Months	.83	.47	.71**	.12
24 Months	.86	.64	.44**	.42**

N = 159–190 for MZ twin pairs and 135–157 for DZ twin pairs.
*$p < .05$; **$p < .01$.

to low correlations with DES fearfulness and CCTI negative emotion, indicating associations with negative emotionality in both measures.

Genetic Analyses

Table 10.6 presents our twin correlations for measures at each age. One can see from the twin correlations that there is a substantial amount of similarity between twins, both MZ and DZ, across measures with higher levels of correlations for MZ twins. This is reflected in our genetic calculations. Calculations for h^2 (for estimates of heritability) and for c^2 (for estimates of common environmental influences) were made according to the regression method of DeFries and Fulker (1985). Significant genetic influences are indicated for the three maternal-report measures at each of the three ages and for observed restraint protest at 24 months. Influences of the common environment were evident for all of our maternal report measures at 24 months and for two of the maternal report measures at 14 months. Only one common environmental influence was indicated for maternal report measures at 20 months—namely, for fights initiated by the child. Influences of the common environment were indicated for observed restraint protest at 20 and 24 months.

The above twin correlations provide estimates of genetic and environmental influences present at each age only. Another question of major interest has to do with influences across age. We know, for example, that genetic influences may change across age. In other words, genetic influences may be unique at each age or they may exert their influences in common across ages. The same is true for environmental influences, which may either be unique at each age or continuous across ages.

Longitudinal model-fitting was done in order to provide quantitative best estimates of genetic and environmental influences across time. The quantitative model-fitting approach, known as the Cholesky decomposition, and introduced in chapter 2, was used (for more details, see also Cherny et al., 1994). The model generated by this approach takes into account three aspects of the data: stability of measures across time, the degree of genetic relatedness of the twins, and cross-twin, cross-age resemblance for the measures (for each twin type).

Although a full model postulates the same number of factors as measures and estimates their influences at given ages and at subsequent ages, parsimonious models can be constructed with analytic procedures that cut down the number of pathways necessary to estimate a "good fit." The most parsimonious models resulting from such model-fitting procedures are presented in figures 10.1 and 10.2.

Figure 10.1 presents the most parsimonious model for observed restraint protest. A genetic factor is present from 14 months, which contributes to the phenotype at all three ages. There are no significant age-specific genetic factors. In regard to common environment, we can see that, in contrast, new influences are indicated at each of the three ages. (These could be socialization influences in which parents respond to age-related changes in the child, for example.) In addition, the model-fitting indicates common environmental contributions to phenotypic stability. In a more technical sense, we can say that the parsimonious model would not allow us, according to its statistical guidelines, to drop the links between factor 1 and measures 2 and 3—or between factor 2 and measure 3.

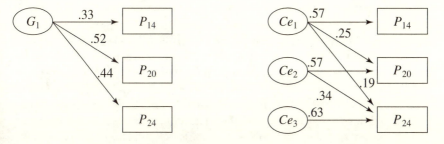

Figure 10.1. Longitudinal model fitting: restraint protest. Genetic (*G*) and common environment (*Ce*) factors contributing to phenotype (*P*) at 14, 20, and 24 months.

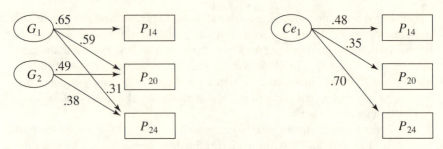

Figure 10.2. Longitudinal model fitting: DES anger. Genetic (G) and common environment (Ce) factors contributing to phenotype (P) at 14, 20, and 24 months.

Figure 10.2 illustrates parsimonious model fitting for one of our maternal-
report measures, DES anger. A single factor accounts for common environmental influences across all three ages. In regard to genetics, both change and
continuity are indicated. New genetic influence on the phenotype is indicated
for 20 months, and both factors influence phenotypes at subsequent ages.

The model fitting for the two other maternal report measures also reveal
distinctively different profiles. (Values are available from the authors on request.) In the case of initiating physical fights, two genetic factors are also
needed for the most parsimonious model. Continuity, as well as change, is
again indicated because genetic influences continue to subsequent ages. The
common environment, in contrast, indicates new influences on the phenotype
at each age with no influences going across age.

For angry outbursts, results of the model fitting are different. We see three
factors of genetic influence. The picture is one of unique genetic influence at
each age on the phenotype and, in addition, of short-term influence across age.
On the common environment side, there are unique influences on the phenotype at each age (three factors are needed), and one influence is indicated
from 14 months that continues to 24 months but not to 20 months.

SUMMARY AND CONCLUSIONS

A striking aspect of our findings concerns the lack of phenotypic correlation
across our primary measures, which were selected as child reactions that have
been associated with the construct of anger. This lack of correlation maintained
in spite of the fact that the separate measures showed coherence in terms of
developmental trends, moderate stability, and genetic and environmental influences on individual differences. This lack of correlation or connectedness
among our measures, in spite of their individual coherence, is intriguing.
Clearly, these response tendencies are not moods but are context specific.
Might their lack of connection also be an example of differentiation of components of a response that would later show connectedness and integration?

The fact that the selected anger-related expressions correlated better with other measures of negative emotion than they did with each other is puzzling. We do know, however, that ratings of restraint protest contained a good deal of general distress, of which anger expressions may have been a part. Perhaps a general propensity for distress could have been included in maternal reports of anger in varying degrees as well.

We are led from the above considerations to an important conclusion. We cannot infer that we have any evidence for a general construct of anger propensity that goes across contexts from our results. Still, the separate response tendencies seen in different contexts remain intriguing, and the behavior genetic modeling analyses also show strikingly different patterns. It may now be useful to review these modeling patterns against the background of the phenotypic developmental patterns we have seen for each measure.

Our observational measure of restraint protest showed a developmental pattern during the second year with an increase from 14 to 20 months and a decrease from 20 to 24 months. There was moderate stability over this time for this measure, and twin correlations indicated significant genetic influence at 24 months with significant common environmental influences at 20 and 24 months. Longitudinal model fitting indicated evidence of genetic continuity, with a genetic factor present at 14 months accounting for phenotypic variance at all three ages. Genetic influences, present early in development, do not seem to change. Common environmental influences showed both continuity and change in that such influences were shared across age, but there were also unique influences at each age. Such results could reflect both age-dependent common socialization influences as well as common socialization influences that are framed early in development and persist.

Our three maternal-report measures showed a similar pattern of increase from 14 to 20 months but had three different patterns from 20 to 24 months— namely, of decline, increase, or no change. Stability was moderate for each of these measures across the second year, and twin correlations indicated significant genetic influence at all three ages; common environmental influences were also prominent at all ages. Longitudinal model fitting indicated a picture of both genetic continuity and change. All three maternal-report measures had influences that were shared across ages, with two (DES anger and initiating fights) having unique genetic influences at 20 months, as well as at 14 months and one (angry outbursts) having unique genetic influences at each of the three ages. Longitudinal model fitting of common environmental influences revealed different patterns for each of the maternal-report measures. These varied from a pattern of strong continuity, with a single factor present at 14 months accounting for phenotypic variance at all ages (DES anger) to a pattern of change with unique factors entering at each age and no shared influences across age (initiating fights). Angry outbursts showed a pattern of continuity and change with respect to common environmental influences. In the case of our maternal-report measures, early-appearing common environmental influences that persist could not only reflect consistent socialization, but could also reflect consistent maternal bias in reporting. Similarly, new influences at subsequent ages could reflect age-dependent maternal perceptual biases as well as age-dependent socialization features.

Overall, we again draw attention to the different developmental patterns of response for our observed measures of reactions to restraint and our maternally reported anger-related expressions during the child's second year. The results were unexpected and do not support a single construct of anger across the separate contexts, of our measures. Perhaps if we had other measurement contexts we would have found more evidence of connectedness. Still, there is a more interesting possibility. Perhaps anger as we usually think of it is less differentiated during the second year. Perhaps the measures we used would be more related to each other later in development. It remains for future research to ascertain if there is evidence for later connectedness and integration and thus evidence for a later-age family of anger-related responses such as we had originally envisioned.

REFERENCES

Averill, J. R. (1979). Anger. In H. E. How & R. A. Dienstbier (Eds.), *Nebraska symposium on motivation: Human emotions* (Vol. 26, pp. 1–80). Lincoln: University of Nebraska Press.

Barrett, K. C., & Campos, J. J. (1987). Perspectives on emotional development II: A functionalist approach to emotions. In J. D. Osofsky (Ed.), *Handbook of infant development* (2nd ed.,) (pp. 555–578). New York: Wiley.

Campos, J. J., Mumme, D. L., Kermoian, R., & Campos, R. G. (1994). A functionalist perspective on the nature of emotion (commentary). In N. A. Fox (Ed.), The development of emotion regulation. *Monographs of the Society for Research in Child Development, 59* (2–3, Serial No. 240).

Camras, L. A., Campos, J. J., Oster, H., Miyake, K., & Bradshaw, D. (1992). Japanese and American infants' responses to arm restraint. *Developmental Psychology, 28,* 578–583.

Cherny, S. S., Fulker, D. W., Emde, R. N., Robinson, J., Corley, R. P., Reznick, J. S., Plomin, R., & DeFries, J. C. (1994). A developmental-genetic analysis of continuity and change in the Bayley Mental Development Index from 14 to 24 months: The MacArthur Longitudinal Twin Study. *Psychological Science, 5,* 354–360.

DeFries, J. D., & Fulker, D. W. (1985). Multiple regression analysis of twin data. *Behavior Genetics, 15,* 467–473.

Dunn, J., & Kendrick, C. (1982). *Siblings: Love, envy, and understanding.* Cambridge, MA: Harvard University Press.

Emde, R. N., Johnson, W. F., & Easterbrooks, M. A. (1987). The do's and don'ts of early moral development: Psychoanalytic tradition and current research. In J. Kagan & S. Lamb (Eds.), *The emergence of morality in young children* (pp. 245–277). Chicago: University of Chicago Press.

Emde, R. N., Izard, C., Huebner, R., Sorce, J. F., & Klinnert, M. D. (1985). Adult judgments of infant emotions: Replication studies within and across laboratories. *Infant Behavior and Development, 8*(1), 79–88.

Fischer, K. W., & Tangney, J. P. (1995). Self-conscious emotions and the affect revolution: Framework and overview. In J. P. Tangney & K. W. Fischer (Eds.), *Self-conscious emotions—The psychology of shame, guilt, embarrassment, and pride* (pp. 3–22). New York: Guilford Press.

Fox, N. A. (1989). Infant response to frustrating and mildly stressful events: A positive look at anger in the first year. *New Directions for Child Development, 45,* 47–64.

Fox, N. A., & Davidson, R. J. (1988). Patterns of brain electrical activity during facial signs of emotion in 10-month-old infants. *Developmental Psychology, 24,* 230–236.

Goldsmith, H., & Campos, J. (1982). Toward a theory of infant temperament. In R. N. Emde & R. J. Harmon (Eds.), *The development of attachment and affiliative systems* (pp. 161–193). New York: Plenum.

Goldsmith, H. H., & Rothbart, M. K. (1988). *The Laboratory Temperament Assessment Battery (LAB-TAB): Locomotor version.* Oregon Center for the Study of Emotions, Technical Report No. 88–01. Eugene, University of Oregon.

Goodenough, F. L. (1931). *Anger in young children.* Minneapolis: University of Minnesota Press.

Izard, C. E., Huebner, R., Risser, D., McGinnes, G. C., & Dougherty, L. (1980). The young infant's ability to produce discrete emotional expressions. *Developmental Psychology, 16*(2), 132–140.

Klinnert, M. D., Sorce, J. F., Emde, R. N., Stenberg, C., & Gaensbauer, T. (1984). Continuities and change in early emotional life. In R. N. Emde & R. J. Harmon (Eds.), *Continuities and discontinuities in development* (pp. 339–354). New York: Plenum.

Lazarus, R. S., (1991). *Emotion and adaptation.* New York: Oxford University Press.

Lewis, M., Alessandri, S. M., & Sullivan, M. W. (1990). Violation of expectancy, loss of control, and anger expressions in young infants. *Developmental Psychology, 26,* 745–751.

Mandler, G. (1990). A constructivist theory of emotion. In N. L. Stein, B. Leventhal, & T. Trabasso (Eds.), *Psychological and biological approaches to emotion* (pp. 21–43). Hillsdale, NJ: Erlbaum.

Miller, P., & Sperry L. L. (1987). The socialization of anger and aggression. *Merrill-Palmer Quarterly, 33,* 1–31.

Plomin, R., Emde, R. N., Braungart, J. M., Campos, J., Corley, R., Fulker, D. W., Kagan, J., Reznick, J. S., Robinson, J., Zahn-Waxler, C., & DeFries, J. C. (1993). Genetic change and continuity from 14 to 20 months: The MacArthur Longitudinal Twin Study. *Child Development, 64*(5), 1354–1376.

Radke-Yarrow, M., & Kochanska, G. (1990) Anger in young children. In N. L. Stein, B. Leventhal, & T. Trabasso (Eds.), *Psychological and biological approaches to emotion* (pp. 297–310). Hillsdale, NJ: Erlbaum.

Stein, N. L., & Levine, L. J. (1990). Making sense out of emotion: The representation and use of goal-structured knowledge. In N. L. Stein, B. Leventhal, & T. Trabasso (Eds.), *Psychological and biological approaches to emotion* (pp. 45–73). Hillsdale, NJ: Erlbaum.

Stenberg, C. R. (1982). The development of anger in infancy. Unpublished doctoral dissertation, University of Denver. Denver, Colorado.

Stenberg, C. R., & Campos, J. J. (1990). The development of anger expressions in infancy. In N. L. Stein, B. Leventhal, & T. Trabasso (Eds.), *Psychological and biological approaches to emotion* (pp. 247–282). Hillsdale, NJ: Erlbaum.

Stenberg, C. R., Campos, J. J., & Emde, R. N. (1983). The facial expression of anger in seven-month-old infants. *Child Development, 54,* 178–184.

11

Empathy and Prosocial Patterns in Young MZ and DZ Twins
Development and Genetic and Environmental Influences

Carolyn Zahn-Waxler
Kimberlea Schiro
JoAnn L. Robinson
Robert N. Emde
Stephanie Schmitz

This chapter focuses on empathic development in the second and third years of life. Patterns of heritability, continuity, and change over this time period are considered, extending earlier twin research on genetic and environmental influences (Zahn-Waxler et al., 1992b) to include later time points. In the MacArthur Longitudinal Twin Study, ages were chosen so as to bracket a major developmental transition from 20 to 24 months. This transition period includes marked changes in cognitive and linguistic abilities, self-regulation, reflective self-awareness and self-differentiation, the appreciation of standards, and a moral sense (Kagan & Lamb, 1987; Plomin et al., 1993; Zahn-Waxler & Radke-Yarrow, 1990). We viewed this as a unique opportunity to consider children's patterns of caring for others in the context of these other pivotal developmental transformations.

Many studies of the development of empathy and prosocial behavior have been conducted in the past few decades. Most of the work has been based on age cross-sectional research designs. In a meta-analysis of age differences in children's expressions of concern for others (Eisenberg & Fabes, 1998), the following age groupings were constructed: (1) infants (less than 3 years of age), (2) preschool (3–6 years of age), (3) childhood (7–12 years of age), (4) adolescents (13–17 years of age), and (5) young adults (18–21 years of age). Comparisons were made both across and within age groups. There was a significant, positive effect size, indicating increases in prosocial behavior with age, when all studies were combined. Although the same direction of effect was found for each of the specific age-group comparisons, it was not significant for the infant and preschool groupings. This could indicate that this aspect of moral development is not seen in early childhood. However, it also could reflect the fact that less developmental research on this topic has been conducted during the early years of life. Hence, there is less established empirical literature from which to derive reliable, valid generalizations. The MacArthur Longitudinal Twin Study is the first to provide an extensive database on empathic development for a large sample of young children at multiple time points.

This investigation of young monozygotic (MZ) and dizygotic (DZ) twins also makes it possible to investigate genetic influence on early expressions of concern for others. Far more research has been devoted to environmental than to early child predispositions or underlying biological bases (Eisenberg & Fabes, 1998; Eisenberg et al., 1990). Socialization practices shown to promote concern for the welfare of others have included modeling, reinforcement, instruction, discussion of others' feelings, role playing, appeals to the feelings of others following rule violations, and also a family atmosphere of nurturance, affection, and low conflict (see reviews by Eisenberg & Fabes, 1998; Grusec & Lytton, 1988; Maccoby & Martin, 1983; Radke-Yarrow et al., 1983). Although socialization plays a significant role in expressions of empathy, there is substantial unexplained variance in these studies, indicating the need to consider constitutional or dispositional factors as well.

Several biological explanations for the presence of empathy have been proposed. Some emphasize the selfish or self-serving nature of altruism (Dawkins, 1989), originating in Wilson's sociobiological speculations about the evolution of altruistic behavior (Wilson, 1975, 1978). In this view, self-sacrificing actions increase an animal's probability that its relatives (who share some of the same genes) will survive and reproduce. Hence, even if prosocial acts lead an animal to perish, shared genes are passed on via the relatives to future generations. Other sociobiologists have argued that altruism could have evolved through reciprocity, even where provider and recipient are not biologically related (Trivers, 1983); that is, histories of benefit build over time through processes of reciprocal exchange.

Not all biological theories begin with the premise that patterns of caring for others ultimately reflect selfish intent. In MacLean's view (1985), empathy emerged with the evolution of mammals in the familial context of extended caregiving, affiliation, and responsibility for others. Both MacLean (1985) and Panksepp (1986) discuss brain processes, focusing particularly on the limbic system, that mediate the formation of social attachments and empathy. MacLean (1973 p. 42) emphasizes the prefrontal cortex as well and the neural interconnections that may enable an individual to "feel one's way into another person in the sense of empathy." Extending this line of reasoning to early human development, Hoffman (1977) suggested that these brain connections integrate primitive emotional responses with higher order cognitive awareness, resulting in empathic arousal that predisposes individuals to altruism.

A biological, evolutionary perspective on empathy, focusing on neurological bases of emotional communication, is also espoused by Brothers (1989). Laboratory studies of nonhuman primates suggest both the presence of an inborn capacity to understand facial expressions and the role of later social experience in responding appropriately to others. Complex facial motor patterns are present at a very early age that allow human infants to match the facial emotion expressions of others (e.g., Field et al., 1982; Haviland & Lelivica, 1987; Meltzoff & Moore, 1977). The capacity for empathy is seen as "present in some precursor form at birth in the normal brain and is elaborated by cognitive maturation and by subsequent experiences in the social milieu" (Brothers, 1989, p. 13).

In Hoffman's (1975, 1982) theory of empathic development in humans, empathy is defined as a vicarious affective response that is more appropriate to someone else's situation than to one's own situation. It is an innate, hard-wired response connecting humans as social beings to the emotional plights of others. It becomes manifest in the first days of life as a precursor pattern termed "global empathy." Here the distress cues of another (e.g., a crying infant) become confounded with unpleasant feelings (i.e., personal distress) aroused in the self. The work of Simner (1971) and Sagi and Hoffman (1976) confirms the uniquely social nature of infants' reflexive crying in infants when they react to other infants' cries; infants do not react as strongly to equally noxious, nonsocial stimuli.

Sometime in the second year of life, global empathy is followed by egocentric empathy. Self-distress begins to diminish as children become aware that another person (rather than the self) is in distress. But children still are more likely to comfort victims in ways that would ameliorate their own distress. With the onset of role taking during the third year of life, children become more aware that other people's needs and feelings may differ from one's own. In this third stage, "empathy for another's feelings," children become able to empathize with a wider range of emotions. They also establish increasingly elaborated prosocial repertoires in their developing capacity to help and comfort others. The fourth level, "empathy for another's life condition," is thought to emerge by late childhood. It extends beyond immediate situations to more distant, abstract circumstances (e.g., orphans). This theory has, in large part, subsequently received empirical confirmation (Zahn-Waxler & Radke-Yarrow, 1982).

Evolutionary theories of empathy and prosocial behavior clearly cannot be tested directly. Nor is it readily possible to distinguish different motives for individuals' caring patterns (e.g., selfish versus selfless), though some have tried (see Batson & Shaw, 1991). One approach to understanding potentially innate individual differences in empathy has been to consider different dispositions, temperament, or personality characteristics that predict caring behaviors (see review by Eisenberg, et al., 1990). Biologically based differences also have been examined in terms of physiological profiles. Psychophysiological manifestations of empathic, other-oriented concern (e.g., heart rate deceleration) and self-distress (e.g., heart rate acceleration) in response to another's distress have been identified (Eisenberg et al., 1988, 1989a, 1991; Fabes et al., 1993).

With regard to genetic influence, adult MZ and DZ twins have been compared on their self-reported empathy and altruism (Davis, et al., 1994; Loehlin & Nichols, 1976; Matthews, et al., 1981; Rushton et al., 1986). A strong heritable component often has been identified, suggesting a disposition toward caring and kindness. Davis et al. (1994) examined the heritability of three facets of empathy: empathic concern, personal distress, and perspective taking. Genetic influence was present for the two affective dimensions but not for perspective taking. This is consistent with the view that aspects of emotionality are part of temperament patterns that may underlie the heritability of empathy. In a study of prosocial patterns in elementary-school–age children (Segal, 1984), greater cooperation was observed between MZ than between DZ twins. In the present study of young children, it was possible to examine genetic influence prior to long socialization histories.

Empathic and prosocial responses to distress in others were assessed in two longitudinal studies in one- to two-and-a-half-year-old children, using both naturalistic observations and standardized distress probes (Zahn-Waxler & Radke-Yarrow, 1982; Zahn-Waxler et al., 1992a). During the second and third years of life, self-distress waned, and a constellation of caring patterns emerged (cognitive exploration of distress, empathic concern, and prosocial behaviors). These components of caring increased with age, with stable individual differences also present. Sharing, cooperation, and conscience also emerge around this point in time. Hence, early empathic development occurs within a broader context of moral internalization (see review by Zahn-Waxler & Radke-Yarrow, 1990). Developmental change also was identified in the present twin sample, between 14 and 20 months of age (Zahn-Waxler et al., 1992b). Modest genetic influence was present for observed empathy, more strongly at 14 months than 20 months. Modest genetic influence also was found for prosocial behavior at both time points, based on maternal report.

In this chapter, developmental and genetic analyses are extended to include two later time points, 24 and 36 months. Patterns of developmental continuity and change, as well as genetic and environmental influences also are examined. Cognitive, affective, and behavioral expressions of concern for others again are considered. Genetically mediated change and continuity as well as environmental sources of change and continuity are examined at multiple time points (see Plomin, et al., 1993 for details).

METHOD

Subjects

Sample characteristics are reported in chapter 3. Sample sizes for the research issues addressed in this chapter were 115–138 MZ twin pairs (66–74 females; 49–64 males) and 95–108 DZ twin pairs (45–49 females; 49–59 males). Sample sizes varied, depending on the age point.

Procedure

At 14, 20, 24, and 36 months, the twins and their mothers were visited at home by two female examiners. One to three weeks later the twin pair visited the laboratory with their mother. During these visits multiple observational measures were obtained within domains of cognition-language, temperament, and emotion. Empathy probes were inserted into these sessions, similarly at each age. Maternal reports of children's prosocial behaviors at each time period were obtained during the course of a maternal interview approximately 45 min in length. Interviews were conducted over the phone, typically between home and laboratory visits.

Empathy probes

Simulation procedures were used in the home and laboratory visits to assess the empathic capabilities of the children. They were done at different points during the session for each twin, when the other twin was not present. Children's reactions were videotaped, with all reliability coding and scoring done from the tapes. A different observer was assigned to each twin in a dyad. The mother and the experimenter simulated distress according to specified scripts. During the home visit, the experimenter pretended to close her finger in a suitcase containing testing materials. The mother pretended to hurt her knee as she got up from the floor. During the laboratory visit, the mother caught her finger in a clipboard, and the examiner bumped into a chair. All simulated actions were accompanied by pain vocalizations at low to moderate volume and pained facial expressions over a 30 s period, with a gradual subsiding of the distress for an additional 30 s. We used these four probes at each of the four time points. For a fifth probe, a tape recording of an infant cry was played over a loudspeaker during the laboratory visit (at the first three time points). For a more in-depth description of these procedures, refer to chapter 3.

The following components of children's responses were examined:

1. *Prosocial acts*: presence versus absence of efforts to help or comfort victim (e.g., gets Band-Aid or pats victim).
2. *Hypothesis testing*: explores and/or attempts to comprehend distress rated on a 4-point scale where 1 = none, 2 = simple nonverbal (e.g., looking from injury to victim's face) or non-verbal (e.g., a single utterance, "Hurt?"), 3 = combinations of nonverbal and verbal exploration (e.g., looking at the injury and its cause and inquiring "Owie?"), 4 = repeated, sophisticated attempts to comprehend the problem (e.g., asking "Does it hurt? How happen? Need kiss?" or looking behind or underneath injury to ascertain cause).
3. *Empathic concern*: expressions of apparent concern to the victim, including facial, vocal, or gestural-postural expressions rated on a 4-point scale, where 1 = absent, 2 = slight (fleeting or slight change of expres-

sion that includes brow furrow), 3 = moderate (sustained sobering of expression that includes brow furrow), 4 = substantial concern (sustained sadness expressed in cooing or sympathetic vocal tones or sympathy face in which the eyebrows are drawn down and brow drawn up over the nose).

4. *Self-distress*: expressions of fear or states of personal distress, rated on a 5-point scale, where 1 = none, 2 = fear present for several seconds (eyes wide and mouth open), 3 = facial grimace with eyes wide, 4 = whimpers, 5 = full blown cries.

5. *Unresponsive actively indifferent*: presence versus absence of uninvolvement or negative involvement in the distress. Presence was scored if the child ignored the victim, withdrew, avoided, or expressed active indifference through callous or aggressive behaviors, such as blaming or hitting the victim.

We used Cohen's kappa to calculate interobserver reliability. Agreement was checked periodically, typically once every 6 months. Each reliability check involved independent scoring of 20–30 distress simulations. For prosocial acts and unresponsive indifferent, agreement was based on presence versus absence. For hypothesis testing, empathic concern, and personal distress, agreement occurred when the same scale point was scored by both observers. Intraclass correlations were used to assess reliability. Eighty-nine paired ratings of children across all four age groups yielded the following coefficients; .90 for prosocial acts, .85 for hypothesis testing, .84 for empathic concern, .82 for self-distress, and .65 for indifference.

Maternal Reports

During the interviews the mothers were asked to rate on a 5-point scale the degree to which each twin showed (1) spontaneous help toward the other twin, (2) sharing and cooperation with the other twin, and (3) sharing and cooperation with another child. They also rated prosocial behaviors toward others in distress on a 3-point scale. Responses were z scored and summed to create a composite measure of prosocial behavior for each twin.

RESULTS

Developmental change, sex of child, and zygosity are considered first, followed by examination of interrelations of measures and continuity across time. Correlations between MZ and DZ pairs are then presented (within and across time periods), followed by heritability estimates. We used six univariate analyses of variance to examine children's responses to distress. Separate ANOVAs were conducted for each of the five observational measures and for the maternal report of prosocial behaviors. Sex and zygosity were between-subjects factors and age a within-subjects factor. Analyses were based on composite scores, averaged across the total number of stimuli to which the child was exposed at each age. We based genetic analyses of maternal interview reports on a score derived from principal components analyses.

Interrelations among variables, within time points, indicated that hypothesis testing, empathic concern, and prosocial behaviors were correlated at each of the four time points. The magnitude of associations was typically about .30 with associated p values of $<.001$. Correlations did not differ as a function of the zygosity or the sex of child. Self-distress was positively correlated with empathic concern at each time point. These associations were low in magnitude but significant at p $<.05$. Analyses conducted separately for boys and girls indicated these correlations were significant for girls but not for boys. There were moderate negative correlations between indifference and the different measures of caring patterns, with correlations averaging about .30. These correlations did not differ as a function of zygosity or sex. Observational and maternal report data were uncorrelated.

Age Changes and Sex Differences

Tables 11.1 and 11.2 provide the means and standard deviations for MZ and DZ twins, respectively, for responses at each of the four time points, separately for boys and girls. There were significant mean increases with age in prosocial behavior ($F_{3,368} = 48.17$, p $<.001$), empathic concern ($F_{3,368} = 57.39$, p $<.001$), and hypothesis testing ($F_{3,368} = 256.06$, p $<.001$), with older children scoring higher than younger children on each of these dimensions. There were significant main effects of sex, with girls scoring higher than boys, for hypothesis testing ($F_{1,370} = 8.83$, p $<.01$), prosocial behavior ($F_{1,370} = 4.53$, p $<.05$), and empathic concern ($F_{1,370} = 21.32$, p $<.001$). There were no significant interactions.

Self-distress and indifference also differed as a function of age and sex of the child. Self-distress decreased with age, leveling off at 24 months, ($F_{3,368} = 7.68$, p $<.001$). Girls showed more self-distress than boys at each age point ($F_{1,370} = 6.21$, p $<.001$). Indifference decreased from 14 to 20 months and then increased again at 24 and 36 months, sometimes exceeding the 14 month level ($F_{3,377} = 28.96$, p $<.001$). Boys showed more indifference to others' distress than did girls ($F_{1,379} = 14.55$, p $<.001$).

Maternal reports of prosocial behavior indicated a significant increase with age ($F_{3,364} = 80.61$, p $<.001$). Girls were reported to show more prosocial behavior than boys ($F_{1,366} = 34.80$, p $<.001$). There were no differences between MZ and DZ twins either on observational or maternal report measures.

Table 11.1. Developmental changes in MZ twins' responses to distress

Type of response to distress	14 Months		20 Months		24 Months		36 Months	
	Girls	Boys	Girls	Boys	Girls	Boys	Girls	Boys
Prosocial acts								
Mean	0.13	0.11	0.14	0.13	0.23	0.18	0.27	0.30
SD	0.18	0.17	0.17	0.18	0.24	0.18	0.27	0.27
Concern for victim								
Mean	2.26	2.08	2.52	2.35	2.55	2.38	2.58	2.54
SD	0.46	0.44	0.41	0.40	0.39	0.35	0.42	0.40
Hypothesis testing								
Mean	2.21	2.03	2.59	2.37	3.17	3.00	3.03	3.18
SD	0.48	0.40	0.56	0.49	0.71	0.68	0.77	0.88
Self-distress								
Mean	1.33	1.25	1.32	1.20	1.18	1.13	1.20	1.15
SD	0.48	0.34	0.44	0.34	0.33	0.29	0.33	0.31
Unresponsive-indifferent								
Mean	0.31	0.45	0.22	0.31	0.33	0.34	0.38	0.41
SD	0.22	0.22	0.26	0.22	0.24	0.26	0.27	0.28
Maternal report of prosocial behavior								
Mean	3.05	2.41	3.82	3.07	3.78	3.27	4.05	3.49
SD	1.01	1.24	0.92	1.02	0.95	1.02	0.80	0.88

Heritability of Responses to Distress

Twin Correlations within Time Periods

A significantly greater correlation between MZ twins (who are genetically identical) and DZ twins (who share only half their segregating genes on average) indicates genetic influence on behavior. Correlations were used first to describe heritability patterns. These correlations are presented in table 11.3, for each measure at each time point. Genetic influence is suggested for the observational measures for each of the components of caring (hypothesis testing, empathic concern, and prosocial behavior). MZ correlations were significant at each of the four time points for empathic concern and prosocial behavior, whereas no DZ correlations were significant. Both MZ and DZ correlations were significant for hypothesis testing, except for one time point, with MZ

Table 11.2. Developmental changes in DZ twins' responses to distress

Type of response to distress	14 Months		20 Months		24 Months		36 Months	
	Girls	Boys	Girls	Boys	Girls	Boys	Girls	Boys
Prosocial acts								
Mean	0.12	0.09	0.14	0.12	0.22	0.19	0.33	0.23
SD	0.15	0.13	0.19	0.18	0.21	0.19	0.30	0.24
Concern for victim								
Mean	2.26	2.18	2.58	2.43	2.55	2.43	2.59	2.55
SD	0.40	0.42	0.40	0.45	0.41	0.40	0.38	0.45
Hypothesis testing								
Mean	2.21	2.12	2.58	2.41	3.16	2.95	3.19	3.14
SD	0.34	0.36	0.56	0.62	0.70	0.66	0.75	0.79
Self-distress								
Mean	1.28	1.31	1.26	1.21	1.21	1.16	1.20	1.19
SD	0.40	0.47	0.35	0.33	0.38	0.38	0.33	0.40
Unresponsive-indifferent								
Mean	0.33	0.35	0.19	0.28	0.31	0.38	0.40	0.42
SD	0.28	0.25	0.21	0.23	0.25	0.25	0.32	0.29
Maternal report of prosocial behavior								
Mean	2.95	2.65	3.62	3.33	3.78	3.46	3.83	3.47
SD	1.12	1.41	0.97	1.23	0.84	1.00	0.90	0.92

Table 11.3. Twin correlations for monozygotic and dizygotic twins, within time periods for individual measures

Responses to distress	14 Months		20 Months		24 Months		36 Months	
	MZ	DZ	MZ	DZ	MZ	DZ	MZ	DZ
Prosocial acts	.36***	−.04	.39***	.19	.20	.14	.43***	.19
Concern for victim	.21*	.02	.33***	.18	.27***	.06	.25***	.08
Hypothesis-testing	.34***	.20*	.14	.20*	.53***	.36***	.43***	.21*
Self-distress	.26**	.16	.08	−.08	−.04	.20*	.15	.08
Unresponsive-indifferent	.28***	.10	.26**	.26**	.22*	.34***	.20*	−.01
Maternal report	.89***	.74***	.87***	.83***	.82***	.74***	.63***	.76***

*p < .05; **p < .01; ***p < .001.

correlations typically exceeding DZ correlations. Self-distress and indifference did not show clear, consistent patterns. For the maternal report measure of prosocial behavior, twin correlations were high, regardless of zygosity, suggesting stronger shared environmental than genetic influences.

Heritability Estimates

The regression model formulated by DeFries and Fulker (1985) was used to provide estimates of heritability (h^2) and shared environmental influences (c^2) (see chapter 2). Table 11.4 provides the model-fitting estimates for genetic and environmental parameters. Dashes in table 11.4 indicate the instances in which constrained estimation procedures were used.

Table 11.4 gives the estimates of genetic and shared environmental influences. There was consistent genetic influence for each of the components of caring, confirming the significance of differences in MZ and DZ twin correlations observed in table 11.3. Prosocial acts, empathic concern, and hypothesis testing each showed significant genetic influence at three out of four time periods. Shared environmental influences did not emerge as a significant factor. There was no evidence for genetic or shared environmental influences for self-distress. An unusual pattern emerged for indifference, with genetic influence (only) present at 14 and 36 months, and shared environmental influence (only) at 20 and 24 months. The maternal-report measure yielded strong effects of the shared environment at all time points and genetic influence only at 14 months.

Twin Correlations across Time

Genetically mediated change and continuity, as well as environmental sources of change and continuity, were examined across time points. Age-to-age genetic continuity was first examined in terms of cross-twin correlations. A composite measure was constructed, summing the scores for children's prosocial acts, empathic concern, and hypothesis testing. This provided a more stable estimate of the principal construct of interest and reduce the number of analyses. Correlations of one twin's score at an earlier time point with the cotwin's score at a later time point are presented in table 11.5 (for the observational measure and in Table 11.6 (for the maternal report measure). Correlations on the diagonal represent the within-time associations. Correlations above the diagonal represent twin 1 at an earlier time period with twin 2 at a later time period; correlations below the diagonal represent twin 2 at an earlier time period with twin 1 at a later time period.

Inspection of table 11.5 indicates five significant off-diagonal correlations for MZ twins, but only one for DZ twins, mainly at immediately adjacent time points, suggesting some genetic influence on continuity across time for observed empathy. Tables 11.6 shows a contrasting pattern for maternal reports of prosocial behavior. Significant correlations are present for both MZ and DZ twins at virtually all time points, suggesting strong, consistent shared environmental influences.

Table 11.4. Genetic and environmental parameter estimates at 14, 20, 24, 36 months

	Model-fitting estimates			
Variable/age	h^2	SE	c^2	SE
Prosocial acts				
14 Months	.30***	.05	—	—
20 Months	.38***	.05	—	—
24 Months	.10	.19	.08	.16
36 Months	.39***	.06	—	—
Concern for victim				
14 Months	.15**	.05	—	—
20 Months	.30	.18	.01	.15
24 Months	.19**	.06	—	—
36 Months	.23***	.06	—	—
Hypothesis testing				
14 Months	.31***	.05	—	—
20 Months	—	—	.14**	.05
24 Months	.35*	.17	.17	.14
36 Months	.42***	.06	—	—
Self-distress				
14 Months	.17	.18	.07	.14
20 Months	.03	.06	—	—
24 Months	—	—	.07	.05
36 Months	.15	.19	.00	.15
Indifference				
14 Months	.25***	.06	—	—
20 Months	—	—	.23**	.05
24 Months	—	—	.27***	.04
36 Months	.15*	.06	—	—
Maternal report of prosocial behavior				
14 Months	.28***	.11	.59***	.08
20 Months	.05	.10	.80***	.08
24 Months	.17	.12	.65***	.09
36 Months	—	—	.62***	.04

h^2 = heritability; c^2 = shared environment. A dash indicates that estimates were constrained, hence yielding modified estimates.
*$p < .05$; **$p < .01$; ***$p < .001$.

Table 11.5. Twin correlations within and across time: composite empathy measure

Twin 1	Twin 2			
	14 Months	20 Months	24 Months	36 Months
MZ twins				
14 Months	.33***	.06	.06	.17
20 Months	.30***	.21*	.11	.12
24 Months	.18*	.20*	.51***	.38***
36 Months	.08	.08	.28**	.43***
DZ twins				
14 Months	.05	.11	.24**	.03
20 Months	−.01	.22*	.17	.03
24 Months	.12	.17	.26**	.07
36 Months	−.07	.12	.13	.21

*$p < .05$; **$p < .01$; ***$p < .001$.

Table 11.6. Twin correlations within and across time: maternal report of prosocial behavior

Twin 1	Twin 2			
	14 Months	20 Months	24 Months	36 Months
MZ twins				
14 Months	.87***	.46***	.36***	.32***
20 Months	.40***	.87*	.44***	.38***
24 Months	.25**	.34***	.82***	.32***
36 Months	.32***	.42***	.37***	.63***
DZ twins				
14 Months	.74***	.47***	.43***	.42***
20 Months	.39***	.83***	.44***	.24*
24 Months	.33***	.42***	.74***	.25*
36 Months	.23*	.19	.32**	.76***

*$p < .05$; **$p < .01$; ***$p < .001$.

Longitudinal model fitting using Mx statistical methods (Neale, 1994) to evaluate the strength of genetic and environmental influences in a multivariate context was conducted. Parameter estimates in the multivariate case may differ from the univariate case, both because of the different modeling approach and because of the observed covariances between the measures that the multivariate model fitting must take into account. In general, for correlated measures, multivariate analyses are more powerful than univariate analyses in detecting significant effects. Model fitting was then conducted for the aggregate measure of observed empathy and for the maternal report of prosocial behavior across the four time points.

The best estimate for observed empathy is the full model for the entire sample; there were no gender differences for parameter estimates for the various time points. Figure 11.1 indicates a moderate genetic effect at 14 months, which contributed to continuity across the later ages. At 20 months, no new

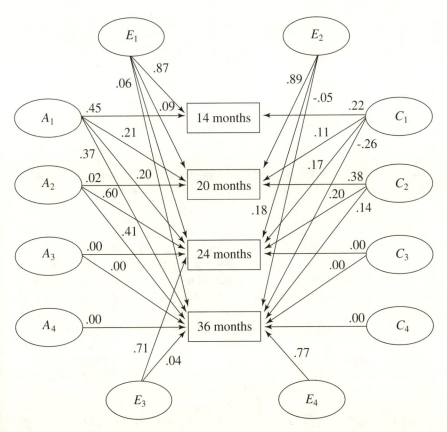

Figure 11.1. Longitudinal model fitting for aggregate measure of observed empathy in MZ and DZ twins.

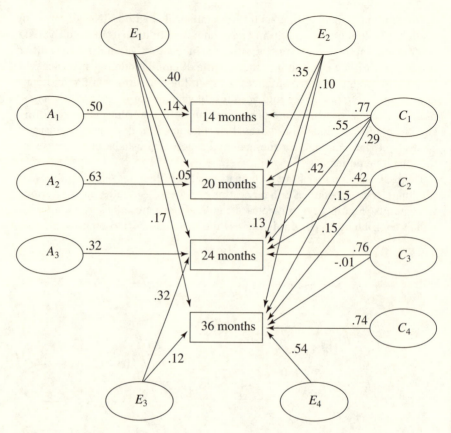

Figure 11.2. Longitudinal model fitting for maternal report of prosocial behavior in MZ and DZ girls.

substantial genetic effect was detected, but continuity from this parameter was observed at 24 and 36 months. No new genetic influences appeared at 24 or 36 months. With regard to common environmental effects, the estimate is weak at 14 months, with weak continuity across time. At 20 months, however, additional variance is explained by a second common environmental parameter, which also shows continuity across later ages. No new common environmental influences appeared at 24 or 36 months. In general, nonshared environment estimates suggest measurement error and little continuity across time.

For the maternal report of prosocial behavior, a model that estimates parameters separately for boys and girls fit the data significantly better than one that estimates parameters for the whole sample. Examination of the best model for girls (see figure 11.2) indicated that genetic effects were age specific for 14, 20, and 24 months and did not contribute to continuity over time. Estimates of common environment, however, showed large age-specific effects at every age and the first common environmental factor also contributed to substantial continuity across the later ages. Nonshared environmental influences were largely age specific, but modest contributions to stability were seen at 14, 20, and 24 months, suggesting that meaningful variance of the nonshared environment is exerting an effect on mothers' reports of girls' prosocial patterns.

For boys, the best model yielded a somewhat different picture (see figure 11.3). Two genetic parameters were estimated, showing substantial genetic influence at 14 and 24 months of age. There were no genetic contributions to continuity. However, four common environmental parameters were estimated, each contributing to age-specific phenotypic variance as well as continuity. The nonshared environmental parameter estimates reflect time-specific influence only. Thus, boys and girls differed somewhat in the degree to which new genetic influences were introduced across the second and third year and in

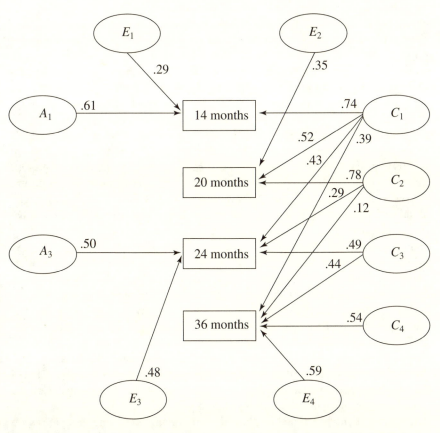

Figure 11.3. Longitudinal model fitting for maternal report of prosocial behavior in MZ and DZ boys.

the role that unique environmental circumstances, events, and relationships played in contributing to continuity over time.

DISCUSSION

The existence of an altruistic personality, a disposition to be compassionate and caring, has often been considered in studies of both children and adults. Research has shown some stability across time, contexts, and different forms of caring, providing evidence for traitlike patterns (e.g., see reviews by Eisenberg & Fabes, 1998; Radke-Yarrow et al., 1983). However, the magnitude of correlations across studies is modest, and patterns are sometimes inconsistent. Hence, the evidence can be used either to support or refute claims of a disposition or trait (also see Batson et al., 1986; Eisenberg, et al., 1989b; Hampson, 1981; Knight et al., 1994; Rehberg & Richman, 1989). Other investigations have focused on associations between empathy/altruism and personality traits such as affiliation, sociability, and shyness (Eisenberg-Berg & Hand, 1979; Eisenberg, et al., 1984; Farver & Branstetter, 1994; Stanhope et al., 1987). Implicit in research focusing on traits and dispositions is the notion of biological underpinnings of caring orientations. Twin and adoption studies, as well as research on autonomic nervous system activity, represent yet other approaches to gain understanding of empathy and altruism as part of our evolutionary heritage.

Most of the research on genetic bases of concern for others has focused on older children, adolescents, and adults. The present study considers the development and unfolding of these patterns in a large sample of MZ and DZ twins in the first years of life. The developmental transitions identified are mainly consistent with earlier research. They indicate a progression from self-distress, conceptualized by Hoffman (1975) as the first primitive form of empathy, to more modulated and multifaceted expressions of caring whereby emotional concern becomes more outer and other directed. Cognitive, affective, and behavioral components of caring increase with age. Moreover, they cohere to form an integrated expression of empathy. A composite of these measures shows continuity over time, indicative of stable individual differences and possible early evidence for empathy as a trait. Visual comparisons of response patterns of the twin samples here with those of singletons seen in earlier normative studies suggest that prosocial behavior may occur less frequently in twins than in nontwins. However, research designs that simultaneously include twins and nontwins have not yet been used. It would be of interest to assess directly whether aspects of self-other differentiation, evidenced by role taking and perspective taking, develop more slowly in twins than in nontwins and contribute to delayed expression of prosocial actions despite apparently similar levels of emotional concern.

Girls scored higher than boys on all components of caring, as well as on self-distress. Such findings have been documented numerous times. What is of interest here is how early the differences emerge. It is worth noting, however, that the differences are not dramatic, indicating that many boys at this young age are just as capable as girls of showing concern for others. Also of interest is the fact that self-distress is higher in girls than in boys. According to Hoffman's (1975) theory and empirical research (Zahn-Waxler et al., 1992a), self-distress wanes during the second year of life and is replaced by more modulated, other-directed expressions of concern. These replacements, however, differ for girls and boys. Although personal distress decreases with age, it does so more slowly and remains at a higher level in girls than in boys.

There are alternative views of the conceptual linkage between personal distress and empathic concern. In one approach, the two constructs are conceptualized as opposites. Empathic concern is outer directed and should facilitate prosocial activity, whereas personal distress leads the individual to turn inward and hence away from the needs of the other person (Batson et al., 1988). There is supportive evidence for this hypothesis in laboratory studies with older children and adults (Batson, et al., 1988; Eisenberg et al., 1989a). Another view (Zahn-Waxler & Radke-Yarrow, 1990) stems from Hoffman's (1975) theory in which self-distress is a primitive form of empathy and hence a precursor (and sometimes correlate) of prosocial behavior. The data on young children support the latter position (Zahn-Waxler et al., 1992a).

Under some circumstances, the ability to focus concern on others, in conjunction with anxiety/self-distress, may result in particularly potent expressions of empathy. A recent study provides supportive evidence (Zahn-Waxler et al., 1995). Girls were more prosocial than boys, but also showed higher autonomic arousal (heart rate and electrodermal activity) during a sadness mood induction, suggesting greater tension. At the same time, girls also showed greater heart rate deceleration than boys at the point of peak intensity of sadness in the mood induction. The very early coexistence of empathy and self-distress, which is more characteristic of young girls than boys, may pave the way to this later capacity to experience considerable internal distress, yet switch more readily and fluidly to an outer-directed focus.

Significant genetic influence was evident for most components of empathy at most time points, based on the observational measures, and genetic continuity across time was present as well. The underlying bases of the phenotypic differences between MZ and DZ twins is less clear. We do not assume that genes code for social emotional behaviors directly (Zahn-Waxler et al., 1992b). Rather, genes code for enzymes, structural proteins, and regulatory factors that, in the context of the environment, influence patterns of brain chemistry and neurohormonal systems. These, then, play a role in individual social and emotional reactions to others in distress. Empathy may be an aspect of temperament or closely linked with particular personality characteristics that have genetic underpinnings.

In the previous report, based on a slightly smaller sample size and fewer time points, genetic influence appeared to be waning by 20 months; it was present only for the affective component—namely, empathic concern (Zahn-

Waxler et al., 1992b). However, generalizations regarding diminution of genetic influence appear premature because significant genetic influences were detected here at 24 and 36 months. Self-distress did not show genetic influence. Children this age display relatively little strong overt distress by the second year of life, in contrast to the reflexive crying of infants. However, others have found genetic influence for self-distress, as well as empathic concern, based on adolescents' self-reports (Davis et al., 1994).

Indifference to others' distress showed an unusual pattern of genetic influence (only) at 14 and 36 months and shared environmental influences (only) at 20 and 24 months. The developmental pattern was similarly atypical, with a decrease in indifference from 14 to 20 months, followed by an increase at the two later time points. The construct of indifference was indexed by a number of variables considered to reflect active lack of concern for the victim, evidenced by withdrawal, avoidance, and/or callous, aggressive behaviors. In retrospect, these varied responses may have different meanings at different ages, and/or be differentially represented in the summary variable at different time points. For example, an avoidant 3 year old might differ markedly from a similarly nonobservant, unengaged 1 year old. Intentionality may be more characteristic of the active lack of caring shown by older than by younger children. Aggression is known to increase during this age period and hence may come to play a relatively greater role in the indifference of older children. Heterogeneity of uncaring actions, including a range of both asocial and antisocial responses to distress, thus may have contributed to the curvilinear pattern identified.

Observational measures and maternal reports of prosocial patterns showed no correspondence and different patterns of genetic and shared environmental influences, both within and across time periods. Observational measures mainly indicated genetic influence at most time points, where maternal report measures mainly indicated shared environmental influences. In longitudinal analyses, genetic influences for observed empathy were similar for boys and girls. They were present early in development and contributed to the patterns of continuity over time. Hence genetic influence seen at the later ages in the univariate analyses were due to the earlier sources of influence (a similar pattern for the weak environmental effects identified). Different models were required to characterize the prosocial behaviors of girls and boys, as perceived by their mothers. Although maternal reports may capture a broader sampling of prosocial behavior, there are also potential biases in maternal perceptions, with social desirability entering into their judgments. Moreover, mothers of young twins may be reluctant to label one child as more kind, caring, and helpful than the other, regardless of zygosity, contributing to the strong shared environmental influences. Boys and girls differed in the degree to which new genetic influences entered in at different time points. Also, nonshared (unique) environmental effects contributed to stability over time for girls but not for boys. Mothers may be attuned to small behavioral changes in daughters that reorganize mothers' perceptions of girls over time; perceptions of their boys may remain more static, suggesting less investment in the empathy of sons. Questions regarding the biological underpinnings of concern for others have not received consistent, sustained attention in the conceptual or empirical literature. Definitional questions remain unresolved, and measurement is often indirect. Adoption and twin studies provide a means to explore further the conditions that underlie our capacity to engage in kind, caring actions. In subsequent behavior genetics research with twins, it will be valuable to examine empathy in relation to temperament patterns and actual socialization practices (as well as inferred shared and nonshared environmental influences) (Robinson et al., 1994). Panksepp (1986) asserts that it is unrealistic to assume that functionally unitary brain circuits will be discovered for global constructs like altruism, sympathy, and prosocial behavior. These are labels for diverse emotional and behavioral patterns that may share outward similarities but are likely to arise from several distinguishable neural systems. A more realistic goal is to identify multiple genetic and environmental sources of influence and to better understand how dispositions interact with socialization experiences to create variation in concern for others.

Acknowledgments: This research was supported by the John D. and Catherine T. MacArthur Foundation through its Research Network in Early Childhood Transitions and by the National Institute of Mental Health Intramural Program. We are grateful to the families who contributed their time and effort, as well as to the research assistants and staff members involved in data collection, coding, and management.

REFERENCES

Batson, C. D., Bolen, M. H., Cross, J. A., & Neuringer-Benefiel, H. E. (1986). Where is the altruism in the altruistic personality? *Journal of Personality & Social Psychology, 50*(1), 212–220.

Batson, C. D., Dyck, J. L., Brandt, J. R., Batson, J. G., Powell, A. L., McMaster, M. R., & Griffitt, C. (1988). Five studies testing new egoistic alternatives to the empathy-altruism hypothesis. *Journal of Personality and Social Psychology, 55*, 52–77.

Brothers, L. (1989). A biological perspective on empathy. *American Journal of Psychiatry, 146*, 10–19.

Davis, M. H., Luce, C., & Kraus, S. J. (1994). The heritability of characteristics associated with dispositional empathy. *Journal of Personality, 62*, 369–391.

Dawkins, R. (1989). *The selfish gene.* Oxford: Oxford University Press.

DeFries, J. C., & Fulker, D. W. (1985). Multiple regression of twin data. *Behavior Genetics, 16*, 1–10.

Eisenberg, N., & Fabes, R. A. (1998). Prosocial Development. In N. Eisenberg (Vol. Ed.) & W. Damon (Series Ed.), *Handbook of child psychology* (5th ed., Vol. 3, pp. 701–778). *Social, emotional, and personality development.* New York: Wiley.

Eisenberg, N., Fabes, R., Bustamante, D., Mathy, R. M., Miller, P. A., & Lindholm, E. (1988). Differentiation of vicariously induced emotional reactions in children. *Developmental Psychology, 24(2),* 237–246.

Eisenberg, N., Fabes, R. A., & Miller, P. A. (1990). The evolutionary and neurological roots of prosocial behavior. In L. Ellis & H. Hoffman (Eds.), *Crime in biological, social, and moral contexts* (pp. 247–260). New York: Praeger.

Eisenberg, N., Fabes, R., Miller, P., Fultz, J., Shell, R., Mathy, R. M., & Reno, R. R. (1989a). Relation of sympathy and personal distress to prosocial behavior: A multimethod study. *Journal of Personality and Social Psychology, 57,* 55–66.

Eisenberg, N., Fabes, R. A., Schaller, M., Carlo, G., & Miller, P. A. (1991). The relations of parental characteristics and practices to children's vicarious emotional responding. *Child Development, 62,* 1393–1408.

Eisenberg, N., Miller, P. A., Schaller, M., Fabes, R. A., Fultz, J., Shell, R., & Shea, C. L. (1989b). The role of sympathy and altruistic personality traits in helping: A reexamination. *Journal of Personality, 57*(1), 41–67.

Eisenberg, N., Pasternack, J. F., Cameron, E., & Tyron, K. (1984). The relation of quantity and mode of prosocial behavior to moral cognitions and social style. *Child Development, 155,* 1479–1485.

Eisenberg-Berg, N., & Hand, M. (1979). The relationship of preschoolers' reasoning about prosocial moral conflicts to prosocial behavior. *Child Development, 50,* 356–363.

Fabes, R. A., Eisenberg, N., & Eisenbud, L. (1993). Behavioral and physiological correlates of children's reactions to others' distress. *Developmental Psychology, 29,* 655–663.

Farver, J. M., & Branstetter, W. H. (1994). Preschoolers' prosocial responses to their peers' distress. *Developmental Psychology, 30,* 334–341.

Field, T., Woodson, T., Greenberg, R., & Cohen, D. (1982). Discrimination and imitation of facial expressions by neonates. *Science, 218,* 179–181.

Grusec, J. E., & Lytton, H. (1988). *Social development: History, theory and research.* New York: Springer-Verlag.

Hampson, R. B. (1981). Helping behavior in children: Addressing the interaction of a person-situation model. *Developmental Review, 1,* 93–112.

Haviland, J. M., & Lelivica, M. (1987). The induced affect response: 10-week-old infants' responses to three emotion expressions. *Developmental Psychology, 23,* 97–104.

Hoffman, M. (1975). Developmental synthesis of affect and cognition and its interplay for altruistic motivation. *Developmental Psychology, 11,* 607–622.

Hoffman, M. (1977). Sex differences in empathy and related behaviors. *Psychological Bulletin, 34(4),* 712–722.

Hoffman, M. L. (1982). Development of prosocial motivation: empathy and guilt. In N. Eisenberg (Ed.), *The development of prosocial behavior* (pp. 281–313). New York: Academic Press.

Kagan, J., & Lamb, S. (1987). *The emergence of morality in young children.* Chicago: University of Chicago Press.

Knight, G. P., Johnson, L. G., Carlo, G., & Eisenberg, N. (1994). A multiplicative model of the dispositional antecedents of a prosocial behavior: Predicting more of the people more of the time. *Journal of Personality & Social Psychology, 66*(1), 178–183.

Loehlin, J. C., & Nichols, R. C. (1976). *Heredity, environment and personality*. Austin: University of Texas Press.

Maccoby, E. E., & Martin, J. A. (1983). Socialization in the context of the family: Parent-child interaction. In E. M. Hetherington (Ed.), *Handbook of child psychology* (Vol. 4), *Socialization, personality, and social development* (4th ed., pp. 1–101). New York: Wiley.

MacLean, P. D. (1973). A triune concept of the brain and behavior. In T. Boag & D. Campbell (Eds.), *The Hincks Memorial Lectures* (pp. 6–66). Toronto: University of Toronto.

MacLean, P. D. (1985). Brain evolution relating to family, play and the separation call. *Archives of General Psychiatry, 42,* 405–417.

Matthews, K. A., Batson, C. D., Horn, J., & Rosenman, R. H. (1981). Principles in his nature which interest him in the fortune of others: The heritability of empathic concern for others. *Journal of Personality, 49,* 237–247.

Meltzoff, A. N., & Moore, M. K. (1977). Imitation of facial and manual gestures by human and neonates. *Science, 198,* 75–78.

Neale, M. C. (1994). *Mx: Statistical Modeling* (2nd. ed.). Richmond, VA: Department of Psychiatry, Medical College of Virginia.

Panksepp, J. (1986). The psychobiology of prosocial behaviors: Separation distress, play, and altruism. In C. Zahn-Waxler, E. M. Cummings, & R. Iannotti (Eds.), *Altruism and aggression: Biological and social origins* (pp. 19–57). Cambridge: Cambridge University Press.

Plomin, R., Emde, R. N., Braungart, J. M., Campos, J., Kagan, J., Reznick, J. S., Robinson, J., Zahn-Waxler, C., & DeFries, J. C. (1993). Genetic change and continuity from fourteen to twenty months: The MacArthur Longitudinal Twin Study. *Child Development, 64,* 1354–1376.

Radke-Yarrow, M., Zahn-Waxler, C., & Chapman, M. (1983). Children's prosocial dispositions and behavior. In E. M. Hetherington (Ed.), *Handbook of child psychology* (Vol. 4), *Socialization, personality, and social development* (4th ed., pp. 469–545). New York: Wiley.

Rehberg, H. R., & Richman, C. L. (1989). Prosocial behavior in preschool children: A look at the interaction of race, gender, and family composition. *International Journal of Behavioral Development, 12,* 385–401.

Robinson, J. L., Zahn-Waxler, C., & Emde, R. N. (1994). Patterns of development in early empathic behavior: Environmental and child constitutional influences. *Social Development, 3*(2), 125–145.

Rushton, J. P., Fulker, D. W., Neale, M. C., Nias, D. K. B., & Eysenck, H. J. (1986). Altruism and aggression: The heritability of individual differences. *Journal of Personality and Social Psychology, 50,* 1192–1198.

Sagi, A., & Hoffman, M. L. (1976). Empathic distress in the newborn. *Developmental Psychology, 12,* 175–176.

Segal, N. L. (1984). Cooperation, competition and altruism within twin sets: A reappraisal. *Ethology and Sociobiology, 5,* 163–177.

Simner, M. L. (1971). Newborn's response to the cry of another infant. *Developmental Psychology, 5,* 136–150.

Stanhope, L., Bell, R. Q., & Parker-Cohen, N. Y. (1987). Temperament and helping behavior in preschool children. *Developmental Psychology, 23,* 347–353.

Trivers, R. L. (1983). The evolution of cooperation. In D. L. Bridgeman (Ed.), *The nature of prosocial development* (pp. 43–60). New York: Academic Press.

Wilson, E. O. (1975). *Sociobiology: The new synthesis.* Cambridge, MA: Harvard University Press.

Wilson, E. O. (1978). *On human nature.* Cambridge, MA: Harvard University Press.

Zahn-Waxler, C., Cole, P. M., Welsh, J. D., & Fox, N. A. (1995). Psychophysiological correlates of empathy and prosocial behaviors in preschool children with behavior problems. (Special Issue on Emotion and Psychopathology.) *Development and Psychopathology, 1,* 27–48.

Zahn-Waxler, C., & Radke-Yarrow, M. (1982). The development of altruism: Alternative research strategies. In N. Eisenberg (Ed.), *The development of prosocial behavior* (pp. 109–139). San Diego: Academic Press.

Zahn-Waxler, C., & Radke-Yarrow, M. (1990). The origins of empathic concern. *Motivation and Emotion, 14,* 107–129.

Zahn-Waxler, C., Radke-Yarrow, M., Wagner, E., & Chapman, M. (1992a). Development of concern for others. *Developmental Psychology, 28,* 126–136.

Zahn-Waxler, C., Robinson, J., & Emde, R. N. (1992b). The development of empathy in twins. *Developmental Psychology, 28,* 1038–1047.

12

Dispositional Cheerfulness
Early Genetic and Environmental Influences

JoAnn L. Robinson
Robert N. Emde
Robin P. Corley

Motivation and communication are two primary functions of emotion. Traditional theories have given more attention to motivation, emphasizing the adaptive functions of emotion which have evolved in our species to structure biological predisposition for arousal and for responding to survival-related features of the environment so as to guide behavior either to approach or withdraw (Darwin, 1872; Ekman, 1971; Izard, 1977; Panksepp, 1985; Schneirla, 1959). Recently, more attention has been directed to the communicative functions of emotion which are also adaptive. Expressing emotion assists the individual in attaining goals, and, by setting in motion feedback for others, it allows for appraisals that can disambiguate environmental events (Campos, et al., 1983; Lazarus, 1991). Both functions, as Darwin (1872) indicated more than a century ago, are interrelated. Individual dispositions in emotional expression reflect both motivational-arousal and communicative functions. Similarly, experience of the developing individual engages both functions with environmental as well as genetic influences over time.

Both functions of emotion are also related to individual differences. Repeated emotional communications with caregivers in everyday routines can contribute to the development of characteristic individual differences in the intensity and temporal parameters of expression, what Goldsmith and Campos (1982, 1990) refer to as a "disposition." This chapter explores the proposition that a disposition toward expressing positive emotions, what we will call "dispositional cheerfulness," is evident in the second year of life. Our investigation considers stability and sources of individual differences in the toddler's expression of positive emotion within the behavioral genetic design of the Mac-Arthur Longitudinal Twin Study.

It is a curious fact that, in comparison with negative emotions, little research has been devoted to the dispositional facets of positive emotions and their expression. Perhaps this imbalance is due to clinical preoccupations with pain and suffering. Perhaps it is due to cultural concerns that have to do with understanding the shared meaning of the many discomforts of life and resolving them. Perhaps the imbalance also has to do with a recognition of the facts of our biology; namely, that evolution has provided more differential incentives and expressions for negative responses in order to survive (involving caution and avoidance) than incentives and expression for positive responses in order to approach and engage. The MacArthur Longitudinal Twin Study provides a unique opportunity to generate empirical data on this topic. But first, it is important to frame our work in the context of regulation.

REGULATION AND POSITIVE EMOTION SYSTEMS

Regulation is a central property of all biological and behavioral systems. It refers to the fact that adaptational functions operate within an optimal range and that outside that range (either from "not enough" or from "too much") corrective influences occur to avoid dysfunction (Emde, 1989; 1994). Correction typically occurs from the activation of other biologically based systems in an interlocking network of regulatory systems. In other words, regulation is interdependent, and any particular system under consideration is both regulating and being regulated.

Not surprisingly, the dysfunctional or dysregulating properties of emotions have occupied more attention from researchers and clinicians than have their regulatory properties (Campos & Barrett, 1984; Emde, 1991; Izard, 1977; Lazarus, 1991; Sroufe, 1979). This has been especially true in the toddler period, when dysregulation and negative emotion features have been emphasized (Kopp, 1989).

Contemporary emotion theorists, however, are now focusing on regulation, as well as dysregulation, and on positive emotion systems, as well as negative emotion systems. Positive emotional states, such as those indicated by infant smiling, are viewed primarily as manifestations of organized and regulated systems, systems that, like those dealing with negative emotions, have evolved because they are organizing and adaptive (Emde, 1991).

Empirical research has shown that expression of negative and positive emotion are independent systems (Emde, 1994; Tellegen, 1985). Working to diminish the expression of negative emotion does not necessarily result in an increase in positive emotion. There is also substantial evidence that positive and negative emotion are organized separately within the central nervous system. Investigations (e.g., Davidson & Fox, 1989; Fox, 1994) have linked positive emotional expression with left frontal EEG activation and negative emotional expression with right frontal EEG activation. As a result of their studies in infancy, Fox and colleagues have speculated that a separate organization of positive and negative affect expression systems in the brain appears to be related to the infant's goals and the contexts of the expressed emotion (Fox, 1994); Fox & Davidson, 1987).

Still, it is unclear if positive as well as negative emotions are organized as temperamental dispositions. Although some investigators, such as Rothbart (1981, 1986) and Goldsmith and Campos (1990), have provided evidence for a smiling/laughter or pleasure temperament dimension, others do not consider positive affect to be temperament related (Rowe & Plomin, 1977). Thomas and Chess's (1977) work on difficult versus easy temperament styles laid the ground for clinical interests, and a great deal of empirical research on early temperament that focused largely "difficult temperament" and on the more negative, difficult style (Bates, 1980). Kagan's well-known research on the development of behaviorally inhibited children is also an example of the general focus in the field of developmental psychology on negative temperament traits (e.g., Kagan, 1994; Kagan et al., 1984). Children displaying the behaviorally inhibited style in the Kagan studies are often contrasted with a group of outgoing/exuberant children about whom little has been studied.

A CHEERFUL DISPOSITION?

Environmental events are often ambiguous, and specific patterns of response are not well established for very young children. During the second year, organized behavior patterns are developing rapidly in children's social-emotional responses to different situations, such as social play opportunities, prohibitions, or situations where others are distressed. In these contexts, emotions are reference points for communication and action. A child with a more cheerful disposition (a tendency to respond to environmental events with positive emotion) may influence others through the use of smiling and positive vocalizations rather than distress. However, positive emotion used in these different situations may express quite different goals. Positive emotion expressed during low-stress situations, such as during social play, may serve a different function than that expressed during high-stress situations, such as prohibitions. In the latter case, positive affect may be used defensively, to ward off the experience of negative affect, as in a buffering function. We therefore arrive at an interesting set of questions. Is there evidence of a cheerful disposition in the toddler age period? To what extent is positive affect generalized across contexts, with individuals who use positive affect more frequently in low-stress situations also using it more frequently in high-stress situations? "Cheerful" individuals may have developed an emotion regulation style that maximizes the utility of positive emotions in engaging others and in coping with challenging situations. Such dispositional differences could therefore lead to different modes of experiencing the environment.

Previous behavioral genetics research on this topic has generated findings that are encouraging with respect to a construct of a cheerful disposition, but uncertainties remain. Some research has demonstrated heritable and shared environmental influences on positive emotions. Goldsmith et al., (1997; 1999) report low heritability and considerable shared environmental influences for infants and toddlers on smiling/laughter based on parental reports using the Infant Behavior Questionnaire and Toddler Behavior Assessment Questionnaire instruments. Matheny (1989) reported increasing heritability during the first and second years for general emotional tone, with positive and negative emotion expression taken as a single dimension. However, these findings may be driven by the negative affect end of the dimension, which consistently shows moderately strong heritable influences on both parental reports and observational measures (Emde et al., 1992; Plomin et al., 1993). Previously, the MacArthur Longitudinal Twin Study reported findings that differed for parental reports and observed emotional expressions. Significant shared environment and low heritability was found for parental reports of positive emotion in the second year. Shared environment and hereditary estimates similar to those reported by Goldsmith et al., (1997; 1999) were found for observed positive tone (Emde et al., 1992; Plomin et al., 1993).

This chapter examines evidence for an organized dimension of positive emotional expressiveness or cheerfulness based on observations made during the child's second year. Individual differences in positive emotion expressions are explored across low-stress and high-stress contexts and over time. Twin analyses provide estimates of genetic and environmental influences on change and continuity of behaviors. A disposition to cheerfulness would be indicated if children who expressed higher levels of positive emotion in low-stress situations also expressed more positive emotion in high-stress situations. Although moderate levels of heritability of positive expressiveness would provide support for a biological basis for a cheerful disposition, environmental factors may also contribute to the emergence of a stable, traitlike behavior.

METHODS

Sample and Procedures

Approximately 100 MZ and 100 DZ twins with unambiguous zygosity were selected for analysis.[1] Behavioral observations were made during home and laboratory situations and parental reports were gathered from questionnaires completed at home and in the lab at three points during the second year: 14, 20, and 24 months. At each age in the home, two low-stress contexts and two high-stress contexts were chosen to analyze children's tendency to express positive emotion, or cheerful disposition. The low-stress situations included administration of the Bayley Scales of Infant Development (mental development items) and approximately 20 min of free play with their cotwin. These two situations sampled behaviors over approximately 1 hr. The two high-stress contexts included response to prohibition and response to the distress of another and were brief in duration, approximately 3 min total.

Contextual features of the low-stress situations differ markedly, which potentially affords the opportunity for different children to experience pleasure in each context. Bayley testing was completed approximately 30 min after the testers arrived at the home. The child was seated in a high chair with a video camera placed on a tripod approximately 6–8 feet away. The testing procedure involved a fairly constant give-and-take between the child and a female tester. A toy was presented and an action modeled, the child responded, the toy was removed, and a new toy was presented. Testers were instructed to engage children playfully and to maintain a positive, encouraging tone. Thus children who are quick to acclimate with the home test situation and who enjoy the sociable exchange with a pleasant tester may express greater positive emotion during this situation than those children who are initially more reserved and are slower to warm up to social play.

The free play was the last procedure in the home visit, occurring approximately 2 hr after the testers arrived at the home. A standard set of 10 toys were arranged on the floor of the family's living room or family room. The mother was asked to sit on a comfortable chair or sofa and was encouraged to respond to but not to initiate the children. The children were allowed to play freely with the toys and each other while the testers (sitting on the floor near the edges of the room) each filmed one child, focusing on the child's torso and head (see chapter 3 for a more detailed description of the procedures). In this situation, children who get pleasure in social toy play, even if they have been slower to warm up, may express greater positive emotion than children who may have tired during the course of the home visit.

In addition to these two low-stress behavioral observations, the children's emotional reactions were observed during two high-stress situations: simulated distress of another, which tended to induce empathy, and a prohibition. Each twin was repeatedly exposed to simulated distress by (1) the examiner, (2) the mother, and (3) a prerecorded infant cry. The first two simulations were performed during home and laboratory visits; the infant cry was presented in the lab. A maximum of five presentations were made at each age. A prohibition was administered just before Bayley testing in the home and again after the inhibition room procedures with the mother. In the home, children were shown a plastic "glitter wand" by the examiner, who then placed it in front of the child, saying "Don't touch." In the lab, the examiner held out a basket of three colorful balls and placed them on the floor in front of the child, repeating "Now, don't touch" (see chapter 3 for additional procedural details).

Measures

We used the Hedonic Tone Scales (Easterbrooks & Emde, 1983) to measure the strength of positive and negative emotion during each of the observed situations. These scales permit time-sampled ratings of peak affect expressions on 4-point scales for both affect dimensions, where 1 = no positive or negative emotion observed, 2 = positive interest or furrowed brow, 3 = smiling or fretting, and 4 = strong laughter or full-blown cry. Because Bayley scales testing frequently lasted for 40–45 min, emotion expressions were rated during 1-min intervals which were alternately sampled in 5-min blocks (i.e., 5 min on, 5 min off). This resulted in 20–25 one-minute ratings which were averaged within the positive and negative dimensions to produce two variables: Bayley-average positive hedonic tone and Bayley-average negative hedonic tone. The 15-min free play was rated continuously, using 30-s time intervals and yielding 30 ratings at 14 months and up to 50 ratings at 20 and 24 months which were averaged to produce two variables: free play-positive hedonic tone and free play-negative hedonic tone. Interobserver reliability was based on multiple pairs of raters independently scoring the same child. Intraclass correlations revealed high levels of agreement (r = .89 for ratings of positive hedonic tone and $r = .85$ for negative hedonic tone).

Positive and negative emotion expressions were rated during the simulated distress and prohibition episodes using 30-s intervals. For simulated distress episodes, a 4-point positive dimension and a 5-point negative dimension were used. The positive dimension was similar to the positive Hedonic Tone Scale, and the negative dimension tapped personal distress, differentiating a wide-eyed fear expression (rating of 2), grimacing (rating of 3), fretting/whining (rating of 4), and full-blown cry (rating of 5). A single rating for each dimension was made for the 30-s simulated distress. These variables are called empathy-positive expression and empathy-distress expression. Interobserver reliability (R) for empathy-positive expression was .87 and for empathy-distress expression was .82. The prohibition situation was rated using the Hedonic Tone Scales. Two ratings were made: the first beginning with the onset of the prohibition and ending when the child touched the toy, and the second beginning with onset of permission to play and lasting 30 s. The two ratings for each dimension were averaged to produce two variables: prohibition-positive hedonic tone and prohibition-negative hedonic tone. Interobserver reliability based on paired, independently rated observations was .67 to .83.

RESULTS

We first examined the pattern of means and correlations of positive hedonic tone across the two situations as a prelude to aggregation of ratings across the Bayley and free play situations. Children tended to express somewhat stronger positive emotion during Bayley testing compared to free play at each age (see table 12.1). In addition, there was a small increase in Bayley-positive hedonic tone across ages. Correlations across Bayley and free play situations within ages generally revealed associations that were moderate and increased over time (see table 12.2). Positive hedonic tone ratings were thus averaged across

Table 12.1. Means and standard deviations for positive hedonic tone across Bayley and free play situations

Age	Bayley			Free play		
	Mean	(SD)	n	Mean	(SD)	n
14 Months	2.41	(.33)	393	2.23	(.27)	394
20 Months	2.44	(.31)	367	2.30	(.28)	369
24 Months	2.50	(.34)	354	2.25	(.22)	354

situations, using the original time-sampled interval ratings, and the resulting variable, positive hedonic tone, is used in subsequent analyses.

If there is a disposition to cheerfulness, we expect that it will have a traitlike quality and will be moderately stable across time and across measurement situations. The stability of positive hedonic tone across ages is shown in Table 12.3. Correlations were low to moderate, with adjacent ages showing somewhat stronger associations than others. Across low- and high-stress measurement situations within ages, we examined the association between measures of positive expressions. Correlations of positive hedonic tone to empathy-positive expression and prohibition-positive hedonic tone within each age are presented in table 12.4. The correlation between positive hedonic tone and empathy-positive expression were low- to-moderate and were stronger at 14 months than at the later ages. Positive hedonic tone and prohibition-positive hedonic tone correlations were consistently moderate across the three ages and were somewhat higher than the correlations with empathy-positive expression. It should be noted that the correlations between positive emotion expressions in the empathy and prohibition situations tended to be lower than the correlations with positive hedonic tone, although they were statistically significant at the $p < .05$ level at each age ($r = .13$, .21, and .16 at 14, 20, and 24 months, respectively). Thus, children's positive emotion expressions showed low-to-moderate levels of stability across contexts, supporting traitlike behavior or disposition.

Table 12.2. Correlations of positive hedonic tone across Bayley and free play situations within age

Age	r	n
14 Months	.34**	391
20 Months	.44**	364
24 Months	.48**	348

**Significant at $p < .01$.

Table 12.3. Correlations of positive
hedonic tone across age

Age	r	n
14–20 Months	.34**	368
20–24 Months	.37**	343
14–24 Months	.25**	356

**Significant at $p < .01$.

Is the disposition to cheerfulness, as indexed by positive hedonic tone, related to expressions of negative emotion? Correlations between positive hedonic tone and negative hedonic tone within age are low, but significant, suggesting a weak inverse relationship between the positive and negative affect systems ($r = -.14, -.12$, and $-.14$). In the high-stress contexts, correlations between positive and negative expressions in the empathy and prohibition situations showed similar pattern of effects at 14 ($r = -.11$ and $-.12$ for empathy and prohibition, respectively) and 20 months ($r = .00$ and $-.13$ for empathy and prohibition, respectively). Although positive and negative expressions in the empathy context were not correlated at 24 months, a stronger negative association was found in the correlation of positive and negative tone during the prohibitions at 24 months ($r = -.24$, p $<.001$). Thus, positive and negative expressiveness were generally independent dimensions in these measurement contexts, as has been found previously in nontwin infancy samples (Easterbrooks & Emde, 1987).

Finally, we computed twin correlations and used the regression method of DeFries and Fulker (1985) to examine the univariate estimates of genetic and common environmental sources of individual differences for positive hedonic tone. As shown in table 12.5, genetic influences are small at 14 and 24 months and moderate at 20 months. Common environmental influences are modest at all three ages, although somewhat stronger at 14 than at 20 or 24 months.

To examine the extent of shared genetic and common environmental variance across ages, longitudinal model fitting, using the Cholesky triangular de-

Table 12.4. Correlations of positive hedonic tone to empathy
and positive hedonic tone to prohibition

Age	Empathy		Prohibition	
	r	n	r	n
14 Months	.30**	396	.34**	394
20 Months	.19**	366	.36**	364
24 Months	.26**	359	.34**	352

**Significant at $p < .01$.

Table 12.5. Twin correlations, heritability (h^2), and common environment (c^2) estimates for positive hedonic tone

Age	MZ	DZ	h^2	SE	c^2	SE
14 Months	.50**	.39**	.22	.26	.28	.20
20 Months	.55**	.34**	.40*	.26	.14	.21
24 Months	.26**	.18*	.16	.30	.10	.24

*Significant at $p < .05$.
**Significant at $p < .01$.

composition as described in Cherny et al. (1994), was applied to the positive hedonic tone measure (see figure 12.1). As can be seen, a single factor accounts for the genetic variance observed at all three ages, suggesting moderate genetic continuity in observed positive expressiveness. Two factors contributed to common environmental influences, suggesting short-term continuity across adjacent ages and some change on common environmental influences at 20 months. Unique environmental influences were only age specific and are not presented. Hence, genetic continuity influences the observed genetic variance in positive expressiveness, whereas both continuity and change were present in the influences from the common environment. Note that the genetic variance explained by genetic continuity is only about 12%, a modest source of influence on behavior.

To determine whether the observed continuity in behavior across contexts (see table 12.3) was explained by shared genetic or common environmental influences, we used the Pedikid Twin-Sibling Pedigree Program (Cardon, 1990) to estimate the best models of shared variances across the three measurement contexts within ages. The next three models were thus multivariate rather than longitudinal. In table 12.6, we see that at 14 months environmental influences, especially common environmental influences, showed strong continuity or shared variance across contexts. Unique environment also showed weak continuity between positive hedonic tone and empathy and prohibition positive

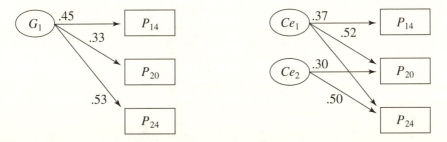

Figure 12.1. Longitudinal model fitting: positive hedonic tone. Genetic (G) and common environment (Ce) factors influencing phenotype (P) at 14, 20, and 24 months.

Table 12.6. Parameter estimates for nonshared environment (*E*), shared environment (*C*), and genetic effects (*G*) from the best-fitting multivariate models for positive expressiveness across contexts at 14, 20, and 24 months

Age/measure	E_1	E_2	E_3	C_1	C_2	C_3	G_1	G_2	G_3
14 Months									
Positive tone	.75			.66			.00		
Positive empathy	.15	.95		.29	.00		.00	.00	
Positive prohibition	.20	.00	.94	.30	.00	.00	.00	.00	.00
20 Months									
Positive tone	.70			.50			.52		
Positive empathy	.19	.78		−.35	.00		.47	.00	
Positive prohibition	.40	.23	.87	.15	.00	.00	.00	.00	.00
24 Months									
Positive tone	.87			.50			.00		
Positive empathy	.15	.91		.25	.30		.00	.00	
Positive prohibition	.15	.14	.86	.38	−.28	.00	.00	.00	.00

expressions in addition to the variance unique to each context. There was no significant genetic variance in the final, best-fit model at 14 months. At 20 months in table 12.6, we see moderate genetic and common environmental continuities across contexts. Moderately strong shared genetic variance was found between positive hedonic tone and empathy positive expression. Continuity was also found across contexts for common environmental influences, with a negative loading for empathy positive expression and much weaker loading for prohibition positive tone. Finally, unique environmental influences exerted continuity as well as change across situations, with noteworthy loadings between the positive hedonic tone during the Bayley-free play situations and the prohibition situation. As can be seen at 24 months, the picture is much the same as at 14 months, with no genetic variance influencing the child's behavior across contexts, and environmental influences showing both consistency and change. Specifically, two common environmental factors were found, with low-to-moderate levels of shared variance across all three situations. Unique environmental variance was largely specific to each context, but a weak consistency across all three contexts was also observed.

DISCUSSION

Our findings provide some evidence for a disposition to cheerfulness observable in individual differences in children's emotional responses during the second year of life. Low-to-moderate levels of consistency were observed in children's positive expressiveness across low- and high-stress situations within each age and over time. Moreover, our longitudinal model-fitting results for positive hedonic tone provided evidence for some continuity in genetic and environmental influences across ages. Across contexts within age, there was also significant shared variance in both common and unique environmental influences in positive expressiveness. However, results for genetic influences across contexts are different. Only at 20 months, the point at which the strongest genetic influence was observed in the univariate analyses, did we also see shared genetic variance in positive expressions across the Bayley-free play and empathy situations.

Our research has been limited by the opportunities we had to observe the children. Each visit was like a snapshot in these children's lives, and the reliability in our sampling of their behavior may limit our findings to low level or modest effects. Furthermore, the observations of cheerfulness in low-stress situations derive from on-going behavior, without explicit pleasure-eliciting incentives. These naturalistic observations may have operated to create a more generalizable set of findings; however, they may also limit the range of behaviors observed. Future research efforts should consider a broader sample of behavior both from the perspective of quantity of behavior observed but also from the types of eliciting events that are sampled.

Our findings suggest that with increasing age, children's positive expressiveness may increase, especially during Bayley testing. Why might such a phenomenon appear? Across the second year, children may develop more stamina in responding in the socially intense testing situation. Alternatively, children may gradually become more expressive across the second year although there are few published data suggesting this to be the case. Bloom (1991) reported no change in emotion expressions across the second year for the expression of positive emotion during dyadic play with mother. Similarly, Robinson, et al. (1993) did not find significant increases in children's positive expressiveness with mothers between 18 and 24 months.

Although we had hypothesized that positive and negative expressions would be independent we observed a weak, inverse association (at the .10 level) between negative expressions and positive expressions in both low- and high-stress contexts. It is particularly understandable that the higher stress contexts might elicit negative responses from some children that might swamp positive expressions and thereby account for a negative correlation. Indeed, some children did show marked responses of this sort. Others, however, responded with a different pattern that was informative. Such children showed a mixture of weak positive and negative expressions as they seemingly attempted to disambiguate these situations and communicate with the examiner or mother.

Children's goals may differ markedly in low- and high-stress situations. In higher stress situations, the child may be actively trying to mitigate the expression of negative emotion in self and others, and positive expressions may buffer the intensity of felt distress for the child. In low stress situations, the expression of positive affect may have shared social goals, and the child may experience mastery pleasure. It is plausible, therefore, that positive affect would have different sources of individual differences in these two situations. The multivariate analysis results showed only weak evidence at best to support this proposition. Low-to-moderate effects were found for the environmental factors across contexts at the younger ages. A more differentiated pattern was seen at 20 months, where the factor loading for shared common environmental variance across the three situations was moderately strong for the Bayley-free play, negatively weighted for empathy, and very weak for prohibition. However, the more general pattern of consistency in the sources of individual differences across contexts does not support the notion that positive expressions served markedly different goals.

The prominence of environmental influences on behavioral continuity across ages and consistency across contexts in this study tends to support the socially constructed nature of emotional communication and regulation in early childhood. A disposition toward cheerfulness is not merely a constitutionally or biologically endowed quality that emerges in development. Still, a constitutional thread was apparent in the modest genetic influences that were found at all three ages and in the longitudinal model fitting, which showed that some genetic continuity was underlying these age-to-age estimates. However, environmental sources of individual differences also contributed to consistencies observed across low- and high-stress contexts which were measured in both home and laboratory settings.

In summary, this investigation considered evidence for a disposition toward cheerfulness during the second of life. Positive expressiveness observed in both low- and high-stress contexts showed low to moderate consistency across these situations and over the three time points measured. Genetic influences were modest but were shared by a single factor over time. At 20 months, a single genetic factor contributed to similarity in positive expressiveness in the Bayley-free play and empathy contexts, but at 14 and 24 months little genetic influence was apparent. Common environmental factors also influenced continuity over time and consistency across contexts. Unique environmental influences did not contribute to continuity over time but did exert a weak effect at each age on behavioral similarities across low- and high-stress situations.

NOTE

1. The same sample has been a basis for different analyses in previous publications. See Emde et al., 1992; Plomin et al., 1993; Robinson et al. 1992; Zahn-Waxler et al., 1992.

REFERENCES

Bates, J. (1980). The concept of difficult temperament. *Merrill-Palmer Quarterly, 26,* 299–319.

Bloom, L. (1991). Developments in expressions: Affect and speech. In N. L. Stein, B. Leventhal & T. Trabasso (Eds.), *Psychological and biological approaches to emotion*. Hillsdale, NJ: Erlbaum.

Campos, J., Barrett, K., Lamb, M. E., Goldsmith, H. H., & Stenberg, C. (1983). Socioemotional development. In M. M. Haith & J. J. Campos (Eds.), *Handbook of Child Psychology* (Vol. 2, pp. 783–915). New York: Wiley.

Campos, J., & Barrett, K. (1984). Toward a new understanding of emotions and their development. In C. Izard, J. Kagan, & R. Zajonc (Eds.), *Emotions, cognition, and behavior*. New York: Cambridge University Press.

Cardon, L. (1990). Pedikid I. D. Twin-Sibling Pedegree Program. Unpublished Fortran program. Institute for Behavioral Genetics, University of Colorado at Boulder.

Cherny, S. S., Fulker, D. W., Emde, R. N., Robinson, J., Corley, R. P., Reznick, J. S., Plomin, R., & DeFries, J. C. (1994). A developmental-genetic analysis of continuity and change in the Bayley Mental Development Index from 14 to 24 months: The MacArthur Longitudinal Twin Study. *Psychological Science, 5*, 354–360.

Darwin, C. (1872). *The expression of emotions in man and animals*. London: John Murray. (Reprinted Chicago: University of Chicago Press, 1965).

Davidson, R. J., & Fox, N. A. (1989). Frontal brain asymmetry predicts infants' response to maternal separation. *Journal of Abnormal Psychology, 98*, 127–131.

DeFries, J. C. & Fulker, D. W. (1985). Multiple regression of twin data. *Behavior Genetics, 22*, 467–473.

Easterbrooks, A., & Emde, R. N. (1983). The Hedonic Tone Scales. Unpublished manuscript, University of Colorado Health Sciences Center, Denver.

Easterbrooks, M. A., & Emde, R. N. (1987). Marital and parent-child relationships: The role of affect in the family system. In R. A. Hinde & J. Stevenson-Hinde (Eds.), *Relations between relationships within families* (pp. 83–103). Oxford: Oxford University Press.

Ekman, P. (1971). *Darwin and facial expression: A century of research in review*. New York: Academic Press.

Emde, R. N. (1989). Toward a psychoanalytic theory of affect: I. The organizational model and its propositions. In S. I. Greenspan & G. H. Pollock (Eds.), *The course of life* (Vol. I), *Infancy* (rev. ed). Madison, CT: International Universities Press.

Emde, R. N. (1991). Positive emotions for psychoanalytic theory: Surprises from infancy research and new directions. *Journal of the American Psychoanalytic Association, 39* (Supplement), 5–44.

Emde, R. N. (1994). Individuality, context, and the search for meaning. *Child Development, 65,* 719–737.

Emde, R., Campos, J., Corley, R., DeFries, J., Fulker, D., Kagan, J., Plomin, R., Reznick, J. S., Robinson, J., & Zahn-Waxler, C. (1992). Temperament, emotion, and cognition at 14 months: The MacArthur Longitudinal Twin Study. *Child Development, 63,* 1437–1455.

Fox, N. A., & Davidson, R. J. (1987). Electroencephalogram asymmetry in response to the approach of a stranger and maternal separation. *Developmental Psychology, 24,* 230–236.

Fox, N. A. (1994). Introduction to Part I. In N. Fox (Ed.), The development of emotion regulation. *Monographs of the Society for Research in Child Development, 59* (2–3, Serial No. 240), 3–6.

Goldsmith, H. H., Buss, K. A., & Lemery, K. S. (1997). Toddler and childhood temperament: Expanded content, stronger genetic evidence, new evidence for the importance of environment. *Developmental Psychology, 33,* 891–905.

Goldsmith, H. H., & Campos, J. J. (1982). Toward a theory of infant temperament. In R. Emde and R. Harmon (Eds.), *Attachment and affiliative systems* (pp. 161–193). New York: Plenum.

Goldsmith, H. H., & Campos, J. J. (1990). The structure of infant temperamental dispositions to experience fear and pleasure: A psychometric perspective. *Child Development, 61,* 1944–1964.

Goldsmith, H. H., Lemery, K. S., Buss, K. A., & Campos, J. J. (1999). Genetic analyses of focal aspects of temperament using the Infant Behavior Questionnaire. *Developmental Psychology, 35,* 972–985.

Izard, C. (1977). *Human emotion.* New York: Plenum.

Kagan, J. (1994). *Galen's professor.* New York: Basic Books.

Kagan, J., Reznick, J. S., Clarke, C., Snidman, N., & Garcia-Coll, C. (1984). Behavioral inhibition to the unfamiliar. *Child Development, 55,* 2212–2225.

Kopp, C. B. (1989). Regulation of distress and negative emotions: A developmental view. *Developmental Psychology, 25,* 343–354.

Lazarus, R. S. (1991). *Emotion and adaptation.* New York: Oxford University Press.

Matheny, A. P., Jr. (1989). Children's behavioral inhibition over age and across situations: Genetic similarity for a trait during change. *Journal of Personality, 57,* 215–236.

Panksepp, J. (1985). Toward a general psychobiological theory of emotions. *Behavioral Brain Sciences, 5,* 407–467.

Plomin, R., Emde, R. N., Braungart, J. M., Campos, J., Corley, R., Fulker, D. W., Kagan, J., Reznick J. S., Robinson, J., Zahn-Waxler, & DeFries, J. C. (1993). Genetic change and continuity from 14 to 20 months: The MacArthur Longitudinal Twin Study. *Child Development, 64,* 1354–1376.

Robinson, J. L., Kagan, J., Reznick, J. S., & Corley, R. (1992). The heritability of inhibited and uninhibited behavior: A twin study. *Developmental Psychology, 31,* 685–696.

Robinson, J. L., Little, C., & Biringen, Z. (1993). Emotional communication in mother-toddler dyads: Evidence for gender differentiation. *Merrill Palmer Quarterly, 39,* 496–517.

Rothbart, M. K. (1981). Measurement of temperament in infancy. *Child Development, 52,* 569–578.

Rothbart, M. K. (1986). Longitudinal observation of infant temperament. *Developmental Psychology, 22,* 356–365.

Rowe, D, & Plomin, R. (1977). Temperament in early childhood. *Journal of Personality Assessment, 41,* 150–156.

Schneirla, T. R. (1959). An evolutionary and developmental theory of biphasic process underlying approach and withdrawal. In M. R. Jones (Ed.), *Nebraska symposium on motivation: 1959.* Lincoln: University of Nebraska Press.

Sroufe, A. (1979). Socioemotional development. In J. Osofsky (Ed.), *Handbook of infant development* (pp. 462–516). New York: Wiley.

Tellegen, A. (1985). Structures of mood and personality and their relevance to assessing anxiety, with an emphasis on self-report. In A. H. Tuma & J. D. Maser (Eds.), *Anxiety and the anxiety disorders* (pp. 681–706). Hillsdale, NJ: Erlbaum.

Thomas, A., & Chess, S. (1977). *Temperament and development.* New York: Brunner/Mazel.

Zahn-Waxler, C., Robinson, J. L., & Emde, R. N. (1992). The development of empathy in twins. *Developmental Psychology, 28,* 1038–1047.

Part IV

Cognition

Section Editor
David W. Fulker

13

Cognitive Development, Cognitive Abilities, and Modularity

David W. Fulker
Robert Plomin

The first twin (Merriman, 1924) and adoption (Theis, 1924) studies investigated cognitive abilities. This topic continues to be a leading edge of genetic research in psychology, both methodologically and substantively (Plomin & McClearn, 1993). For this reason, it is surprising that psychological reviews of cognitive development rarely mention genetic research. For example, the 1983 edition of the *Handbook of Child Psychology* (Mussen, 1983) includes an entire volume of more than 1000 pages on cognitive development, yet genetics is barely mentioned. One reason for this gulf lies in differing levels of analysis. Genetic research focuses on individual differences, asking why some children develop more quickly and attain higher levels on tests of cognitive abilities. Most theories and research on cognitive development, on the other hand, ignore differences among children and focus instead on species-typical development, describing what children, on average, can do at particular ages. Piaget (1952), for example, explicitly excluded individual differences from his work, most of which involved very small samples of children. Small samples are appropriate if each child is thought to represent a species-typical pattern of development.

These are two views of development: normative development (means) and individual differences (variances). Such perspectives are never right or wrong, merely useful or not useful for particular purposes. For example, currently a major area of interest in cognition is modularity. In the strict sense, modularity refers to innate and invariant information-processing units (Fodor, 1983). Theoretical discussions of the issue of modularity focus on the invariance of species universals rather than on the variance of differences among individuals within a species. In contrast, empirical data brought to bear on the issue often involve individual differences of a sort. For example, rare genetic variants such as Williams syndrome and rare environmental variants such as brain damage are typically used to support the argument for domain specificity, although these are considered as odd exceptions to an otherwise invariant species-typical development.

Using modularity as an example, genetic research on the normal range of genetic and environmental sources of individual differences in cognitive processes can bring a powerful empirical approach to bear because it can investigate the extent to which genetic effects on one cognitive process covary with genetic effects on other cognitive processes. From a genetic perspective, modularity implies genetically distinct abilities—genetic effects on one cognitive ability should be independent of genetic effects on other cognitive abilities. The issue of genetic overlap among traits can be addressed by multivariate genetic analysis, which focuses on the covariance among traits rather than on the variance of each trait considered separately.

Genetic overlap is the topic of two of the chapters in this section. Chapter 14 uses multivariate genetic analyses to elucidate specific cognitive abilities. Most tests of infant development, including the Bayley Scales of Infant Development used in the MacArthur Longitudinal Twin Study, were developed many decades ago. They rely on commonplace tasks such as drawing, naming objects, and building with blocks and were not designed to assess specific cognitive abilities. Although these are reasonable tasks to assess cognitive development, it is also reasonable to take advantage of experimental procedures that have been devised in recent decades and to assess specific cognitive abilities more systematically.

As a step in this direction, the MacArthur Longitudinal Twin Study included two tasks that go beyond standard infant test items. One is a word comprehension task that does not require a verbal response from the infants in order to obtain a purer assessment of receptive language abilities unconfounded by expressive abilities. This task takes advantage of work in experimental psychology showing that children's gaze fixation can be used as a sensitive index of their interest. Children were shown pairs of slides of objects such as a cat and a shoe and asked if they saw one of the objects. Increased fixation to the target slide was scored as a correct response. The second task attempted to assess memory for locations because memory is not assessed on standard tests like the Bayley. This task was like the shell game. A small object with which the child was playing was put under one of several cups, with a shield blocking the child's view of the cups for a delay interval. A correct response was scored if the child removed the appropriate cup after the delay interval. Task difficulty was varied with the number of cups and the delay interval.

Phenotypically, the results indicate the importance of molarity as well as modularity. That is, the word comprehension and memory tasks correlate about .4 and .2, respectively, with the Bayley Mental Development Index (MDI) at 24 months. However, some specificity is suggested because the two tasks correlate only .1 with each other, although it should be noted that the stability of the measures from 20 to 24 months is only .3 and .2, respectively (in contrast to the MDI stability of .65). Multivariate genetic analyses made a stronger case early in the second year of life for modularity by showing that genetic influences on each task are largely independent of genetic influences on the other task and independent of the Bayley MDI. At 24 months, however, the situation was reversed. Genetic influences on memory were largely the same genetic influences that affected the Bayley MDI; oddly, word comprehension showed no genetic influence at this age. The results for memory at 24 months suggest genetic molarity rather than modularity. If replicated, this finding would be interesting because the major developmental theory of modularity assumes that development proceeds in the opposite direction, from molarity to modularity (Karmiloff-Smith, 1992). Perhaps the findings at 24 months reflect integrative processes that coordinate functioning and come into play as a consequence of developmental transitions between 20 and 24 months and are in place thereafter. Although more research is needed to substantiate these findings, they are given greater credence by similar results suggesting substantial genetic overlap among tests of diverse cognitive abilities that have been reported in sibling and parent–offspring adoption analyses from the Colorado Adoption Project (CAP; e.g., DeFries et al., 1994) and in combined CAP and twin analyses (Cardon & Fulker, 1993). Other multivariate genetic analyses in adulthood also show that genetic effects on specific cognitive abilities overlap with general cognitive ability to a surprising extent (Plomin et al., 1997a).

Chapter 16 in this part presents even more striking results relevant to the issue of modularity. It addresses the key issue of the relationship between thought and language, studied more prosaically in terms of tests of mental development (Bayley and Stanford-Binet) and language development (Sequenced Inventory of Communication Development; SICD). The multivariate genetic approach to the issue of modularity asks to what extent genetic effects on general mental development overlap with genetic effects on language development. The answer that emerges from these analyses is similar to the results reported in chapter 14. It is a developmental story of early genetic modularity in the second year of life that becomes genetic molarity by 36 months. That is, at 14 months, genetic effects on the Bayley measure of general mental development are independent of genetic effects on expressive and receptive language as assessed by the SICD. However, by 36 months, the situation is reversed: Genetic effects on the Standford-Binet completely overlap with genetic effects on expressive language and also overlap, albeit not completely, with genetic effects on receptive language. One caveat is that tests of mental development at 36 months, but not at 14 months, incorporate many items about language development. This raises the possibility that genetic overlap at 36 months might be an artifact of the inclusion of language items in tests of mental development. However, the results of other multivariate genetic analyses, mentioned above, support the hypothesis of some slight genetic modularity in the face of substantial genetic molarity.

Chapters 14 and 16 also provide interesting results of longitudinal genetic analyses, which suggest some hints of genetic change from age to age during this fast-moving development period but in the context of general genetic continuity. Chapter 15 focuses on longitudinal genetic analysis of general cognitive ability as indexed by the MDI of the Bayley measure and reaches a similar conclusion. Moderate genetic influence, accounting for 30–40% of the total variance, was found at 14, 20, 24, and 36 months of age. Other research suggests that genetic influence increases in early childhood (Fulker et al., 1988), in middle childhood and adolescence (Plomin et al., 1997b), and throughout the life span (McCartney et al., 1990; McGue et al., 1993). Longitudinal genetic analysis indicates that this genetic influence is largely stable from age to age, although there are hints of new genetic influences emerging at 24 and 36 months, but not at 20 months.

The chapters in this part thus provide dramatic illustrations of the power of genetic research on individual differences to address fundamental issues about cognitive development. In the case of modularity, the results are fascinating because they go against the grain of current thinking, suggesting that genetics favors molarity rather than modularity. The longitudinal results support an emerging principle that genetic influence on cognitive development is largely responsible for continuity, although there are tantalizing hints of change in genetic effects from age to age. In addition, these analyses uncover some interesting findings concerning environmental influences on cognitive development. Most notably, shared environmental influences account for some significant variance for the various measures of cognitive and language development, but these effects are very general, spanning all of the measures and all of the ages. In our view, these are some of the most important discoveries made in recent years about cognitive development and underline the need to bridge the gap between the two worlds of research on cognitive development.

REFERENCES

Cardon, L. R., & Fulker, D. W. (1993). Genetics of specific cognitive abilities. In R. Plomin & G. E. McClearn (Eds.), *Nature, nurture, and psychology* (pp. 99–120). Washington, DC: American Psychological Association.

DeFries, J. C., Plomin, R., & Fulker, D. W. (1994). *Nature and nurture during middle childhood.* Cambridge, MA: Blackwell.

Fodor, J. A. (1983). *The modularity of mind.* Cambridge, MA: MIT Press.

Fulker, D. W., DeFries, J. C., & Plomin, R. (1988). Genetic influence on general mental ability increases between infancy and middle childhood. *Nature, 336,* 767–769.

Karmiloff-Smith, A. (1992). *Beyond modularity: A developmental perspective on cognitive science.* Cambridge, MA: MIT Press.

McCartney, K., Harris, M. J., & Bernieri, F. (1990). Growing up and growing apart: A developmental meta-analysis of twin studies. *Psychological Bulletin, 107,* 226–237.

McGue, M., Bouchard, T. J., Iacono, W. G., & Lykken, D. T. (1993). Behavioral genetics of cognitive ability: A life-span perspective. In R. Plomin & G. E. McClearn (Eds.), *Nature, nurture, and psychology* (pp. 59–76). Washington, DC: American Psychological Association.

Merriman, C. (1924). The intellectual resemblance of twins. *Psychological Monographs, 33,* 1–58.

Mussen, P. H. (1983). *Handbook of child psychology.* New York: Wiley.

Piaget, J. (1952). *The origins of intelligence in children.* New York: International University Press.

Plomin, R., DeFries, J. C., McClearn, G. E., & Rutter, M. (1997a). *Behavioral genetics.* New York: W. H. Freeman.

Plomin, R., Fulker, D. W., Corley, R., & DeFries, J. C. (1997b). Nature, nurture and cognitive development from 1 to 16 years: A parent-offspring adoption study. *Psychological Science, 8,* 442–447.

Plomin, R., & McClearn, G. E. (1993). *Nature, nurture, and psychology.* Washington, DC: American Psychological Association.

Theis, S. V. S. (1924). *How foster children turn out.* New York: State Charities Aid Association, Publication no. 165.

14

Experimental Assessment of Specific Cognitive Abilities during the Second Year of Life

Steven M. Wilson
Robin P. Corley
David W. Fulker
J. Steven Reznick

Many researchers in the field of intelligence have accepted the notion of a hierarchical structure of adult intelligence, consisting of a number of specific ability groups as well as a general g factor (Cardon, 1994; Humphreys & Davey, 1988). Psychometrically, specific cognitive abilities are often defined either as the results (following rotation) of a factor analysis of test scores, either without first extracting a g factor (e.g., Guilford, 1967), or as the remaining systematic variance from a diverse test battery after g has been removed (Jensen, 1972, 1979). The nature and number of these abilities, however, varies widely. Thurstone (1938), for example, described six ability groups which he called verbal, space, number, word fluency, memory, and perceptual speed. Conversely, Guilford (1967) suggested the existence of more than 120 distinctly measurable abilities. In this chapter we focus on the genetic and environmental etiology of individual differences in two specific abilities that can be assessed in young children, word comprehension and memory for location.

The measurement of specific cognitive abilities in infants and young children can be a difficult undertaking, as reliability of the measures tends to be low. Indeed, based on the work of Rice et al. (1986a) and Singer et al. (1984), Cardon (1994) noted that specific cognitive abilities do not seem to emerge as measurable entities until 3 years of age. Further, Rice et al. (1986b) found more validity for their measures of specific abilities in 4 year olds than in 3 year olds.

There have been some encouraging findings of specific cognitive abilities in 3-year-old children. For example, previous research by Singer et al. (1984) with 3 year olds found that a battery of eight tests yielded four major specific abilities, which they identified as verbal, memory, perceptual speed, and spatial abilities. Similar work by Rice et al. (1986a) with 4-year-olds found the same four abilities which, while intercorrelated (thus implying a g factor), were still distinctly measurable.

Psychometrically, the structure of intelligence in infants and young children is not well understood, and questions remain as to the nature of general and specific cognitive abilities at this time of life (Reznick et al., 1996). Developmentally, little is known about influences—genetic and environmental—on the development of specific abilities in younger children, particularly during the second year of life (Plomin et al., 1990).

Thus, two major questions are presented with respect to intelligence and infants: first, can measures be constructed to provide useful and reliable information about specific cognitive abilities during the second year of life? Second, to the extent that these abilities can be reliably measured, how do genes and environment contribute to the continuity and change of these abilities throughout the second year?

To provide preliminary information with regard to these issues, the MacArthur Longitudinal Twin Study included measures designed to evaluate two specific cognitive abilities: word comprehension and memory for location. A sorting task, in which the child was given a tray containing toys of two different categories, was also administered; however, the purpose of this task was largely to provide a break from the focused attention of the word comprehension task (Plomin et al., 1990). These measures, in addition to a measure of general cognitive ability, the Bayley Mental Development Index (MDI; Bayley, 1969), which is described in detail in chapter 16, were administered at three time points during the second year of life: 14, 20, and 24 months of age. The tests are described in greater detail below.

To test for the presence of measurable specific abilities, measures were subjected to analysis at a phenotypic level to detect reliable variance in the specific measures apart from general cognitive ability. Behavioral genetic analyses using covariance structure models were also applied at each time point to test for reliable variability in the specific ability measures at latent genetic and environmental levels. Finally, longitudinal behavioral genetic analyses were performed within each specific ability measure to assess how genes and environmental effects influenced continuity and change in these measures.

A major advantage of behavioral genetic analyses as compared to pure phenotypic analysis is that they can separate the various influences that may be acting on the continuity and change of the phenotype in question into genetic and environmental sources of variation. Thus, genetic and environmental influences that may be obscured in phenotypic analyses might be elucidated through behavioral genetic analyses. For example, Cardon et al. (1992) found distinct patterns of influence for additive genetic effects (a^2), shared environmental effects (c^2), and unique environmental effects (e^2) on the development of general intelligence in children. In the present context, it might be expected that genetic and environmental influences may have distinct patterns of influence, either as a common factor influencing all of the measures, or as influences acting uniquely on each individual measure.

Behavioral genetic analyses can also provide useful information with respect to the question of test reliability in infants. Specifically, the unique environmental (e^2) parameter of behavioral genetic analyses can indicate the reliability of the measures, as it provides an upper bound for measurement error. More important, if the e^2 parameter can be fit with non-zero estimates only on the main diagonal of a multivariate matrix (see equation 6), then this implies that environmental factors which made the twins' scores dissimilar seemed largely to affect only one measure or one time point because the off-diagonal elements, which represent cross-measure or cross-time point effects, were shown to be nonsignificant. Psychometrically, this pattern of results suggests that measurement error rather than persistent idiosyncratic influences or systematic biases is the cause of dissimilarity among the scores of the twins. (One example to elucidate this point: if the tester of one twin were systematically more stringent or lenient in scoring than another, then there would be correlated influences causing one sibling to score differently from another. This type of systematic influence would likely present as significant off-diagonal loadings in the unique environmental parameter.)

SPECIFIC ABILITIES MEASURES

Word Comprehension

For the word comprehension task, the child was seated in a high chair in front of two rear-projection screens which were placed side by side 9 inches apart; a different object was displayed on each screen. An experimenter observed the focus of attention of the child through a small window above the screens and pressed a toggle switch to indicate on which screen the child was fixating. A computer tallied the total time spent fixating on each object.

Each word comprehension trial had two phases: a preprompt phase, followed by a prompted phase. In the preprompt phase one item was shown on each screen simultaneously for 8 s. The experimenter recorded fixation to each item. The slides went off for 1 s and then reappeared in exactly the same location for the commencement of the prompted phase. The experimenter, who was visible through the opening above the projection screens, but was blind to the location of the specific slides, asked the child: "Do you see the (name of object)?" "Where is the (name of object)?" "Look at the (name of object)?", where the target item shown on one of the slides and the word to be comprehended was spoken. The slides remained visible for 8 s, during which time the experimenter prompted the child and recorded fixation. There was a 3-s interval between trials, after which the procedure was repeated with a new pair of slides and a new word prompt.

Table 14.1. Easy, moderately difficult, and extremely difficult words administered to test word comprehension at 14, 20, and 24 months of age

| Age | Difficulty level | | |
	Easy	Moderately difficult	Extremely difficult
14 months	Cat	Boat	Purse
	Shoe	Lamp	Tree
	Ball	Balloon	Bee
	Dog	Socks	Broom
	Car	Blocks	Bread
20 months	Bike	Pillow	Drawer
	Fork	Jacket	Mop
	Cake	Pencil	Penguin
	Frog	Elephant	Map
	Hand	Crib	Camera
24 months	Zebra	Cowboy	Slipper
	Glove	Wolf	Jar
	Birdhouse	Refrigerator	Flag
	Roof	Snowman	Alligator
	Flag	Ladder	Scarf

Each child performed 15 trials with target words selected to be either easy, moderately difficult, or extremely difficult, based on findings from previous research (Reznick, 1990). The trials were ordered so that each block of three trials contained one word from each difficulty level. The words tested at each level, along with their level of difficulty, are shown in table 14.1.

The task was scored by subtracting percent fixation to the target slide in the preprompt phase (the baseline level of fixation to the target item) from the percent fixation to the target slide in the prompted phase. Word comprehension was inferred if the child demonstrated a 15% increase over the baseline level in fixation to the target slide following the prompt.

The overall word comprehension score was computed in two different and theoretically distinct ways. In keeping with prior reports (Emde et al., 1992; Plomin et al., 1990, 1993), one measure was simply the total number of words comprehended. However, other recent research has suggested that this total score might be confounded with attentiveness (Reznick et al., 1996). To compensate for this confounding, word comprehension was scored as the percentage of words comprehended of the total attended to by the infant. Thus, if the child attended to all 15 words, then the percentage would be computed as the number correct out of 15; if the child attended to, say, 8 words, then 8 would serve as the denominator from which the percent comprehended was computed. Analyses were performed on the data scored in each fashion.

Memory for Location

Like the word comprehension task, the memory for location task was conducted during the laboratory testing session. The child was seated in a high chair with two small cups placed in an inverted position, centered on a board in front of the child. The child was given a small attractive toy to play with. The toy was later taken away by the experimenter, and as the child watched it was placed under one of the inverted cups. A shield was then placed between the child and the cups to block the child's view of the cups for a delay interval. After the prescribed interval, the shield was removed, and the child was encouraged to find the toy. A trial was considered a success if the child removed the appropriate cup to reveal the toy. Practice trials were given in which the shield was not placed between the child and the cups until the child demonstrated comprehension of the task by locating the hidden toy.

Task difficulty varied both by the number of cups under which the toy could be hidden and also by the time for which the cups were occluded by the shield; for each correct response the task difficulty was increased. The first test trial was conducted with two cups and a 1-s delay. If the child successfully located the toy, the delay was increased to 5 s, and following another correct response, to 10 s. After a correct response at a 10-s delay, the number of cups was increased to four, and the delay was again started at 1 s. Success at this level led to a final circuit with an increase to six cups from which the child could choose. Correct performance with six cups at a 10-s delay led to a final trial with six cups at 15-s delay.

An incorrect response at any level led to a second trial at that level; a second failure on a level led to the termination of the procedure. The total number of trials successfully completed constituted the score for this measure.

PHENOTYPIC CORRELATIONS

The phenotypic correlations among the specific abilities measures and the Bayley MDI at each of the time points are presented in table 14.2. Examination of table 14.2 reveals that the MDI is significantly correlated with each measure at each time point. This pattern of correlations seems to indicate that, at least at this crude level, there is indeed a general factor of intelligence at a phenotypic level that is measurable during the second year of life.

The pattern of correlations among the measures of specific abilities is not as clear. One would expect to see positive correlations within each measure across the three time points; such positive correlations would also suggest some amount of stability for the measures. Further, were these abilities distinct from each other, their intercorrelations should be smaller than intracorrelations.

In looking at the data, for each time point (i.e., 14, 20, and 24 months), there is a significant correlation ($.104 \leq r \leq .127$) between the memory for location

Table 14.2. Phenotypic correlations (number of cases) among the Bayley Mental Development Index (MDI), memory for location (Mem), and word comprehension (WC) tests, each measured at 14, 20, and 24 months of age

	MDI 14	MDI 20	MDI 24	Mem 14	Mem 20	Mem 24	WC 14	WC 20
MDI 20	.475** (663)							
MDI 24	.432** (668)	.669** (637)						
Mem 14	.149** (656)	.214** (590)	.169** (593)					
Mem 20	.117** (553)	.147** (551)	.199** (546)	.043 (512)				
Mem 24	.136** (604)	.168** (584)	.217** (607)	.052 (546)	.175** (515)			
WC 14	.188** (639)	.289** (571)	.278** (574)	.104** (626)	.060 (496)	.087* (527)		
WC 20	.223** (496)	.377** (490)	.417** (484)	.132** (455)	.127** (478)	.000 (459)	.299** (457)	
WC 24	.175** (486)	.322** (461)	.380** (485)	.070 (435)	.083 (414)	.104* (487)	.165** (435)	.286** (431)

Two-tailed probability significant at *p ≤ .05; **p ≤ .01.

task and the word comprehension task. At the very least, this pattern indicates that the abilities measured by these tasks are not orthogonal to each other.

Looking at each measure individually, the word comprehension task appears to behave somewhat as a reliable measure, with statistically significant correlations at each time point and temporally adjacent measurements having larger correlations than the correlation between 14 and 24 months. However, it does not necessarily seem to act as a specific ability measure, as there are a number of significant correlations between word comprehension and memory for location (see table 14.2.) The memory for location task, on the other hand, did not show stability between the 14-month time point and either of the later measurements, while the 20- and 24-month time points were correlated, perhaps suggesting that this test does begin to assess a specific cognitive ability starting at about 20 months. Overall, at least at a phenotypic level, we cannot assert with certainty that there are specific abilities of memory for location and word comprehension that can be measured during the second year of life.

A principal components analyses of the phenotypic data formalize the patterns of correlations observed in table 14.2. Table 14.3 summarizes the results of orthogonal rotation of the factors, and two factors with eigenvalues > 1 were extracted. As examination of table 14.3 reveals, a general factor accounts for 27.4% of the variance in these measures; however, the memory for location task at 20 and 24 months does not load on this factor. Rather, these two measurements seem to form their own second factor. These results may indicate that the memory for location task used in this study may serve as a stable

Table 14.3. Orthogonally rotated factor loadings and percentage of variance accounted for by principal components analysis of the Bayley Mental Development Index (MDI), memory for location (Mem), and word comprehension (WC) at 14, 20, and 24 months

Measure	Factor 1	Factor 2
MDI 14	.565	
MDI 20	.778	
MDI 24	.769	.312
Mem 14	.332	
Mem 20		.670
Mem 24		.785
WC 14	.542	
WC 20	.692	
WC 24	.541	
Percentage of variance	32.3	12.2

Only factors with eigenvalues $\geq$ 1 and loadings $\geq$ .30 are given.

measure of that specific cognitive ability starting in the latter part of the second year.

Although some interesting results are suggested by phenotypic analyses, at this level ambiguity still remains as to whether infant intelligence may be viewed as a general factor, a constellation of specific abilities, or both. However, as mentioned above, analysis at a strictly phenotypic level may obscure the nature of latent processes that influence infant intelligence. That is, some processes may exert a general influence, while others may affect specific abilities; further, these influences may change over time (Cardon et al., 1992). Behavioral genetic analyses, in the form of multivariate covariance structure models of latent genetic and environmental influences, can help disentangle the components of the phenotype and may shed some light on the nature of these underlying structures.

BEHAVIORAL GENETIC ANALYSES

Model

The present analysis models phenotypic scores as the sum of latent genetic, common environmental, and specific environmental influences. Figure 14.1 depicts this relationship in a path diagram where **P** is a vector of phenotypic scores; $\mathbf{F}_G$, $\mathbf{F}_C$, and $\mathbf{F}_E$ are latent vectors of genetic, common environmental, and specific environmental influences, respectively; and $\mathbf{\Lambda}_G$, $\mathbf{\Lambda}_C$, $\mathbf{\Lambda}_E$ are factor loadings from the phenotypes to the respective factors. Genetic and environmental

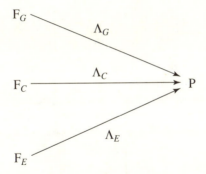

Figure 14.1. Structural model of the latent sources of phenotypic scores. F_G, F_C, and F_E are vectors of additive genetic, common environmental, and specific environmental influences, respectively; Λ_G, Λ_C, and Λ_E are factor loadings from the phenotypes to the respective factors; P is a vector of phenotypic scores.

factors are assumed to be uncorrelated in this model, as previous research has shown such correlations to be negligible for mental ability measures (Cardon et al., 1992; Cyphers et al., 1989).

The structural equation for the phenotypic scores corresponding to the path diagram in figure 14.1 is as follows:

$$\mathbf{P} = \Lambda_G \mathbf{F}_G + \Lambda_C \mathbf{F}_C + \Lambda_E \mathbf{F}_E. \tag{1}$$

This model of the phenotypic scores of an individual is extended to derive the expected covariances of a multivariate model both within and between twins by assuming correlations among the components. Monozygotic (MZ) and dizygotic (DZ) twins would be expected to have correlations of 1.0 and .5, respectively, for additive genetic influences, and common environmental influences are unity for siblings raised together.

Equation 1 can be expanded to represent the expected within-subject covariance ($\mathbf{CP}_W$) for the multivariate model as follows:

$$\mathbf{CP}_W = \Lambda_A \Lambda'_A + \Lambda_C \Lambda'_C + \Lambda_E \Lambda'_E, \tag{2}$$

where each Λ' is the transpose of the corresponding Λ matrix of factor loadings.

Expected between-subject covariances ($\mathbf{CP}_B$) are given by:

$$\mathbf{CP}_B = k\Lambda_A \Lambda'_A + \Lambda_C \Lambda'_C, \tag{3}$$

where k takes on the value of 1.0 or 0.5 in accordance with the genetic relatedness of MZ and DZ twins, respectively.

For the present analyses, the Λ matrices are 3×3 lower-triangular matrices with the factor loadings of each of the three measures, the MDI, memory for location, and word comprehension, at each time point. Developmental patterns were also explored for each measure across the three time points.

Thus, the general pattern of the lambda matrices is

$$\Lambda = \begin{bmatrix} b_{11} & & \\ b_{21} & b_{22} & \\ b_{31} & b_{32} & b_{33} \end{bmatrix}, \tag{4}$$

where there are as many potential factors as there are measures. (The letter b simply refers to a factor loading.)

The questions raised in this chapter suggest specific patterns for the Λ matrices. For example, if infant IQ can be best described as a general factor, then the expected loadings in the Λ matrices would be as follows:

$$\Lambda = \begin{bmatrix} b_{11} & & \\ b_{21} & 0 & \\ b_{31} & 0 & 0 \end{bmatrix}, \tag{5}$$

with all three measures having loadings on the first factor, and negligible (shown theoretically as zero above) loadings on the latter factors.

Conversely, if specific cognitive abilities are being reliably measured, then genetic or environmental influences for each of the measures would be expected to show significant unique variance only. In this case the Λ matrices would be expected to take the following form:

$$\Lambda = \begin{bmatrix} b_{11} & & \\ 0 & b_{22} & \\ 0 & 0 & b_{33} \end{bmatrix}, \tag{6}$$

with significant loadings for each variable on its own factor, and trivial, if non-zero, loadings on other factors.

Longitudinal analyses within the same measure across time points would be expected to have Λ matrices with similar forms. Continuity within a given measure, where the same influences that were present at 14 months continued to have an effect at the later time points, would be demonstrated by a Λ matrix in the form shown in equation 5. Change in the influences on a given measure from time point to time point should show up as specific loadings similar to those shown in equation 6.

Pedigree Analysis

In the present data set there are different patterns of observations resulting from missing data. For a given family, one or both twins might have one or more missing observations; however, the pattern of missing data does not appear to be systematic. Accordingly, the data were analyzed using a maximum-likelihood pedigree technique (Lange et al., 1976) via the Mx Statistical Modeling package (Neale, 1994). The advantage of the pedigree technique is that it allows every available data point to be used, rather than deleting families with missing data. For a more detailed explanation of pedigree analysis, see chapter 15 Appendix Note 3.

At each time point, various parameters of the model were constrained to be zero to address the question of whether infant intelligence shows evidence of a general factor or specific abilities at a genetic and environmental level. In each case, specific environmental effects were evaluated first to find a more parsimonious model which fit the data. The same procedure was then followed for shared environmental effects, and finally additive genetic effects in order to determine the relative pattern of their influence.

RESULTS

14 Months

Table 14.4 shows the results of model fitting at 14 months. As can be seen from this table, specific environmental effects (e^2) could be diagonalized, indicating that the components that make up this parameter test error and unique environmental factors were not correlated across the measures. Next we see that the common environmental influences (c^2) could be fit to a single factor, which suggests that common environmental effects serve as a constant background affecting all the measures in a similar manner. The common environmental factor could actually be dropped from the model without a significant deterioration in fit (model 4), but in so doing, no additive genetic parameters (a^2) could be dropped; thus c^2 was left as a single common factor to try to ascertain the mode of influence of additive genetic factors. Tests of these models (models 5–7) revealed that genetic effects could be diagonalized without loss of fit. This pattern suggests that genetic effects seem to be working independently on each measure, and that, at least at an additive genetic level, there are specific cognitive abilities measurable as early as 14 months. The additive genetic component, however, could not be made to fit as a single common

Table 14.4. Model testing results for genetic and environmental components at 14 months across the MDI, memory for location, and word comprehension

Model[a]	Parameters estimated	−2 LL	Models compared	χ^2 Difference	df	p
1. ACE	18	7068.352				
2. ACE_d	15	7068.441	2 vs. 1	0.089	3	.993
3. AC_1E_d	12	7068.441	3 vs. 2	0.0	3	1.000
4. AE_d	9	7074.946	4 vs. 3	6.505	3	.089
5. $A_dC_1E_d$ [b]	9	7069.793	5 vs. 3	1.352	3	.717
6. $A_1C_1E_d$	9	7080.132	6 vs. 3	11.691	3	.009
7. C_1E_d	6	7100.419	7 vs. 6	30.626	3	.000

[a] Model parameters are additive genetic (A), common environment (C), and unique environment (E). A 1 following a parameter indicates that the parameter was modeled as a single common factor; a d following a parameter indicates that it was modeled as separate abilities, with loadings only along the main diagonal.
[b] Model retained. LL, log-likelihood.

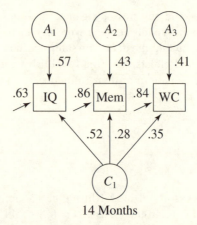

14 Months

Figure 14.2. Standardized loadings of additive genetic (A), common environmental (C), and unique environmental factors on the Bayley MDI (IQ), the memory for location measure (Mem), and the word comprehension (WC) measure at 14 months.

factor, nor could it be dropped altogether, which provides additional support for this finding.

Figure 14.2 shows the most parsimonious model that explained the data well. This figure gives the factor loadings for each parameter and demonstrates the unique genetic effects on each measure separately, while common environmental effects act as a single common factor. As can be observed in this figure, especially for the measures of specific abilities most of the variance is accounted for by unique environmental factors, which include test (un)reliability, or measurement error. However, even at this early age, additive genetic effects seem to account for about 16–18% of the variance, and common environmental effects seem to account for about 8–12% of the variance in the measures of specific abilities.

20 Months

The picture at 20 months was similar to that at 14 months. Additive genetic effects seemed to be influencing each measure individually, suggesting again that at least at this level, these specific cognitive abilities can be measured during the second year of life. As table 14.5 shows, the most parsimonious model again showed that additive genetic effects described the data adequately only as unique abilities (model 5); the common factor model for genetic effects (model 6) was rejected. Once again, common environmental effects acted as a single factor.

Figure 14.3 provides the magnitude of the loadings estimated in model 5, the most parsimonious model to adequately fit the data. As can be seen from this figure, the unique environmental parameter accounts for a majority of

Table 14.5. Model testing results for genetic and environmental components at 20 months across the MDI, memory for location, and word comprehension

Model[a]	Parameters estimated	−2 LL	Models compared	χ^2 Difference	df	p
1. ACE	18	6082.065				
2. ACE_d	15	6088.594	2 vs. 1	6.529	3	.089
3. AC_1E_d	12	6090.386	3 vs. 2	1.792	3	.617
4. AE_d	9	6109.994	4 vs. 3	19.608	3	.000
5. $A_dC_1E_d$ [b]	9	6094.774	5 vs. 3	4.388	3	.225
6. A_1C_1E	9	6104.266	6 vs. 3	13.880	3	.003
7. C_1E_d	6	6130.410	7 vs. 6	35.636	3	.000

[a] Model parameters are additive genetic (A), common environment (C), and unique environment (E). A 1 following a parameter indicates that the parameter was modeled as a single common factor; a d following a parameter indicates that it was modeled as separate abilities, with loadings only along the main diagonal.
[b] Model retained. LL, log-likelihood.

the variance in the specific abilities measures. In comparison with the loadings estimated at 14 months (see figure 14.2), there is an increase in the common environmental effects and a decrease in additive genetic effects for the word comprehension task, and the opposite pattern is observed for the memory for location task, where there was a larger loading on the additive genetic factor and a reduction in the common environmental influences. This pattern is further demonstrated in the 24-month analyses, as shown below.

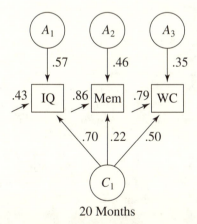

20 Months

Figure 14.3. Standardized loadings of additive genetic (A), common environmental (C), and unique environmental factors on the Bayley MDI (IQ), the memory for location measure (Mem), and the word comprehension (WC) measure at 20 months.

Table 14.6. Model testing results for genetic and environmental components at 24 months across the MDI, memory for location, and word comprehension

Model[a]	Parameters estimated	−2 LL	Models compared	χ^2 Difference	df	p
1. ACE	18	6595.416				
2. ACE_d	15	6602.314	2 vs. 1	6.898	3	.075
3. AC_1E_d	12	6603.932	3 vs. 2	1.618	3	.655
4. AE_d	9	6619.077	4 vs. 3	15.145	3	.002
5. $A_dC_1E_d$ [b]	9	6612.680	5 vs. 3	8.748	3	.033
6. $A_1C_1E_d$	9	6608.002	6 vs. 3	4.070	3	.254
7. C_1E_d	6	6642.512	7 vs. 6	34.510	3	.000

[a] Model parameters are additive genetic (A), common environment (C), and unique environment (E). A 1 following a parameter indicates that the parameter was modeled as a single common factor; a d following a parameter indicates that it was modeled as separate abilities, with loadings only along the main diagonal.
[b] Model retained. LL, log-likelihood.

24 Months

At 24 months the picture for additive genetic effects on the three measures appeared to change. Although common environmental influences could still be modeled as a single common factor, as table 14.6 shows, the model with genes acting independently on each measure (model 5) no longer fit. Rather, the most parsimonious model that still adequately explained the data showed genetic effects to be acting as a common factor (model 6). These results suggest that additive genetic effects may initially act on each ability separately and then form a more generalized ability toward the end of the second year.

Figure 14.4 shows this emerging pattern; both genetic and shared environmental effects are best modeled as a common factor. However, it is interesting to note from the magnitude of these loadings that the pattern observed in the 20-month data became markedly more prominent at 24 months. Specifically, the loading from the genetic factor to word comprehension dropped to null, as did the loading from the common environmental factor to to the memory for location task. For both of these measures, approximately 75% of the variance is due to unique environmental factors and measurement error, with the other 25% being contributed by additive genetic factors for the memory for location task and common environmental effects for the word comprehension task. Thus, although the results do show that additive genetic effects cannot be modeled as independent abilities at this time point, a general ability factor includes only the MDI and the memory for location task; it does not include the word comprehension measure at a genetic level.

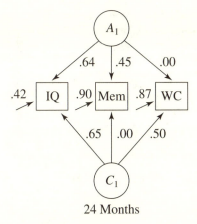

24 Months

Figure 14.4. Standardized loadings of additive genetic (A), common environmental (C), and unique environmental factors on the Bayley MDI (IQ), the memory for location measure (Mem), and the word comprehension (WC) measure at 24 months.

Word Comprehension across Time Points

Longitudinal behavioral genetic analyses within a measure can provide additional information concerning test reliability and how genes and environment contribute to continuity and change within that measure. Table 14.7 shows the results of model fitting for the word comprehension measure at 14, 20, and 24 months; scoring method was the total number of words comprehended. Preliminary analyses of these data showed considerable evidence of continuity (a single common factor) for the measure from time point, to time point, while little evidence for change (loadings along the main diagonal) was found. Thus the focus of these analyses became whether it was genetic or environmental influences that were responsible for this continuity. As table 14.7 demonstrates, there is no reliable distinction between modeling a genetic common factor (model 5) or a shared environmental common factor (model 8). Thus no model could be retained as being the most parsimonious descriptor for this measure.

Table 14.8 shows the same analyses performed on the word comprehension measure using the percent correct of total words attended to by the infant as the means of scoring. The results for this scoring method were identical to those for the total number correct; there is evidence of familiality and evidence of continuity, but there is insufficient power to detect whether these traits are due to genetic or shared environmental influences.

Memory for Location across Time Points

Preliminary longitudinal analyses of the memory for location measure helped elucidate the finding of a separate factor uniting the 20- and 24-month time

Table 14.7. Model testing results for genetic and environmental components of continuity and change across 12, 20, and 24 months for word comprehension (scoring method is number correct)

Model[a]	Parameters estimated	−2 LL	Models compared	χ^2 Difference	df	p
1. ACE	18	6797.691				
2. ACE_d	15	6802.067	2 vs. 1	4.376	3	.224

Test of genetic parameters with shared environment dropped

3. AE_d	9	6806.050	3 vs. 2	3.983	3	.679
4. A_dE_d	6	6890.620	4 vs. 3	88.553	3	.000
5. A_1E_d	6	6811.138	5 vs. 3	5.088	3	.165

Test of shared environment parameters with genetic influences dropped

6. CE_d	9	6806.851	6 vs. 2	4.784	3	.572
7. C_dE_d	6	6885.306	7 vs. 6	85.306	3	.000
8. C_1E_d	6	6811.704	8 vs. 6	4.853	3	.183

[a] Model parameters are additive genetic (A), common environment (C), and unique environment (E). A 1 following a parameter indicates that the parameter was modeled as a single common factor; a d following a parameter indicates that it was modeled as separate influences for each time point, with loadings only along the main diagonal.

Table 14.8. Model testing results for genetic and environmental components of continuity and change across 12, 20, and 24 months for word comprehension (scoring method is percent correct)

Model[a]	Parameters estimated	−2 LL	Models compared	χ^2 Difference	df	p
1. ACE	18	5740.829				
2. ACE_d	15	5743.145	2 vs. 1	2.316	3	.509

Test of genetic parameters with shared environment dropped

3. AE_d	9	5745.427	3 vs. 2	2.282	3	.892
4. A_dE_d	6	5801.749	4 vs. 3	56.322	3	.000
5. A_1E_d	6	5747.425	5 vs. 3	1.998	3	.573

Test of shared environment parameters with genetic influences dropped

6. CE_d	9	5745.420	6 vs. 2	2.275	3	.893
7. C_dE_d	6	5800.788	7 vs. 6	55.368	3	.000
8. C_1E_d	6	5747.132	8 vs. 6	1.712	3	.634

[a] Model parameters are additive genetic (A), common environment (C), and unique environment (E). A 1 following a parameter indicates that the parameter was modeled as a single common factor; a d following a parameter indicates that it was modeled as separate influences for each time point, with loadings only along the main diagonal.

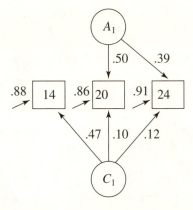

Figure 14.5. Standardized loadings of additive genetic (A), common environmental (C), and unique environmental factors on the memory for location measure at 14, 20, and 24 months of age.

points. Specifically, these analyses suggested that there should be a single loading for this measure at 14 months, followed by a separate factor influencing both of the later time points. As is often the case in factor analysis studies, there are multiple paths (all mathematically valid) which can be followed in interpreting the data.[1]

In keeping with the format chosen for the present chapter (i.e., continuity and change across all three time points), these data were analyzed looking for either a single common factor or a main diagonal of unique loadings for each time point. As a result of this chosen constraint, the most parsimonious model to describe the data had both shared environmental and additive genetic factors. However, the pattern of these loadings shows that there needs to be a separate loading for the 14-month time point than the 20- and 24-month time points. As can be seen from figure 14.5, there was a larger loading of shared environment on the 14-month time point, whereas the loadings at the 20-and 24-month time points were negligible. Conversely, the additive genetic factor had moderate loadings linking the 20- and 24-month time points.

Inspection of table 14.9 reveals that given a common factor for the shared environment, the genetic factors could be modeled either as separate factors at each time point (model 5) or as a single factor (model 6) and still provide an adequate fit of the model to the data. However, using a single common factor resulted in one less parameter being estimated and thus was chosen as the preferred model.[2]

DISCUSSION

The questions proposed in this chapter were (1) can measures be constructed to provide useful and reliable information about specific cognitive abilities

Table 14.9. Model testing results for genetic and environmental components of continuity and change across 12, 20, and 24 months for memory for location

Model[a]	Parameters estimated	−2 LL	Models compared	χ^2 Difference	df	p
1. ACE	18	7693.930				
2. ACE_d	15	7694.719	2 vs. 1	0.789	3	.852
3. AC_1E_d	12	7694.868	3 vs. 2	.149	3	.985
4. AE_d	9	7695.852	4 vs. 3	0.984	3	.805
5. $A_dC_1E_d$	9	7698.132	5 vs. 3	3.264	3	.353
6. $A_1C_1E_d$ [b]	8[c]	7695.904	6 vs. 3	1.322	4	.724
7. C_1E_d	6	7711.899	7 vs. 6	13.571	2	.004
8. A_dE_d	6	7728.792	8 vs. 5	30.660	3	.000
9. A_1E_d	6	7710.054	9 vs. 6	11.726	2	.008

[a] Model parameters are additive genetic (A), common environment (C), and unique environment (E). A 1 following a parameter indicates that the parameter was modeled as a single common factor; a d following a parameter indicates that it was modeled as separate influences at each time point, with loadings only along the main diagonal.
[b] Model retained.
[c] Loading for 14 months was constrained to be zero, because when it was estimated it took on a small negative loading (see figure 5).
LL, log-likelihood.

during the second year of life? and (2) to the extent that these abilities can be reliably measured, how do genes and the environment contribute to the continuity and change of these abilities throughout the second year? The short answer to the first question seems to be a guarded "yes." Given the difficulties inherent in measuring psychological variables in infants, it does appear that some reliable information can be gleaned about specific cognitive abilities during the second year. The short answer to the second question seems to be that both additive genetic and shared environmental influences contribute to continuity in the measurement of specific abilities, and unique environmental influences contribute only to change in these abilities. These findings are expounded below.

With respect to the first question, results from the MacArthur Longitudinal Twin Study provide some interesting findings concerning the measurement of the specific cognitive abilities of word comprehension and memory for location during the second year of life. Although phenotypically these specific abilities are difficult to distinguish from general cognitive abilities, results from the memory for location measure provide evidence to make us guardedly optimistic that this ability may be measurable as early as 20 months of age even at a phenotypic level. Further, behavioral genetic results suggest that there may be reliable genetic effects that influence each measure individually, providing evidence for measurable specific cognitive abilities during the second year. These results suggest that at least at an additive genetic level, the specific cognitive abilities of word comprehension and memory for location may be measurable as early as 14 months of age.

It is apparent that reliability for these measures of infant intelligence is not high (as indicated by the magnitude of the unique environmental loadings, which include measurement error, especially on the specific ability measures). However, of that variance which is reliable, both genetic and environmental influences contribute to the phenotype of infant intelligence at each time point. Shared environmental effects seems to act as a general factor, whereas genetic effects seem to provide specific influences early on.

To expound on this point, phenotypically it was difficult to ascertain whether infant intelligence can be modeled as a common factor, a constellation of specific abilities, or both. However, at a behavioral genetic level, additive genetic effects seem to influence infant intelligence as specific abilities at the initial measurements, and a general factor may possibly emerge starting at the end of the second year. Common environmental influences, in contrast, seem to contribute to the development of intelligence as a general factor throughout the second year of life. Thus the proposition that even an infant's intelligence is best modeled as both a general factor and specific abilities seems to have some merit genetically.

Another interesting finding from the present data concerns the second question. Within the specific ability measures there seems to be continuity for both measures throughout the second year. With respect to the memory for location measure, whether genes, environment, or both influence this continuity could not be differentiated with these data. The present data showed a distinct pattern in which there was change from 14 to 20 months, followed by continuity in the measure from 20 to 24 months. However, to avoid the risk of overfitting models, this finding was not explored in the present chapter; for a more complete analysis of this measure, see Reznick et al. (1996).

For the word comprehension measure, there was clear evidence for continuity from time point to time point, but there was insufficient power to ascertain whether this continuity was due to genetic or environmental influences. Theoretically relevant changes in the scoring method did not provide additional power to elucidate the source of continuity evident in the measure.

Overall, although the results of the present study do not provide conclusive answers to the questions proposed in this chapter, they do indicate that behavioral genetic analyses can provide sufficient power to detect at least the specific cognitive abilities of word comprehension and memory for location at latent genetic and environmental levels.

NOTES

1. An alternative method of analysis performed on these data (Reznick et al., 1996) used a two-factor model, with a single loading for shared environmental effects at 14 months, and then a factor for continuity between the 20- and 24-month time points; all additive genetic parameters were dropped. Al-

though this model does adequately describe the data, the Akaike Information Criterion (AIC) for the shared environmental model is −16.069 as compared to −18.886 for the same pattern of loadings in an additive genetic model with all of the shared environmental parameters dropped. The AIC for the model retained was −17.917. Thus, although each model adequately describes the data, the AIC empirically suggests that the additive genetic model is the preferable one unless a specific theory drives one to reject genetic influences in favor of environmental ones—as was the case in Reznick et al. (1996)—or unless there is some other theoretical question, as is the case in the present chapter.

2. The reason that the single factor resulted in one less parameter merits some explanation. In estimating the loadings for the common factor model (model 6), Mx (Neale, 1994) attached a very small positive loading to the 14-month time point. This first loading then caused the loadings at 20 and 24 months to become negative with moderate values. (The sign of loadings in any factor analysis are arbitrary, so all of the loadings on a given factor can be reversed without consequence.) The implication of this pattern of loadings is that there appear to be two genetic factors: one for the first time point and a second factor for the remaining time points. Thus, if the loading for the first time point is constrained to be zero, it is equivalent to dropping the factor for that time point. Constraining the loading at 14 months to be zero, the loadings for the remaining time points are free to take on their optimal values in their factor. As this could be done without significant loss of fit (see model 6, table 14.9), this model was accepted as the most parsimonious. (Note that the factor for the 20- and 24-month time points could not be dropped; model 7.) This pattern of results suggests that the second factor from the principal components analysis, which was dedicated to the memory for location measure at 20 and 24 months, may be due largely to additive genetic influences.

REFERENCES

Bayley, N. (1969). *Manual for the Bayley Scales of Infant Development.* New York: Psychological Corporation.

Cardon, L. R. (1994). Specific cognitive abilities. In J. C. DeFries, R. Plomin, & D. W. Fulker (Eds.), *Nature and nurture during middle childhood* (p. 57–76). Oxford: Blackwell.

Cardon, L. R., Fulker, D. W., DeFries, J. C., & Plomin, R. (1992). Continuity and change in general cognitive ability from 1 to 7 years of age. *Developmental Psychology, 28,* 64–73.

Cyphers, L. H., Fulker, D. W., Plomin, R., & DeFries, J. C. (1989). Cognitive abilities in the early school years: No effects of shared environment between parents and offspring. *Intelligence, 13,* 369–386.

Emde, R. N., Plomin, R., Robinson, J. A., Reznick, J. S., Campos, J., Corley, R., DeFries, J. C., Fulker, D. W., Kagan, J., & Zahn-Waxler, C. (1992). Temperament, emotion, and cognition at 14 months: The MacArthur Longitudinal Twin Study. *Child Development, 63,* 1437–1455.

Guilford, J. P. (1967). *The nature of human intelligence.* New York: McGraw-Hill.

Humphreys, L. G., & Davey, T. C. (1988). Continuity in intellectual growth from 12 months to 9 years. *Intelligence, 12* 183–197.

Jensen, A. R. (1972). *Genetics and education.* New York: Harper & Row.

Jensen, A. R. (1979). g: Outmoded theory or unconquered frontier? *Creative Science & Technology, 2,* 16–29.

Lange, K., Westlake, J., & Spence, M. A. (1976). Extensions to pedigree analysis: III. Variance components by the scoring method. *Annals of Human Genetics, 39,* 485–491.

Neale, M. C. (1994). *Mx: Statistical modeling* (2nd ed.). Richmond, VA: Department of Psychiatry, Medical College of Virginia.

Plomin, R., Campos, J., Corley, R., Emde, R. N., Fulker, D. W., Kagan, J., Reznick, J. S., Robinson, J., Zahn-Waxler, C., & DeFries, J. C. (1990). Individual differences during the second year of life: The MacArthur Longitudinal Twin Study. In J. Columbo & J. Fagan (Eds.), *Individual differences in infancy: Reliability, stability, & predictability* (pp. 431–455). Hillsdale, NJ: Erlbaum.

Plomin, R., Emde, R. N., Braungart, J. M., Campos, J., Corley, R., Fulker, D. W., Kagan, J., Reznick, J. S., Robinson, J., Zahn-Waxler, C., & DeFries, J. C. (1993). Genetic change and continuity from fourteen to twenty months: The MacArthur Longitudinal Twin Study. *Child Development, 64,* 1354–1376.

Reznick, J. S. (1990). Visual preference as a test of infant word comprehension. *Applied Psycholinguistics, 11,* 145–166.

Reznick, J. S., Corley, R., & Robinson, J. (1997). A longitudinal twin study of intelligence in the second year. *Monographs of the Society for Research in Child Development, 62*(1), i–vi, 1–154.

Rice, T., Corley, R., Fulker, D. W., & Plomin, R. (1986a). The development and validation of a test battery measuring specific cognitive abilities in four-year-old children. *Educational and Psychological Measurement, 46,* 699–708.

Rice, T., Fulker, D. W., & DeFries, J. C. (1986b). Multivariate path analysis of specific cognitive abilities in the Colorado Adoption Project. *Behavior Genetics, 16,* 107–125.

Singer, S., Corley, R., Guiffrida, C., & Plomin, R. (1984). The development and validation of a test battery to measure differentiated cognitive abilities in three-year-old children. *Educational and Psychological Measurement, 44,* 703–713.

Thurstone, L. L. (1938). *Primary mental abilities.* Chicago: University of Chicago Press.

15

Continuity and Change in General Cognitive Ability from 14 to 36 Months

Stacey S. Cherny
David W. Fulker
Robert N. Emde
Robert Plomin
Robin P. Corley
John C. DeFries

The Bayley Scales of Infant Development (BSID; Bayley, 1969) are the most widely used measurement scales of general intelligence in infancy and were the first commonly used measures nominated for a marker variable for purposes of making comparisons among longitudinal studies in early development (Bell & Hertz, 1976). However, early in the second year of life, tests of general cognitive ability, such as the BSID, are only moderately predictive of later general intelligence (e.g., Cherny et al., 1995; Wilson, 1983). Thus, there are both continuities and discontinuities from infancy onward in the development of general cognitive ability.

The testing of mental abilities in infancy presents special issues regarding the organization of competencies over time (McCall, 1979; Sternberg & Powell, 1983). Major organizational changes in infancy have been documented in a number of longitudinal studies. Bell and co-workers (1981) noted a longitudinal inversion of intensity from the newborn period to the beginning of the child's third year in which overly intense behaviors across a number of early measures were linked to a later outcome of low intensity of functioning. The separate longitudinal studies of Emde et al. (1976) and of Kagan et al. (1978) also found major changes in developmental organization across infancy, with qualitative transformations in both cognitive and emotional development. The basis for discontinuities as well as for continuities from infancy was reviewed by Rutter (1987), who pointed to the need for more extensive empirical approaches. Qualitative transformations in mental development were a basis for earlier "stage" theories of development (e.g., Erikson, 1950; Piaget, 1952; Spitz, 1959). Changes in developmental organization do not, however, have necessary implications with respect to the stability of individual differences. In a penetrating analysis of longitudinal data from the Berkeley growth study, McCall and co-workers (1977) found periods of developmental change that were accompanied by increased variability in individual differences and speculated that these periods of change represented major shifts in organization.

The McCall et al. (1977) analyses used a factor analytic approach to examine whether children's performance on the predecessor of the BSID (California first year and California preschool scales) fit a model of mental development that was either continuous and stable or discontinuous and unstable. Their analyses supported a stage model of development because of significant changes in the rank order of individual performance and in the content of the first principal component. In a subsequent review, McCall (1979) proposed a theoretical model that took into account existing longitudinal information. A major change in the structuring of mental function was seen to begin around 18 months of age. This age marks the beginning of symbolic relation of thought, both nonverbal and verbal. Multiword speech and language comprehension become increasingly important in mental assessment at the end of the child's second year. Moreover, beginning at this age, correlations increase between infant mental test scores and childhood IQ tests; thus there is a tendency for greater prediction to later IQ when infant testing becomes more verbal in character. Similarly, it is at this age that the level of prediction to later IQ can be increased by adding parental socioeconomic status in a multiple regression equation (McCall, 1979; McCall et al., 1977).

McCall's (1979) model included both genetic and environmental influences as causes of individual differences in mental performance during infancy. Summarizing cognitive data with respect to parent–child similarity, sibling similarity, and similarity of twins from the longitudinal Louisville Twin Study (e.g., Matheny et al., 1976; Wilson, 1983), McCall (1979) concluded that the role of heritable influences in early infancy was either modest or small and increased toward the end of the child's second year. Using Waddington's conceptual metaphor of canalization (Waddington, 1957) as modified by Scarr-Salapatek (1975), McCall (1979) proposed that there is a major organizational change at 18–24 months that accounts for this shift. Mental development is strongly canalized before that age and less so afterward. Behaviors that characterize mental performance before 18–24 months are accomplished by all infants brought up in a species-typical environment, and environmental and biological perturbations are buffered by self-righting tendencies. Beginning at 18–24 months of age, concomitant with language and symbolic relations becoming more important in mental development, individual differences due to both genetic and environmental influences become more apparent.

Although McCall based his proposal on an observed increase in genetic variance across ages 18–24 months, his theory can really only be tested by examining genetic covariation over time. If organizational changes are occurring during this time period, we would expect some new genetic variation to appear at the later ages that was not present at earlier ages. In other words, organizational changes should produce correlations less than unity between genetic influences across time. Advances in quantitative behavioral genetics allow us to document and study such developmental change as the McCall model implies. Twin and adoption methods, when used in longitudinal studies, permit estimates of genetic and environmental influences on individual differences at each age, as well as separable influences of each as contributors to stability and change in development. Furthermore, advances in molecular genetics have given added incentive to document organizational change in early mental development in terms of both genetic and environmental sources of influence with respect to change. Genes "turn on" and "turn off" across development, and genetic expression is known to be dependent on interactions with both the cellular and supercellular environment. For example, although some of the same genes may be influencing general cognitive ability across ages 18–24 months, there may also be different genes operating at 24 months that were not exerting an influence earlier in development. Moreover, genes can regulate change itself (e.g., Plomin et al, 1993).

The present chapter presents data on monozygotic (MZ) and dizygotic (DZ) twin pairs from the MacArthur Longitudinal Twin Study that were subjected to longitudinal genetic analysis. Comparisons of correlations for MZ and DZ twins allow us to assess heritable and environmental influences over time. We now recognize that similar behaviors can reflect different influences at different times in development. Does the change in the structure of mental performance as reflected in the Bayley Mental Development Index (MDI) in the latter part of the second year of life mean that the genetic and environmental sources of influence are also different across this transition? The McCall model suggests that this might be so, and this hypothesis can now be tested.

Advances in genetic model fitting, used in a longitudinal twin study such as the current one, now permit us to assess the sources of variation across multiple ages. When applied to twin data, a longitudinal genetic model can partition observed phenotypic variance and covariance into parts due to genetic, shared family environmental influences, and nonshared environmental influences unique to individuals. The extent to which there is continuity and change in each of these three separate components can then be estimated and tested.

SUBJECTS AND MEASURES

Subjects were identical (MZ) and same-sex fraternal (DZ) twin pairs participating in the ongoing MacArthur Longitudinal Twin Study (Emde et al., 1992; Plomin et al., 1990) at the University of Colorado, Boulder. Longitudinal data from more than 350 MZ and DZ twin pairs tested at 14, 20, 24, and 36 months are available on what is now the complete sample for these ages.

The mental development component of the BSID (Bayley, 1969) was administered during the home visit at each age. The first home visit occurred between

13.5 and 15.5 months, the second between 19.5 and 21.5 months, and the third between 23.5 and 26 months. Children generally were seated in a high chair, although a small number sat in booster seats at a table. The twins were tested at the same time by different examiners and were situated in adjoining rooms in the home, with the mother seated within view of each twin. The test was administered according to standardized procedures, with basal and ceiling levels established to derive the MDI. At age 36 months, the Stanford-Binet IQ test (Terman & Merrill, 1973) was given using standard procedures.

Developmental Model

The developmental model that was fitted to these data was first proposed by Eaves et al. (1986) and represents a combination of a single general factor present at all ages and a quasi-simplex model of specific effects arising at each age point and their subsequent transmission to later ages. The general factor implies a static developmental process where an influence present at an early age persists across the entire period. The simplex implies a more dynamic process in which new variation arises at each age, persists to the next age, and is of progressively decreasing importance at subsequent ages. The two processes are illustrated in figure 15.1 for four age points. When models of this kind have been applied to longitudinal data at the phenotypic level (Humphreys and Davey, 1988), a simplex was found to provide a better account of the data than a common factor from ages 1–9 years.

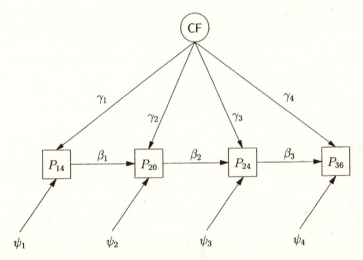

Figure 15.1. Simplex and common factor model of cognitive development from age 14 months through 36 months. The β_i and γ_i are path coefficients and the ψ_i are variances to be estimated.

The path diagram of the longitudinal developmental model, shown in figure 15.1, implies specific mathematical expectations for the covariance structure among measures of cognitive ability across ages. These expectations, given in Appendix Note 1, may either be derived from the path diagram using path tracing rules (Li, 1975) or obtained using matrix algebra and the LISREL model (Jöreskog and Sörbom, 1989).

Genetic Extension of the Developmental Model

With genetically informative data, such as those available on MZ and DZ twin pairs in the MacArthur Longitudinal Twin Study, we can examine developmental structures at both the environmental and the genetic levels to test if similar or different developmental processes are operating at these different levels. Three sources of variation may be observed at each time point. The expectation for the correlation between MZ twin pairs is h^2, the proportion of variation ascribable to genetic influence, plus c^2, the proportion of variance due to environmental influences shared by siblings. For DZ twin pairs, the expected correlation is c^2 plus one half the genetic variance, h^2. A third component, e^2, ascribable to the effects of the nonshared environment, or influences unique to the individual, is $1 - h^2 - c^2$. These are the same components of variance derivable from the classical twin study. The multivariate development simplex and common factor model used in the present chapter is readily extended for use with twin data, as shown in Appendix Note 2.

In summary, the model had three types of parameters to explain continuity and change: (1) the loadings (γ_i) of the common set of genes or environmental influences (CF) influencing the measures (P_i) at all ages; (2) new genetic or environmental influences or innovations appearing at each age (ψ_i); and (3) age-to-age transmission of genetic and environmental influences (β_i). The transmission of genetic (or environmental) influences from age to age is what developmentalists generally refer to as continuity. The new variation at each age reflects developmental change. Changes in the structure of cognitive functioning (e.g., McCall, 1979) might be expected to manifest themselves as the appearance of such new genetic or environmental variation during development.

Fitting the Models

To test the adequacy of our models and to estimate the importance, or magnitude, of each type of influence, we compared the observed patterns of tests scores, both among family members (e.g., pair of twins) and across ages, with those expected from the model. The parameters of the model can then be estimated by minimizing the discrepancy between the expected and the observed covariance matrix, using optimization routines on the computer. Details of our model-fitting procedures are given in Appendix Note 3.

Before model fitting, the data were first standardized within each age, across all individuals as a single group. This standardization procedure effectively

Table 15.1. Twin correlations at each age

Measure	Age (months)	MZ		DZ	
		r	N	r	N
Bayley MDI	14	.56	196	.39	170
Bayley MDI	20	.80	173	.65	149
Bayley MDI	24	.84	174	.61	153
Sanford-Binet	36	.78	159	.51	151

$p < .0001$ for all correlations.

eliminates age differences in variances, which most likely are merely a result of using different tests at different ages, while preserving MZ, DZ, twin 1, and twin 2 variance differences. Resulting parameter estimates presented below were standardized to imply a phenotypic variance of unity, using an extension of the procedures employed by LISREL (Jöreskog and Sörbom, 1989).

Twin Correlations

Before doing any developmental modeling, examining the twin correlations is a necessary first step. These correlations are presented in table 15.1 for the Bayley MDI data at 14, 20, and 24 months and for the Stanford-Binet data at 36 months, along with the number of twin pairs on which these correlations were computed. It is clear that in all cases, the MZ correlations are higher than those of the DZs, implying genetic influence. These correlations are relatively uniform for both MZ and DZ pairs at ages 20–36 months, but lower for both types of twins at 14 months. Of course, simply examining twin correlations at each age does not yield an optimal analysis of longitudinal data but is just a starting point. Of more interest are the cross-sibling, cross-time correlations, and developmental models of them, such as the one described above, which allow further dissection of the developmental process.

Application of the Development Model

In Table 15.2, we present estimates of h^2, c^2, and e^2 obtained from fitting the full simplex and common factor model to the twin data (see also Appendix Note 4). Heritability appears relatively stable across all ages, with the exception of 20 months, when it is a bit lower. The proportions of variance due to shared family environmental influences are relatively small at 14 months but increase dramatically at 20 months, then decrease over the next two age points. It should be noted that the genetic and shared environmental variance components estimated from fitting the full model to these data are somewhat different from what univariate analyses of the twin correlations at each age might yield. These differences arise because the multivariate models take into ac-

Table 15.2. Estimates of h^2, c^2, and e^2 for general cognitive ability at each age

	Age (months)			
Variance component	14	20	24	36
h^2	.36	.29	.39	.40
c^2	.21	.51	.44	.37
e^2	.43	.19	.17	.23

count the cross-twin, cross-time covariance structure, which can affect the within-time parameter estimates. Use of the cross-twin, cross-time information is an inherent advantage of a multivariate longitudinal analysis. When sample sizes are small, the standard errors of the twin covariances are relatively large. However, the covariances between one twin at an earlier age and the other member of the pair at a later age also contribute information to the within-age twin covariances. This information would not be used in a univariate analysis at each age; thus, resulting variance component estimates would have larger standard errors and might differ from those obtained from the multivariate analysis.

We next performed tests of the various components of the full model to determine which aspects of the model are essential for explaining these data. Tests of the nonshared environmental parameters were conducted first, as shown in table 15.3. The nonshared environmental common factor (model 2) was necessary to explain these data. This single degree of freedom test is quite powerful. The transmission of nonshared environmental variance (modeled in the $\mathbf{B}_E$ matrix) was not necessary (model 3), however, for an adequate model

Table 15.3. Tests of developmental patterns

Model	Form	-2 LL[a]	N_{Par}[b]	χ^2	df	p
1	Full model	21248.767	40			
2	Model 1, drop Γ_E	21255.052	39	6.285	1	< .02
3	Model 1, drop $\mathbf{B}_E$	21256.054	37	7.287	3	> .05
4	Model 3, drop $\mathbf{B}_C$	21285.308	34	29.254	3	< .001
5	Model 3, drop Λ_C	21256.063	36	.009	1	> .90
6	Model 5, drop Ψ_C	21257.466	33	1.403	3	> .70
7	Model 6, drop $\mathbf{B}_G$	21274.616	30	18.553	3	< .001
8	Model 6, drop Λ_G	21257.466	32	0.000	3	> .999
9	Model 8, drop Ψ_G	21374.181	29	16.715	3	< .001
10	Model 8, drop ψ_{G2}	21259.772	31	2.306	1	> .10
11	Model 8, drop ψ_{G3}	21275.753	31	18.287	1	< .001
12	Model 8, drop ψ_{G4}	21274.963	31	17.497	1	< .001

[a] $-2 \times$ log-likelihood of the data.
[b] Number of free parameters.

fit. The specific factors were not tested because measurement error contained in those parameters is essential for the model. The model with specific factors and a single common factor at the nonshared environmental level (model 2) was used as the base model for the next set of tests, those of the shared environmental developmental processses.

At the shared environmental level, the transmission parameters were necessary to explain these data (model 4); the common factor, however, was not (model 5). In addition, the age-specific shared environmental variances arising at each of the later ages were also not necessary (model 6). Therefore, the model that had only a single initial shared environmental variance which transmitted to later ages was used as the base model for tests of the genetic components of the model. It should be noted that this model is strictly equivalent to a common factor model, with no specific variances and a separate parameter estimated for each loading. Such a model has previously been found to explain the shared environmental component of cognitive development data from twins and adoptive and nonadoptive sibling pairs (Cherny et al., 1997). Given this equivalence, we chose to present the model in this equivalent form in the path diagram of the final reduced model, shown below.

At the genetic level, the transmission parameters were essential in accounting for the covariance structure (model 7), while the common factor accounted for essentially none of the variance and covariance (model 8). Finally, age-specific genetic influences were also necessary (model 9). When testing each genetic time-specific loading individually, all but the new variation at 20 months were statistically significant (models 10–12).

The final reduced model of the development of general cognitive ability across ages 14–36 months is presented in figure 15.2. The observed variables and the shared and nonshared environmental factors are standardized to unit variance, and the genetic factors are standardized to have a variance equal to the heritability at each respective age. This model shows a small common factor at the nonshared environmental level, a stronger shared environmental common factor, and strong age-to-age transmission of genetic variance. In addition, new genetic variance appears at 24 and 36 months, but not at 20 months.

DISCUSSION

McCall (1979) proposed a model of early development which includes a major organizational change in the structure of mental functioning at about 18 months of age that is indicative of the incorporation of symbolic thought. McCall's model included the proposal that heritable variation is relatively limited during early infancy but increases by the end of the second year. Similarly, he proposed that environmental influences also increase their effect on individual differences after the shift to symbolic thought. He further suggested a reorganization of cognitive functioning during the second year. The present analysis, however, indicates that heritability is relatively constant across ages

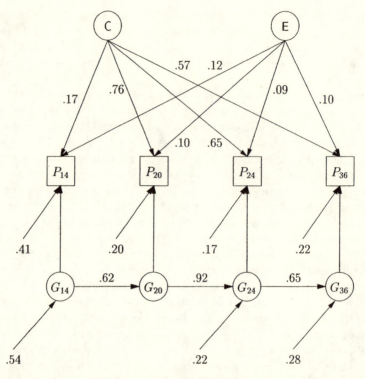

Figure 15.2. Reduced model of cognitive development. Residual variances on the G_i factors are parameters in Ψ_G. Residuals variances loading directly on the phenotypes are parameters in Ψ_E.

14–36 months, although shared environmental variance does increase after 14 months. Both genetic and shared environmental influences contribute to the observed continuity in performance, and, to a much lesser extent, nonshared environmental influences also contribute to continuity. Furthermore, new genetic variation is expressed at 24 and 36 months, indicating that general cognitive ability undergoes developmental change from 14 to 36 months. This is consistent with McCall et al.'s (1977) finding of discontinuity in individual differences at 21 months of age and his proposal of a reorganization occurring at this time. Although the magnitude of the shared environmental influences increases from 14 to 20 months and slightly decreases thereafter, those influences present at 14 months are also present at the later ages, suggesting that more global environmental sources of variation operate on individual differences from at least the beginning of the second year. Nonshared environmental influences contribute only modestly to continuity in general cognitive ability. Measurement error is likely the primary source of variation in this component.

A limitation on inferences stems from the possibility of longitudinal change due to the 36-month assessment (Stanford-Binet) using a different instrument from that used in earlier assessments. A second possible limitation concerns the use of these assessments as measures of general intelligence. There is long-standing debate as to whether intelligence is best viewed as a unitary general capacity (Spearman, 1904; Stern, 1914) or as a set of separable abilities (Gardner, 1983; Guilford, 1967; Thorndike, 1914). The likely resolution of this question is that intelligence can be viewed from either perspective. Recent attempts to explore genetic and environmental influences on separate abilities suggest a complex pattern of effects, but one major influence is the increasing importance of language during the second year (Reznick et al., 1993). The present study is based on a unitary measure of intelligence, but modeling of developmental effects also suggests discontinuity late in the second year, which may be due to the increasing influence of language on MDI scores.

The strength and utility of the longitudinal behavior genetic design used here should also be noted. The present analysis makes use of all available data points using maximum-likelihood methods, making the analysis optimally powerful. Analyzing genetically informative data obtained at multiple points in development allowed us to assess not only the extent of continuity and developmental change present in general cognitive ability from ages 14–36 months, but allowed us to assess the extent to which the observed continuity and change is due to genetic differences among individuals or due to differences in the environments in which the children are raised. Furthermore, environmental differences were partitioned into two mutually exclusive types: those differences shared by members of a twin pair and those differences unique to the individual. In addition, application of these powerful behavior genetic methods to our relatively large sample of more than 350 twin pairs gives us confidence in the substantive conclusions and well as in the reliability of the estimated parameters.

In conclusion, the observed continuity in general cognitive ability from 14 to 36 months can be attributed to both genetic and shared environmental influences. Furthermore, new genetic variation appears at 24 and 36 months, but not at 20 months, indicating that general cognitive ability undergoes developmental change from 14 to 36 months. This is consistent with McCall et al.'s (1977) finding of discontinuity at 21 months. Shared environmental influences do not contribute to change; their effects are monolithic in the sense that they span all three ages. That is, no shared environmental influences unique to a particular age were observed. Finally, nonshared environmental influences do not contribute importantly to continuity and are likely due, at least in part, to measurement error.

APPENDIX

Note 1

Using the LISREL model, the developmental model can be specified using three matrices, **B**, **Γ**, and **Ψ**. As shown on the path diagram, the **B** matrices contain the age-to-age transmission parameters, the Γ matrices contain the common factor loadings, and the Ψ matrices contain the time-specific variances or the new variation at each age. The three parameter matrices are as follows:

$$\mathbf{B} = \begin{bmatrix} 0 & 0 & 0 & 0 \\ \beta_1 & 0 & 0 & 0 \\ 0 & \beta_2 & 0 & 0 \\ 0 & 0 & \beta_3 & 0 \end{bmatrix}, \tag{1}$$

$$\Gamma = \begin{bmatrix} \gamma_1 \\ \gamma_2 \\ \gamma_3 \\ \gamma_4 \end{bmatrix}, \tag{2}$$

and

$$\Psi = \begin{bmatrix} \psi_1 & 0 & 0 & 0 \\ 0 & \psi_2 & 0 & 0 \\ 0 & 0 & \psi_3 & 0 \\ 0 & 0 & 0 & \psi_4 \end{bmatrix}. \tag{3}$$

The free parameters are indicated by the appropriate Greek letter, and all other elements of the above matrices are fixed. The parameters in Ψ were constrained to be positive. The expectations for the phenotypic covariance structure, Σ, would be:

$$\Sigma = (\mathbf{I} - \mathbf{B})^{-1}(\Gamma\Gamma' + \Psi)(\mathbf{I} - \mathbf{B}')^{-1}, \tag{4}$$

where $\mathbf{I}$ is an identity matrix.

In the present case, we have only four time points of measurement, yielding an observed covariance matrix with 10 observed statistics, so we cannot estimate all 11 parameters shown above. To identify the model, we must impose some constraint on the parameters. We chose to equate the common factor loadings, contained in the Γ matrix to identify the model. Previous use of this model has shown a single parameter in Γ has been sufficient in explaining longitudinal general cognitive ability data (Cardon et al., 1992; Cherny and Cardon, 1994; Cherny et al., 1997).

Note 2

The extension of the developmental simplex and common factor model to genetically informative twin data involves increasing the number of parameter matrices and parameters threefold from the phenotypic model described above. Each parameter matrix has a genetic, shared environmental, and non-shared environmental analogue, with three corresponding expected covariance matrices for each of these three sources of individual differences. The MZ and DZ expected covariance matrices take a special form whereby they are partitioned into four equal quadrants. The top-left and bottom-right quadrants contain the within-twin variances and covariances and therefore contain the phenotypic variances and covariances. The other two quadrants contain the cross-sibling variances and covariances. These are expected to differ across the four groups of varying genetic and environmental similarity. The expected co-variance matrices take the following form:

$$\Sigma = \begin{bmatrix} \Sigma_G + \Sigma_C + \Sigma_E & r\Sigma_G + \Sigma_C \\ r\Sigma_G + \Sigma_C & \Sigma_G + \Sigma_C + \Sigma_E \end{bmatrix}, \tag{5}$$

where r is the genetic correlation; $r = 1$ for MZ pairs and .5 for DZ pairs, and the Σ_G, Σ_C, and Σ_E are the expected genetic, shared environmental, and non-shared environmental covariance matrices obtained from equation 4 computed using the component parameter matrices for each of these three sources of individual differences. When suitably standardized, these expected covariance matrices are analogous to the univariate h^2, c^2, and e^2 quantities discussed above.

Note 3

As in any longitudinal study, we do not have data on every single subject at each time point. To make optimal use of the data we have, we must fit the model directly to the raw data rather than to observed covariance matrices. A maximum-likelihood (ML) pedigree approach (Lange et al., 1976) was used to analyze the data in an optimal manner, allowing us to accommodate any pattern of missing data. We maximized the following ML pedigree log-likelihood (LL) function, using the freely available and highly flexible Mx modeling package (Neale, 1994):

$$LL = \sum_{i=1}^{N} [-\frac{1}{2} \ln |\Sigma_i| - \frac{1}{2}(\mathbf{x}_i - \boldsymbol{\mu}_i)'\Sigma_i^{-1}(\mathbf{x}_i - \boldsymbol{\mu}_i) + constant], \tag{6}$$

where $\mathbf{x}_i$ = vector of scores for twin pair i; Σ_i = appropriate MZ or DZ expected covariance matrix; N = total number of twin pairs; $\boldsymbol{\mu}_i$ = vector of expected MZ or DZ means; and

$$2(LL_1 - LL_2) = \chi^2 \tag{7}$$

for testing the difference between two alternative models. The vector of means is modeled as a set of free parameters, allowing estimation of different means in MZs and DZs and twin 1 and twin 2. Mean structure is not of primary interest here, and so no theoretical model was imposed. However, use of the present fit function allows such a possibility, if modeling mean structure is of interest. Use of this fit function, as opposed to the more common ML function used by such programs as LISREL (Jöreskog and Sörbom, 1989) and EQS (Bentler, 1989) allows all the data to be analyzed. This method will recover population parameters provided that the missing data are "missing at random" (MAR), in the terminology of Little and Rubin (1987). That is, the data that are missing may be either a random subset of those observed (or "missing completely at random"; MCAR) or missing based on scores which are observed (MAR). Use of the more common fit function for complete data would necessitate elimination of those sibling pairs from the analysis who were not measured at all time points, which would mean losing information unnecessarily. Furthermore, had this been done, we would need to make the more stringent assumption that data not obtained were MCAR and not merely MAR. In the case where there are no missing data, this pedigree function yields the same results as the ML function for covariance matrices.

Note 4

One may be concerned about whether the highly restrictive simplex model plus a common factor with a single parameter can adequately explain these data. Because the pedigree fit function we use does not yield a test statistic for the adequacy of the model, we fit a Cholesky decomposition model to the genetic, shared environmental, and nonshared environmental covariance structure. The Cholesky decomposition fits as many parameters as there are unique elements in the observed covariance matrix, thereby completely recovering the observed covariance structure. This Cholesky model yielded a function value of 21242.949, which is equal to $-2LL$, with 46 estimated parameters (which includes 16 means). The full simplex and common factor model yielded a function value of 21248.767 for 40 parameters. These two models can be considered nested because the Cholesky completely decomposes each of the three estimated covariance structures. We can then take the difference between these doubled log-likelihoods, which yields 5.818, which is asymptotically distributed as a χ^2 statistic with 6 degrees of freedom. The probability of such a χ^2 occurring by chance given that the two models are really not different is $> .40$. We can therefore conclude that the simplex plus common factor model can adequately explain these data.

REFERENCES

Bayley, N. (1969). *Manual for the Bayley Scales of Infant Development*. New York: Psychological Corporation.

Bell, R. Q., & Hertz, T. W. (1976). Toward more comparability and generalizability of developmental research. *Child Development, 47,* 6–13.

Bell, R. Q., Weller, G. M., & Waldrop, M. D. (1981). Newborn and preschooler: Organization of behavior and relations between periods. *Monographs of the Society for Research in Child Development, 36*(1, 2).

Bentler, P. M. (1989). *EQS Structural Equations Program Manual*. Los Angeles: BMDP Statistical Software.

Cardon, L. R., Fulker, D. W., DeFries, J. C., & Plomin, R. (1992). Continuity and change in general cognitive ability from 1 to 7 years of age. *Developmental Psychology, 28,* 64–73.

Cherny, S. S., & Cardon, L. R. (1994). General cognitive ability. In J. C. DeFries, R. Plomin, & D. W. Fulker (Eds.), *Nature and nurture during middle childhood* (pp. 45–56). Cambridge, MA: Blackwell.

Cherny, S. S., Fulker, D. W., & Hewitt, J. K. (1997). Cognitive development from infancy to middle childhood. In R. J. Sternberg & E. L. Grigorenko (Eds.), *Intelligence, heredity, and environment* (pp. 463–482). Cambridge: Cambridge University Press.

Eaves, L. J., Long, J., & Heath, A. C. (1986). A theory of developmental change in quantitative phenotypes applied to cognitive development. *Behavior Genetics 16,* 143–162.

Emde, R. N., Gaensbauer, T., & Harmon, R. J. (1976). *Emotional expression in infancy.* New York: International University Press.

Emde, R. N., Plomin, R., Robinson, J., Reznick, J. S., Campos, J., Corley, R., DeFries, J. C., Fulker, D. W., Kagan, J., & Zahn-Waxler, C. (1992). Temperament, emotion, and cognition at 14 months: The MacArthur Longitudinal Twin Study. *Child Development, 63,* 1437–1455.

Erikson, E. (1950). *Childhood and society.* New York: Norton.

Gardner, H. (1983). *Frames of mind: The theory of multiple intelligences.* New York: Basic Books.

Guilford, J. P. (1967). *The nature of human intelligence.* New York: McGraw-Hill.

Humphreys, L. G., & Davey, T. C. (1988). Continuity in intellectual growth from 12 months to 9 years. *Intelligence 12,* 183–197.

Jöreskog, K. G., & Sörbom, D. (1989). *LISREL 7: A Guide to the Program and Applications* (2nd ed.). Chicago: SPSS, Inc.

Kagan, J., Kearsley, R., & Zelazo, P. (1978). *Infancy: Its place in human development.* Cambridge, MA: Harvard University Press.

Lange, K., Westlake, J., & Spence, M. A. (1976). Extensions to pedigree analysis: III. Variance components by the scoring method. *Annals of Human Genetics, 39,* 485–491.

Li, C. C. (1975). *Path analysis: A primer.* Pacific Grove, CA: Boxwood Press.

Little, R. J. A., & Rubin, D. B. (1987). *Statistical analysis with missing data.* New York: Wiley.

Matheny Jr., A. P., Dolan, A. B., & Wilson, R. S. (1976). Within-pair similarity on Bailey's Infant Behavior Record. *Journal of Research in Personality, 14,* 224–234.

McCall, R. B. (1979). The development of intellectual functioning in infancy and the prediction of later IQ. In J. D. Osofsky (Ed.), *The handbook of infant development* (pp. 707–741). New York: Wiley.

McCall, R. B., Eichorn, D. J., & Hogarty, P. S. (1977). Transitions in early mental development. *Monographs of the Society for Research in Child Development,* No. 171.

Neale, M. C. (1994). *Mx: Statistical Modeling* (2nd ed.). Richmond, VA: Department of Psychiatry, Medical College of Virginia.

Piaget, J. (1952). *The origins of intelligence in children* (2nd ed.). New York: International University Press.

Plomin, R., Campos, J., Corley, R., Emde, R. N., Fulker, D. W., Kagan, J., Reznick, J. S., Robinson, J., Zahn-Waxler, C., & DeFries, J. C. (1990). Individual differences during the second year of life: The MacArthur Longitudinal Twin Study. In J. Columbo & J. Fagan (Eds.), *Individual differences in infancy: Reliability, stability, & predictability* (pp. 431–455). Hillsdale, NJ: Lawrence Erlbaum.

Plomin, R., Emde, R. N., Braugart, J. M., Campos, J., Corley, R., Fulker, D. W., Kagan, J., Reznick, J. S., Robinson, J., Zahn-Waxler, C., & DeFries, J. C. (1993). Genetic change and continuity from fourteen to twenty months.

The MacArthur Longitudinal Twin Study. *Child Development, 64,* 1354–1376.

Reznick, J. S., Robinson, J., & Corley, R. (1993). Genetic and environmental effects on intelligence in the second year. Unpublished manuscript.

Rutter, M. (1987). Continuities and discontinuities in infancy. In J. D. Osofsky (Ed.), *The handbook of infant development* (pp. 1256–1296). New York: Wiley.

Scarr-Salapatek, S. (1975). Genetics and the development of intelligence. In F. D. Horowitz (Ed.), *Review of child development research.* Chicago: University of Chicago Press.

Spearman, C. (1904). General intelligence, objectively determined and measured. *American Journal of Psychology, 14,* 201–293.

Spitz, R. (1959). *A genetic field theory of ego formation.* New York: International Universities Press.

Stern, W. (1914). The psychological methods of testing intelligence. In *Educational psychological monographs,* Vol. 13. Baltimore, MD: Warwick & York.

Sternberg, R. J., & Powell, J. S. (1983). *The development of intelligence.* In P. H. Mussen (Ed.), *Handbook of child psychology* (4th ed., pp. 341–419). New York: Wiley.

Terman, L. H., & Merrill, M. A. (1973). *Stanford-Binet Intelligence Scale: 1972 Norms Edition.* Boston: Houghton-Mifflin.

Thorndike, E. (1914). *Educational psychology,* Vol. 3. New York: Columbia University Press.

Waddington, C. H. (1957). *The strategy of the genes.* London: Allen & Unwin.

Wilson, R. S. (1983). The Louisville Twin Study: Developmental synchronies in behavior. *Child Development, 54,* 298–316.

16

Language and Cognition

Susan Young
Stephanie Schmitz
Robin P. Corley
David W. Fulker

Language can be characterized as a variety of interactive, communicative behaviors which demonstrate an exchange of information, including those which are nonverbal. Studies of the development of prelinguistic communicative behaviors such as speech perception, cooing, babbling, gesturing, and first words (Bates, et al, 1987) show that these emerge in a child's language repertoire during the first 2 years of life.

Much of the research examining how and why children acquire language has been carried out from a theoretical perspective. Social learning theorists, or *contextualists* (Zimmerman, 1983), focus on children's apparent imitation of and interactions with parents and siblings. Further, they purport that as utterances are differentially reinforced by others, they increase in frequency and complexity. Ethologists argue that communicative skills develop as an adaptive response to the child's physical, psychological, and environmental demands (Charlesworth, 1983). However, those from a more biobehavioral perspective point to the predictability of the developmental stages of language acquisition. The nativists' view (e.g., Chomsky, 1972) proposes that children are born with an innate set of correlated rules that set a relatively universal course for language development. While the rate at which children make the transition from stage to stage varies, the pattern or order of these transitions is consistent (Menyuk, 1971). However, this position has been challenged by suggestions that there may be differing developmental pathways that children follow in language acquisition (Bates, et al., 1988; Bloom, 1970).

Substantial evidence is available for what makes children similar in the way they transition from prelinguistic communication to grammatical and semantic mastery. Despite these explanations for progressive stages of communication development, little is known about why some children advance through those stages more quickly than others or why some develop specific language deficits or delays. These theoretical models focus on the developmental patterns commonly seen in children and minimize the importance of individual differences.

More contemporary research examines language as a problem-solving component of cognitive ability or as a part of meaningful social exchange. These perspectives approach the study of individual differences as a mechanism for answering the question, What is (isn't) innate or learned? rather than simply, Is (isn't) it innate or learned? (Shore, 1995). These issues are related to a long-standing debate regarding the conceptualization of cognitive ability. Numerous factor analytic studies have examined the construct of general ability versus a set of specific, related abilities such as linguistics, spatial abilities, reasoning, and mathematical skills. Although many studies have demonstrated the presence of a general cognitive factor that can be a reliable measure for predicting future cognitive ability, modularity theorists (e.g., Gardner, 1983) view language as one of several individual mental units. Developmental models of specific abilities can be applied to examine differential etiology at various ages and possible gender differences that may be domain specific. A model developed by Dale et al., (1989) proposed that measures of infant intelligence can be scaled with three factors representing expressive language, receptive language, and nonverbal items (e.g., memory, object relations). Although expressive language such as verbalizations and gestures can be observed more directly, receptive language is often inferred when the child responds appropriately to questions or instructions. These factors are often studied as separate domains because children may understand many more words than they can articulate, particularly at early stages in development.

THE STUDY

In a behavioral genetic design, longitudinal data on related individuals can be examined to uncover the causes of these individual differences throughout development. Specifically, these models use familial resemblance to tease apart the influences of genetic and environmental factors contributing to this variability. Previous studies using genetically informative data to address the etiology of individual differences in communicative competence have shown consistent evidence for both familial and nonfamilial factors. Adoption studies, which use phenotypic correlations between biological parents and their children given up for adoption as well as those between unrelated siblings, have demonstrated substantial family resemblance for early language development (Baker, 1983; Casler, 1976; Plomin & DeFries, 1985; Thompson & Plomin, 1988). In particular, they have suggested that communication development is both genetically and environmentally mediated. Estimates of the heritability of early language skills in children participating in the Colorado

Adoption Project (Thompson & Plomin, 1988) have ranged from .22 to .42, suggesting that the proportion of observed variability in language due to genetic influences is between 22% and 42%.

Twin studies, the most common design in behavioral genetic research, use within-pair correlations to test etiological models of genetic and environmental mediation. Phenotypic associations suggest genetic and/or environmental effects are responsible for behavior because twins share both their genetic makeup (estimated at 100% for monozygote twins or 50% for dizygotic twins) as well as their home environment (estimated as 100% for all twins). Although twin studies of individual differences in normal language development are relatively absent from the literature, a number of small twin studies have demonstrated significant genetic influences on developmental language delay and specific language impairment (Bishop, 1997, Bishop, et al., 1995; Olson et al., 1989; Sandbank & Brown, 1990). Additionally, a recent report on a sample of more than 3,000 twin pairs suggests that heritability is greater in twins selected for language delay than in unselected twins (Dale et al., 1998). In contrast, there is a relatively large literature on the genetic and environmental influences on general cognitive ability in childhood (see chapter 15).

Preliminary reports on the MacArthur Longitudinal Twin Study have been published on language and related cognitive factors including univariate and longitudinal results from a subsample of nearly 200 twin pairs during assessment periods at 14 and 20 months (Emde et al., 1992; Plomin et al., 1993). Early findings in these reports suggest that there may be different etiological patterns for expressive and receptive language at these early ages, despite the fact that these skills are highly correlated in individuals. The purpose of the present longitudinal twin analyses is to explore the following questions regarding the etiology of individual differences in communicative development with an enlarged sample at four ages:

1. *What are the source(s) and magnitude of twin resemblance for expressive and receptive language skills in infancy and early childhood?* A central aim of this study is to estimate the relative influences of genetic and environmental factors on individual differences in language skills during an early period of rapid development. To investigate the sources of individual differences in a behavioral genetic model, substantial family resemblance for the traits of interest must be determined. In twin studies, intrapair correlations provide the preliminary information necessary for fitting causal models, which provide estimates of the relative contributions of genetic and environmental influences on early communicative abilities. Further, these analyses can compare the etiological structure for two highly correlated phenotypes, expressive versus receptive language skills.

2. *Do genetic and environmental influences change or remain stable over time?* We are not only interested in the consistency of language skills from infancy to early childhood, but also in the stability of the etiology of these

behaviors as children develop. Longitudinal genetic models provide the leverage to examine stability and change in familial and nonfamilial factors over time. If the shared environment has a strong influence on language, do the same environmental influences affect communicative skills at ages two and three? For example, if maternal responses contribute to shared environmental influences, do these responses develop and change as the children move from providing nonverbal to verbal cues?

Similar questions can be posed for potential genetic effects. Longitudinal twin models allow us to compare the magnitude of heritability at various stages of development. Further, these models provide information regarding genetic correlations across time. Strong associations would suggest that the genes at work at one age continue to have an influence at subsequent ages. However, a nonsignificant genetic correlation would suggest that, rather than a cumulative effect, new genetic factors are "turning on" across time.

3. *Are there substantial gender differences in early language acquisition, and do these differences affect our estimates of genetic and environmental influences?* Investigations of gender differences in cognitive abilities have repeatedly shown a female advantage for language acquisition and verbal ability in infancy, childhood, and adulthood (Fenson et al., 1994). This contrast has also been demonstrated in samples of adoptees (Hardy-Brown & Plomin, 1985; Plomin & DeFries, 1985) and preschool-aged twins (Hay et al., 1987). However, what is not well established is at what age(s) these differences emerge, and the nature of the factors driving these differences. We examined possible gender differences (i.e., mean differences in performance) in expressive and receptive language scores. Second, we investigated gender differences in our estimates of the relative influences of genetic and environmental factors on language. These models allow us to ask, for example, whether girls (or boys) show greater heritability for language during infancy or early childhood, or conversely, whether they are differentially impacted by environmental influences acting on the pair (e.g., exposure to *Sesame Street*).

4. *Are there factors unique to the development of language or general cognitive ability?* Longitudinal data on multiple traits provide another dimension to these analyses. Multivariate twin models can be applied using associated phenotypes (e.g., language and general intelligence) to examine possible common causal factors at a single age as well as longitudinally.

Previous twin and adoption studies have demonstrated that early communication skills predict later performance on the same tasks, as well as concurrent and future performance on tests of general cognitive ability (Casler, 1976; DiLalla et al., 1990; Hardy-Brown, et al., 1981; Thompson & Plomin, 1988). In the present study, we examined the extent to which individual differences in language and general cognitive ability are influenced by common and independent factors during the first 3 years of life. In other words, these models provide an explanation for what drives the phenotypic covariation between early communication abilities and general cognitive ability.

METHODS

Sample and Procedures

Depending on age point and measure, sample sizes (in the analyses presented in this chapter) for MZ twins ranged from 147 to 193 pairs and for DZ pairs from 133 to 170 pairs. See chapter 3 for a detailed description of sample ascertainment and demographic characteristics. Twins were assessed at 14, 20, 24, and 36 months of age. These ages were chosen to bracket periods of accelerated development for the major phenotypes in the MacArthur study (Plomin et al., 1990). The language and general mental ability assessments were conducted during the home visit component of the study. Two female examiners conducted the battery of assessments, including a number of cognitive, emotional, behavioral, and physical measures. The focus of this chapter is communicative skills assessed by the Sequenced Inventory of Communication Development (SICD; Hedrick, et al. 1975). The SICD was designed to identify small progressive steps in communication development using two major scales: (1) expressive, composed of items requiring imitation or production of sounds and words, such as What do you wear on your feet? and (2) receptive, containing items that assess the ability to discriminate or understand words and commands; "give me the cup and the ball" requires the child to understand what is being requested and to respond by handing the objects to the examiner. The raw score represents the sum of items performed by the child supplemented by information solicited from the mother. Because of the possible performance bias that could result from the unfamiliar examiner eliciting children's responses, these supplemental data (i.e., parental endorsements) may represent skills that have been demonstrated in their natural home environment. To shorten the length of administration time, a subset of items was designated at each age point which represented items above and below the children's abilities. At each testing session following the 14-month assessment, items the child had failed previously were readministered and age-appropriate items were added. This change required a scoring system that differed slightly from the standardized procedure.

Measures of general cognitive ability (GCA) came from the Bayley Scales of Infant Development Mental Development Index (MDI; Bayley, 1969), administered at the first three time points, and the Stanford-Binet Intelligence Scale (Terman & Merill, 1973), used at age 36 months. For a detailed description of the administration procedure for the GCA assessments, see chapter 3. Assessment of language skills and GCA were integrated into one test session which

Table 16.1. Means and standard deviations for the Sequenced Inventory of Communication Development (SICD) and general cognitive ability (GCA)

Age (months)	SICD expressive	SICD receptive	GCA
14	20.6 ± 3.8	17.5 ± 6.0	104.5 ± 14.0
	(777)[a]	(779)	(783)
20	30.1 ± 5.9	35.0 ± 6.5	104.4 ± 17.3
	(697)	(692)	(692)
24	37.7 ± 6.8	44.7 ± 4.7	107.7 ± 19.0
	(683)	(665)	(695)
36	8.9 ± 4.2	11.3 ± 5.0	103.1 ± 17.9
	(624)	(636)	(639)

[a] Sample sizes are in parentheses.

required approximately 40 min to complete. Scores containing an excess (> 15%) of invalid or missing items were not used in the analyses. Missing data were demonstrated to be nonsystematic in that mean scores from children who completed all assessments did not differ significantly from those from children who missed one or more of the scheduled assessment procedures. Thus, pairwise deletion was applied, yielding only those scores available for complete pairs at each time point.

RESULT

Descriptive Analyses

Table 16.1 outlines the mean scores for the twins' performance on the SICD and the measures of general intelligence. Although there were slight differences in the means between the MZ and DZ groups, none was significant. Note that the mean scores for expressive and receptive language are considerably smaller at the 36 month assessment. Items that are no longer age appropriate are assumed to be mastered by age three and thus are not readministered.

Longitudinal within-pair correlations for the two language components and GCA are presented in table 16.2. With the exception of three values, each correlation across domain and age is statistically significant. Age-to-age correlations within each domain for language and GCA are shown in bold. These data suggest moderate predictability of future performance from previous performance. The correlations are stronger for adjacent ages than those more distant in time (moving away from the diagonal), and range from .29 to .67.

Correlations within each time point and across measures are those contained in the sections along the main diagonal (table 16.2). These associations are also statistically significant, ranging from .43 to .71. Similarity in performance for

Table 16.2. Longitudinal correlations across domain

Age		14 Months			20 Months			24 Months			36 Months		
		E	R	GCA	E	R	GCA	E	R	GCA	E	R	GCA
14 Months	E	1.0	.47	.43	**.42**	.35	.28	**.34**	.25	.24	.11[a]	.07[a]	.12[a]
	R		1.0	.43	.40	**.54**	.43	.34	**.46**	.44	.34	**.34**	.35
	GCA			1.0	.35	.39	**.48**	.30	.37	**.43**	.25	.25	**.29**
20 Months	E				1.0	.60	.59	**.64**	.46	.51	**.36**	.33	.37
	R					1.0	.64	.55	**.67**	.60	.55	**.50**	.55
	GCA						1.0	.52	.56	**.67**	.54	.50	.53
24 Months	E							1.0	.59	.63	**.47**	.36	.44
	R								1.0	.68	.57	**.51**	.57
	GCA									1.0	.59	.53	**.61**
36 Months	E										1.0	.71	.69
	R											1.0	.65
	GCA												1.0

E = Sequenced Inventory of Communication Development (SICD) expressive; R = SICD receptive; general cognitive ability (GCA) = Bayley Mental Development Index at ages 14, 20, and 24 months, Stanford-Binet at 36 months. Age-to-age correlations within each domain are boldface.
[a] Nonsignificant; all other values are statistically significant.

communicative ability and GCA gradually increases with age. Further, no consistent pattern emerges that would indicate that one of the two language components is more associated with GCA than the other.

Within-pair twin correlations for expressive and receptive language and GCA at each age are shown in table 16.3. In each instance, correlations for monozygotes (MZ) exceeded those for dizygotes (DZ), but this difference is more pronounced for expressive language than for receptive language. These patterns suggest that modeling these associations will reveal a lower estimate of heritability for receptive language than for the other two measures. As with language, the pattern of correlations supports the idea that some heritable factors are operating on general cognitive ability.

Table 16.3. Within-pair twin correlations for language and general cognitive ability (GCA)

Age (months)	Expressive		Receptive		GCA	
	MZ	DZ	MZ	DZ	MZ	DZ
14	.64	.53	.82	.80	.57	.39
20	.76	.56	.78	.69	.80	.64
24	.86	.57	.71	.67	.84	.58
36	.70	.56	.73	.54	.78	.57

All correlations are statistically significant at the $p < .01$ level.

Longitudinal Analyses

These twin correlations were used in structural equation modeling to estimate the contributions of genetic and environmental factors to the phenotypic (observable) variance. A series of longitudinal twin models was fit to data on expressive and receptive language scores separately.

The models assume that the observed scores (phenotypes) are the sum of the underlying latent factors representing additive genetic influences (A_i), shared environmental influences (C_i) and unique environmental influences (E_i). (Note: Genetic influences are referred to as A rather than G or H because these models are testing only additive genetic effects and thus do not include the possible dominance effects.) To model our longitudinal data, we used the Cholesky triangular decomposition (Neale & Cardon, 1992) for the ACE model. The Cholesky model allows for a first factor that influences each of the four time points, a second factor that influences time points 2–4, a third factor that influences points 3 and 4, and a fourth that influences the last time point only. Longitudinal data are compatible with the Cholesky model in that it provides a natural ordering of the variables.

The decomposition of the variance–covariance matrix among the observed scores can be obtained by a regression of the observed measures on the latent factors.[1] To correspond to the expected additive genetic relationships, intrapair correlations among the additive factors (A_i) are constrained to be 1.0 and .5 for MZ and DZ twins, respectively. Similarly, shared environmental factors are constrained to be correlated at 1.0, and the unique environmental factors are correlated 0, by definition, for both twin groups. Factor loadings were estimated by maximum likelihood (ML) methods using Mx (Neale, 1994). This program carries out a numerical search for the parameter values which minimizes a ML function. The function is distributed as a χ^2 with degrees of freedom equal to the number of observed statistics minus the number of estimated parameters. Submodels of the full decomposition were tested by constraining certain coefficients to zero and recalculating the likelihood ratio for the fit of the reduced (nested) model. To determine whether a significant loss of fit resulted from the reduction of the model χ^2 difference tests were used. See chapter 15 for further details of these analyses.

Longitudinal models for the SICD expressive language component show that genetic influences are statistically significant at all four time points (table 16.4). The magnitude of the effect increases up to age 24 months when nearly half of the observed variance is explained by genetic factors and then drops off somewhat at age three. Environmental influences remain strong at all four ages. However, the etiological pattern looks somewhat different for the receptive language scores from the SICD. Although heritability is nonsignificant for receptive language at 14, 20, and 24 months, it accounts for 27% of the observed variance at 36 months. Early home (shared) environment has a substantial effect throughout this period, but decreases with age.

Table 16.4. Sequenced Inventory of Communication Development parameter estimates

	Expressive			Receptive		
Age	h^2	c^2	e^2	h^2	c^2	e^2
14 Months	.24*	.39*	.37	.06	.75*	.19
20 Months	.31*	.43*	.26	.17	.59*	.24
24 Months	.47*	.37*	.16	.10	.58*	.32
36 Months	.24*	.45*	.31	.27*	.44*	.29

$*p < .05$.

To search for a sufficiently parsimonious model that best fits these data, we carried out a series of model comparisons. Results from fitting a full model and several nested models for expressive language are described in table 16.5. The full model, which estimates both time-specific and cumulative effects, fit reasonably well (model 1), with a χ^2 of 49.40 ($p = .20$). Constraining the values on the off-diagonal for the unique environmental component (e^2) to zero does not reduce the fit of the model substantially ($p > .20$, model 1 versus 2). This suggests that these influences were specific to each time point and that unique experiences, accounting for variance at each age, do not have lasting effects. Further, it assures us that measurement error, which is also part of the e^2 component, is not systematic, but specific to each measurement. Because the magnitude of the shared environmental factors was strong and consistent across time, a single common C factor was tested (model 3). The results showed that this simplified model of shared environmental effects fit the data adequately. This suggests that a common set of environmental characteristics contribute to twin resemblance in expressive language at these four ages.

Model 4, which constrains loadings that represent the continuity of genetic effects (off-diagonal elements) to zero resulted in a significant increase in the χ^2 ($p < .01$), suggesting that there is some genetic correlation for expressive language from infancy to early childhood. In other words, a portion of the genetic factors operating at early ages continue to influence behavior at later

Table 16.5. Model comparison: expressive language

Model	χ^2	df	p^a	Test	Diff χ^2	df	p^b
1. Full	49.40	42	.20				
2. ACE_d	57.14	48	.17	1 vs. 2	7.74	6	>.20
3. AC_1E_d	64.97	54	.15	2 vs. 3	7.83	6	>.20
4. $A_dC_1E_d$	184.32	60	.00	3 vs. 4	119.35	6	<.01

[a] p values for model.
[b] p vales for χ^2 tests of model comparisons.

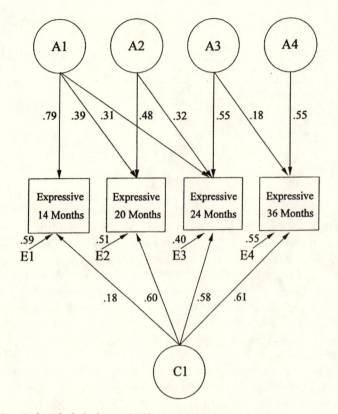

Figure 16.1. Reduced Cholesky model for expressive language.

ages, suggesting these influences are making a meaningful contribution to the model fit.

Figure 16.1 illustrates the genetic and environmental loadings of the reduced model for expressive language. The time-specific genetic influences (e.g., A_1 loading on 14 months, A_2 loading on 20 months) range from .48 to .79. In contrast, those genetic influences carried over from previous ages (e.g., A_1 loading on 20 months) are lower in magnitude, ranging from .18 to .39. A single shared environmental factor (represented as C_1) suggests that the same family environmental influences are affecting expressive language at all four ages. The magnitude of the loadings suggest that these influences have the smallest impact at 14 months. Loadings from the first and second genetic factors on the 36-month measure were estimated at zero and thus are not shown as paths in this diagram.

Table 16.6 shows the magnitude of the genetic correlations for the reduced model. These associations suggest strong genetic continuity among the first three ages. The estimates of .00 for r imply that genetic influences on expressive language at 36 months are independent of those acting at 14 and 20

Table 16.6. Genetic correlations: expressive language

	14 Months	20 Months	24 Months	36 Months
14 Months	1.00			
20 Months	.63	1.00		
24 Months	.44	.62	1.00	
36 Months	.00	.00	.25	1.00

months. Because the reduced model for expressive language suggests that a single common environmental factor adequately fit the data, environmental correlations would be estimated at unity.

Table 16.7 presents the model comparisons for receptive language. Again, the unique environmental variance components could be diagonalized with a nonsignificant change in the model fit (model 2). Model 3 shows that a single factor can explain all of the genetic variance, suggesting that, although the genetic influence on receptive language is small, it is consistent across the four ages. In other words, new genetic influences, independent of those contributing to the variance in receptive language at 14 months, are small and nonsignificant in our sample. However, the shared environmental factors showed a different pattern. A model that simplifies the C factors by dropping elements that represent the continuity of shared environmental influences over time (off-diagonal elements of C) shows that these effects are significant ($p < .01$). Further, a test of a single common environmental factor also showed a significant degradation of the fit of the model ($p < .01$), suggesting that the environmental influences that are age specific are also important in receptive language development.

Figure 16.2 shows the factor loadings of the parsimonious model (model 3) for receptive language. Although the same genetic effects may be influencing receptive language from age 14 months to 36 months, these effects vary in magnitude. Shared environmental factors show the largest loadings for age-specific effects. The shared environmental correlations for receptive language (Table 16.8) were substantial, ranging from .45 to .85. As expected, the relative

Table 16.7. Model comparisons: receptive language

Model	χ^2	df	p^a	Test	Diff χ^2	df	p^b
1. Full	58.34	42	.05				
2. ACE_d	60.34	48	11	1 vs. 2	2	6	>.90
3. A_1CE_d	63.51	54	.18	2 vs. 3	3.17	6	>.75
4. $A_1C_dE_d$	113.44	60	.00	3 vs. 4	49.94	6	<.01
5. $A_1C_1E_d$	161.04	60	.00	3 vs. 5	97.53	6	<.01

$^a p$ values for model.
$^b p$ values for χ^2 tests of nested model comparisons.

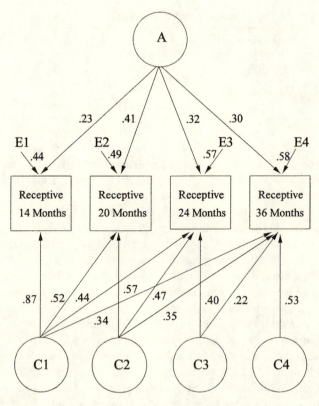

Figure 16.2. Reduced Cholesky model for receptive language.

Table 16.8. Shared environmental correlations: receptive language

	14 Months	20 Months	24 Months	36 Months
14 Months	1.00			
20 Months	.68	1.00		
24 Months	.58	.85	1.00	
36 Months	.45	.64	.70	1.00

magnitudes of these correlations suggest greater consistency for adjacent ages than those more distant in time. For example, the correlation between 24 and 36 months ($r = .70$) suggests that half of the shared environmental variance at these two ages is common to both.

Gender Differences

Gender differences were examined for each of the communication measures and GCA. These results (table 16.9) confirm an expected difference with mean scores for girls significantly higher than mean scores for boys with respect to language at each age (Fenson et al., 1994, Hyde, 1981). Further, girls show significantly higher means for GCA at 20, 24, and 36 months ($p = .07$ at 14 months). Mean differences in performance do not, by themselves, imply differences in causal factors for boys and girls. However, we can investigate whether there is, for example, a greater genetic influence on expressive language for girls than for boys. In subsequent analyses, which allowed parameters to differ for boys and girls, we investigated possible differences in genetic and environmental effects at each age. Despite significant and consistent gender differences in phenotypic means, there were no significant differences in etiological patterns for the two groups. There was a general trend for estimates of genetic influences to be slightly higher in males, and for estimates of shared environmental influences to be somewhat higher in females. Subsequent analyses were carried out using a model with the male and female twins combined into a single group.

As described in Methods, supplemental information on the child's SICD performance (provided by the mother) is part of the SICD scores. To address the

Table 16.9. Gender differences in languages and general cognitive ability (GCA)

Age		Males			Females			
		Mean	SD	N	Mean	SD	N	p
14 Months	E	20.18	3.69	391	21.11	3.86	386	<.01
	R	16.02	5.81	393	19.03	5.88	386	<.01
	GCA	103.61	14.35	396	105.40	13.67	387	>.05
20 Months	E	28.97	5.75	353	31.20	5.81	326	<.01
	R	33.63	6.93	347	36.41	5.60	327	<.01
	GCA	100.42	17.33	352	108.60	16.34	321	<.01
24 Months	E	36.34	6.65	336	39.20	6.58	325	<.01
	R	43.69	5.01	325	45.66	3.86	319	<.01
	GCA	103.93	18.50	344	111.57	18.23	329	<.01
36 Months	E	8.21	4.29	298	9.51	3.91	303	<.01
	R	10.17	4.95	302	12.11	4.71	311	<.01
	GCA	99.80	17.51	308	105.53	17.76	311	<.01

E = Sequenced Inventory of Communication Development (SICD) expressive; R = SICD receptive; GCA = Bayley Mental Development Index at ages 14, 20, and 24 months, Stanford-Binet at 36 months.

concern that parental endorsement items on the SICD may bias these estimates, we removed items not reflecting a direct response from the child and recomputed scores. Refitting these models to the modified scores produced little change in the parameter estimates at 20–36 months. However, an increase in e^2 (the proportion of phenotypic variance due to specific environment and measurement error) at 14 months occurred under these conditions, suggesting that there may be more measurement error at this age. Overall, the patterns among the parameter estimates across time remained the same.

Language and GCA

Early language ability has been shown to correlate with concurrent and future performance on tests of cognitive ability. As reported earlier, within-person correlations among the SICD scores and measures of GCA replicate that finding in our sample (table 16.2). To address the question of whether the correlations between language and GCA can be explained by common genetic or environmental factors, we tested a model that estimates both the common factors and the measure-specific influences for GCA, receptive, and expressive language. These trivariate analyses were done for each age individually and thus did not directly examine longitudinal characteristics of the data.

In the common factor model, the phenotypic variance is partitioned into genetic and environmental influences that are common among the three measures and the residual genetic and environmental variance which is specific to each measure.[2]

Table 16.10 shows the parameter estimates for the full model at 14 months. The effects of the genetic common factor are small and nonsignificant and thus can be dropped from the model. However, there are significant genetic effects unique to the Bayley MDI and those unique to the SICD expressive language measure. Shared environmental influences common to all three measures are significant, accounting for more than half of the variance in the SICD receptive language and about 20% in the other two measures. There are also significant shared environmental influences for receptive language that are independent of those for expressive language and GCA. Figure 16.3 represents the best fit-

Table 16.10. Variance components: full model for 14 months

	A		C		E	
	CF	S	CF	S	CF	S
Bayley MDI	.05	.32*	.20*	.00	.16	.27
Expressive	.12	.15*	.22*	.14	.16	.21
Receptive	.03	.00	.51*	.26*	.05	.15

CF = common factor; S = specific factor.
*$p < .05$.

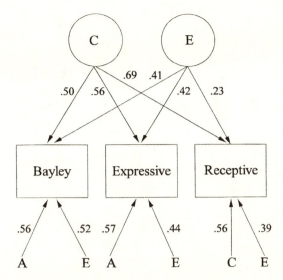

Figure 16.3. Reduced common factor model for 14 months.

ting model for the 14-month data and provides the parameter estimates for descriptive purposes. All paths in the reduced model represent significant effects.

In contrast, at 36 months (table 16.11) there is a significant common genetic factor influencing all three abilities, explaining more than a third of the total variance in GCA, as measured by the Stanford-Binet. Interestingly, specific genetic variance emerges for receptive language only. All significant shared environmental effects that influence these three measures at 36 months come from the common factor. In contrast, the measure-specific shared environmental influences are estimated at (or near) zero. The ratio of e_{CF}^2 to e_S^2 suggests that nonshared environmental influences (along with measurement error) contributing to the observed scores are generally measure specific.

Table 16.11. Variance components: full model for 36 months

	A		C		E	
	CF	S	CF	S	CF	S
Binet	.37*	.06	.30*	.02	.03	.22
Expressive	.22*	.00	.47*	.00	.05	.26
Receptive	.15*	.15*	.42*	.00	.08	.20

CF = common factor; S = specific factor.
*$p < .05$.

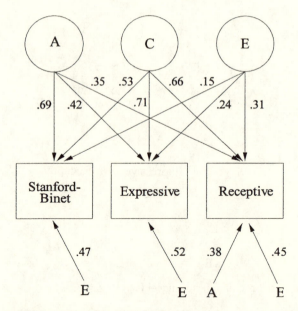

Figure 16.4. Reduced common factor model for 36 months.

Figure 16.4 shows the simplified model that best fits the 36-month twin data. These results suggest that as children develop from 14 and 36 months of age, the genetic influences on language and GCA become less domain specific. Shared environmental influences also lose their specificity across this developmental period, as the common factor accounts for all of these effects by the final time point in the analyses. The reduced models for the 20- and 24-month assessments show intermediate patterns of this transition from specific to common etiology among the language and general cognitive measures. These findings are more consistent with the view that specific and general cognitive abilities are highly related both phenotypically and etiologically as early as age three.

DISCUSSION

We set out to answer a number of questions regarding individual differences in language development and GCA during infancy and early childhood. These ages are ideal in that they span the period of the development of single words and the "vocabulary burst" (Fenson et al., 1994). Behavioral genetic analyses of data from MZ and DZ twins provided several interesting findings. First, within-pair correlations suggest that the magnitude of twin resemblance is substantial for expressive and receptive language skills in infancy and early childhood. A similar pattern is seen for measures of GCA at each of the four ages (14, 20, 24, and 36 months). Multivariate analyses allow us to simultaneously estimate the relative contributions of genetic and environmental factors to language variability at each age and across time, as well as to describe the sources of covariation between language and GCA.

The patterns of phenotypic correlations between language and GCA or between measures at time A and time B may lead researchers to believe that these factors share a common etiology. Longitudinal analyses estimating the influences of genetic and environmental factors show a more complex pattern. Early communicative behavior appears to be influenced by both genetic and environmental factors. However, these analyses suggest that expressive language may be more genetically mediated at the earlier ages than receptive language, which is largely influenced by shared environmental factors in infancy. The genetic etiology for receptive language from 14 to 36 months can be represented by a single factor, suggesting strong continuity in these influences early in development. However, four factors are needed to explain the genetic influences in expressive language. Thus, while some genetic influences are consistent from 14 to 36 months, there are also significant new genetic effects being expressed across time.

The shared environmental influences are substantial and correlated across time for both expressive and receptive language. Assuming that this primarily represents the home environment during infancy and early childhood, it could be argued that these familial characteristics affecting the development of language remain quite consistent across time. Shared environmental influences such as parental responsiveness during the early months of life may be related to verbal performance in infancy, whereas verbal modeling and demand for child verbalization might influence performance at later ages.

Our examination of gender differences in language skills and GCA shows that females are consistently advanced in expressive and receptive language, as measured by the SICD, as well as in general cognitive ability (with the exception of GCA at 14 months) during this developmental period. A gender comparison of the etiological structure of these measures shows that although estimates of heritability are slightly higher for boys than girls, the differences are not statistically significant. In general, model comparisons suggest that the differences in early language development in boys and girls (i.e., differences in mean scores) are not due to significant differences in the magnitude of genetic or environmental influences underlying these abilities.

The analysis of the covariation among GCA, expressive language, and receptive language skills suggests that as the children acquire more language skills, there is more convergence in etiology of these abilities. That is, the specificity of the causal factors seems to decrease with age. Expressive language seems to share more of the genetic influence with GCA than receptive language does, but all seem to be affected by the same persistent environmental influences.

Although a composite score that combines expressive and receptive language scores has been used in previous studies of the SICD (Thompson & Plomin, 1988), results from these twin studies suggest that individual differences in these two areas of early language development are not entirely overlapping. This distinction may be less important at older ages, as the patterns of etiological structure become more similar. The dramatic rate of change for some cognitive skills observed in early developmental periods (Cherny et al., 1994) may begin to decelerate (developmental stabilization) around 3 years of age. These infant/child measures may relate to an emerging general cognitive factor (Humphreys & Davey, 1988), supporting previous evidence that early communicative behavior is useful in predicting cognitive abilities later in development.

NOTES

1. The regressions of the observed scores on the latent variables for the Cholesky decomposition model are represented as follows:

$$SICD_{14} = a_{11}A_1 + c_{11}C_1 + e_{11}E_1$$
$$SICD_{20} = a_{21}A_1 + a_{22}A_2 + c_{21}C_1 + c_{22}C_2 + e_{21}E_1 + e_{22}E_2$$
$$SICD_{24} = a_{31}A_1 + a_{32}A_2 + a_{33}A_3 + c_{31}C_1 + c_{32}C_2 + c_{33}C_3 + e_{31}E_1 + e_{32}E_2 + e_{33}E_3$$
$$SICD_{36} = a_{41}A_1 + a_{42}A_2 + a_{43}A_3 + a_{44}A_4 + c_{41}C_1 + c_{42}C_2 + c_{43}C_3 + c_{44}C_4$$
$$+ e_{41}E_1 + e_{42}E_2 + e_{43}E_3 + e_{44}E_4,$$

where $SICD_i$ represents each of the two language scores (expressive and receptive) at the four age points.

2. The regressions of the observed scores on the latent variables for the common factor model are now represented as follows:

$$GCA_{AGE} = a_{c1}A_c + a_{s1}A_{s1} + c_{c1}C_c + c_{s1}C_{s1} + e_{c1}E_c + e_{s1}E_{s1}$$
$$SICD\text{-}E_{AGE} = a_{c2}A_c + a_{s2}A_{s2} + c_{c2}C_c + c_{s2}C_{s2} + e_{c2}E_c + e_{s2}E_{s2}$$
$$SICD\text{-}R_{AGE} = a_{c3}A_c + a_{s3}A_{s3} + c_{c3}C_c + c_{s3}C_{s3} + e_{c3}E_c + e_{s3}E_{s3}$$

REFERENCES

Baker, L. A. (1983). Bivariate path analysis of verbal and nonverbal abilities in the Colorado Adoption Project. Unpublished doctoral dissertation. University of Colorado, Boulder.

Bates, E., Bretherton, I., & Snyder, L. (1988). *From first words to grammar: Individual differences and dissociable mechanisms.* New York: Academic Press.

Bates, E., O'Connell, B., & Shore, C. (1987) Language and communication in infancy. In J. D. Osofsky (Ed.), *Handbook of infant development* (2nd ed.). New York: Wiley.

Bayley, N. (1969). *Manual for the Bayley Scales of Infant Development.* New York: Psychological Corporation.

Bishop, D. V. (1997). Pre- and perinatal hazards and family background in children with specific language impairments: A study of twins. *Brain and Language, 56,* 1–26.

Bishop, D. V., North, T., & Donlan, C. (1995). Genetic basis of specific language impairment: Evidence from a twin study. *Developmental Medicine and Child Neurology, 37,* 56–71.

Bloom, L. (1970). *Language development.* Cambridge: MIT Press.

Casler, L. (1976). Maternal intelligence and institutionalized children's developmental quotients: A correlational study. *Developmental Psychology, 12,* 64–67.

Charlesworth, W. R. (1983). An ethological approach to cognitive development. In C. J. Brainerd (Ed.), *Recent advances in cognitive-developmental theory: Progress in cognitive development research.* New York: Springer-Verlag.

Cherny, S. S., Fulker, D. W., Emde, R. N., Robinson, J., Corley, R. P., Reznick, J. S., Plomin, R., & DeFries, J. C. (1994). A developmental-genetic analysis of continuity and change in the Bayley Mental Development Index from 14 to 24 months: The MacArthur Longitudinal Twin Study. *Psychological Science, 5,* 354–360.

Chomsky, N. (1972). *Language and mind.* New York: Harcourt, Brace & World.

Dale, P. S., Bates, E., Reznick, J. S., & Morisett, C. (1989). The validity of a parent report instrument of child language at 20 months. *Journal of Child Language, 16,* 239–249.

Dale, P. S., Simonoff, E., Bishop, D. V., Eley, T. C., Oliver, B., Price, T. S., Purcell, S., Stevenson, J., & Plomin, R. (1998). Genetic influences on language delay in two-year-old children, *Nature Neuroscience, 1,* 324–328.

DiLalla, L. F., Plomin, R., Fagan, J. F., Thompson, L. A., Phillips, K., Haith, M. M., Cyphers, L. H., Fulker, D. W. (1990). Infant predictors of preschool and adult IQ: A study of infant twins and their parents. *Developmental Psychology, 26,* 759–769.

Emde, R. N., Plomin, R., Robinson, J., Corley, R. P., DeFries, J. C., Fulker, D. W., Reznick, J. S., Campos, J., Kagan, J., & Zahn-Waxler, C. (1992). Temperament, emotion, and cognition at fourteen months: The MacArthur Longitudinal Twin Study. *Child Development, 63,* 1437–1455.

Fenson, L., Dale, P., Reznick, J. S., Thal, D., Bates, E., Thal, D. J., & Pethick, S. J. (1994). Variability in early communicative development. *Monographs of the Society for Research in Child Development, 59,* 1–173.

Gardner, H. (1983). *Frames of mind: The theory of multiple intelligences.* New York: Basic Books.

Hardy-Brown, K., & Plomin, R. (1985). Infant communicative development: Evidence from adoptive and biological families for genetic and environmental influences on rate differences. *Developmental Psychology, 21,* 378–385.

Hardy-Brown, K., Plomin, R., & DeFries, J. C. (1981). Genetic and environmental influences on the rate of communicative development in the first year of life. *Developmental Psychology, 17,* 704–717.

Hay, D. A., Prior, M., Collett, S., & Williams, M. (1987). Speech and language development in preschool twins. *Acta Geneticae Medicae et Gemellologiae, 36,* 213–223.

Hedrick, D. L., Prather, E. M., & Tobin, A. R. (1975). *Sequenced Inventory of Communication Development*. Seattle: University of Washington Press.

Humphreys, L. G., & Davey, T. C. (1988). Continuity in intellectual growth from 12 months to 9 years. *Intelligence, 12,* 183–197.

Hyde, J. S. (1981). How large are cognitive gender differences? A meta-analysis using w^2 and *d. American Psychologist, 36,* 892–901.

Menyuk, P. (1971). *The acquisition and development of language.* Englewood Cliffs, NJ: Prentice-Hall.

Neale, M. C. (1994). *MX: Statistical modeling.* Richmond, VA: Department of Psychiatry, Medical College of Virginia.

Neale, M. C., & Cardon, L. R. (1992). *Methodology for genetic studies of twins and families.* Dordrecht, Netherlands: Kluwer.

Olson, R., Wise, B., Conners, F., Rack J., & Fulker D. W. (1989). Specific deficits in component reading and language skills: Genetic and environmental influences. *Journal of Learning Disabilities, 22,* 339–348.

Plomin, R., & DeFries, J. C. (1985). *Origins of individual differences in infancy: The Colorado Adoption Project.* San Diego, CA: Academic Press.

Plomin, R., Campos, J., Corley, R., Emde, R. N., Fulker, D. W., Kagan, J., Reznick, J. S., Robinson, J., Zahn-Waxler, C., & DeFries, J. C. (1990). Individual differences during the second year of life: The MacArthur Longitudinal Twin Study. In J. Colombo & J. Fagan (Eds.), *Individual differences in infancy: Reliability, stability, prediction* (pp. 431–455). Hillsdale, NJ: Erlbaum.

Plomin, R. Emde, R. N., Braungart, J. M., Campos, J., Corley, R. P., Fulker, D. W., Kagan J., Reznick, J. S., Robinson, J. Zahn-Waxler, C., & DeFries, J. C. (1993). Genetic change and continuity from fourteen to twenty months: The MacArthur Longitudinal Twin Study. *Child Development, 64,* 1354–1376.

Sandbank, A. C., & Brown, G. A. (1990). An examination of the psychological and behavioral factors in the development of language retardation in twins. *Acta Geneticae Medicae et Gemellologiae, 39,* 497–500.

Shore, C. M. (1995). *Individual differences in language development.* Thousand Oaks, CA: Sage.

Terman, L. M., & Merrill, M. A. (1973). *Standford-Binet Intelligence Scale: 1972 norms edition.* Manual for the 3rd revision form L-M. Boston: Houghton Mifflin.

Thompson, L. A., & Plomin, R. (1988). The Sequenced Inventory of Communication Development: An adoptions study of two- and three-year olds. *International Journal of Behavioral Development, 11,* 219–231.

Zimmerman, B. J. (1983) Social learning theory: A contextualist account of cognitive functioning. In C. J. Brainerd (Ed.), *Recent advances in cognitive-developmental theory: Progress in cognitive development research.* New York: Springer-Verlag.

Part V

Environmental Contributions and Cross-Domain Integrations

Section Editors
Robert N. Emde
John K. Hewitt

17

Context and Specificity of Individual Differences and Their Origins

Robert N. Emde

Chapters 18 and 19 in this part concern aspects of the maternal environment and the context of our observations in the MacArthur Longitudinal Twin Study. The picture that emerges is a dynamic one of mutual influences. For example, one question that is asked is, To what extent do mothers behave differently with monozygotic (MZ) and dizygotic (DZ) twins? McGuire and Roch-Levecq found evidence that maternal perceptions of differential attention, discipline, and affection differ for the two classes of twins across four ages in that mothers more often saw themselves as behaving similarly with MZ twins. Although this finding seems to violate the common environment assumption for genetic comparisons of twin types (see chapter 2), it is also consistent with findings reported by others in the literature; namely, that parents respond to early behavioral differences in their children. Considering the fact that MZ twins are genetically identical (and DZ twins are not), this finding could also serve as an example of a gene–environment correlation.

The context of our observations is also important for thinking about measurement. Again, the picture is a dynamic one, especially where the socioemotional behaviors of children are involved. Robinson, Zahn-Waxler, and Emde found that the child's empathic responding occurs regularly in the context of a familiarized tester who feigns distress; such empathic responding also occurs regularly in the context of a mother who feigns distress. Although empathic responding occurs to each in the home, as well as in the laboratory, twin analyses reveal context differences with respect to genetic and environmental influences. For empathy, hereditary influences predominate in the context of tester distress, whereas common environmental influences predominate in the context of maternal distress. The authors speculate that a central ingredient of the latter context may be the intimate parenting relationship, with its shared history of meaning (including socialization experiences). Common experiences in this relationship, because of their salience in this domain, could swamp genetic influences on individual differences.

Cherny et al. present findings of shyness observed in the separate contexts of home and laboratory. The findings indicate both situational and cross-situational influences. Although observations of home and laboratory shyness are not highly correlated and although shyness is not highly stable from 14 to 20 months, what is stable across ages and situations is genetically mediated. Moreover, there are significant genetic influences in each setting that are independent of each other.

Perinatal and anthropometric variables are also relevant to our thinking about development. Martin and Hershberger found that birth weight is correlated with general cognitive development at the early ages of our study, but it is not correlated at 36 months, suggesting that the influence of this variable sharply diminishes or disappears by then. Chambers et al. performed twin analyses at the four ages of our study for height, weight, and body mass index, the latter being calculated because of its common clinical use as an indicator of body fat. Findings indicate that genetic influence accounts for most of the variation at each age and that new genetic influence does come in at later time points. Environmental influence accounts for a minority of variance at all ages. Both of these chapters provide evidence for cross-domain specificity of individual differences rather than an interconnectedness. Martin and Hershberger found no association of perinatal variables with temperament, and Chambers et al. found no correlation of body mass index with temperament at any of the four ages. Additionally, Chambers et al. found no evidence that anthropometric indices are correlated with other measures of emotional and cognitive development.

The theme of specificity of individual differences, rather than connectedness across domains of the child's behavior, continues in chapter 23 by Kubicek, Emde, and Schmitz. Correlations at the four ages between maternal reports of temperament and general cognitive functioning were either nonsignificant or low, as were correlations between maternal reports and temperament and language. Phenotypic correlations between observed ratings of temperament and language indicated meaningful associations involving the ratings of affect-extraversion and of task orientation. Twin analyses were therefore carried out to examine possible common genetic and environmental origins of these two domains. The chapter details the results of a series of bivariate modeling analyses. Similar patterns from modeling were seen for the temperament ratings of affect-extraversion and task orientation in relation to general cognitive functioning. The nonshared environment had significant influence at 14 and 20 months, and genetic factors had significant influence only at 24 months. For bivariate modeling with respect to the temperament ratings and language, significant influences were largely confined to expressive language, and again significant genetic influences were limited to 24 months. It is tempting to speculate that the pattern of bivariate genetic influences at 24 months may be evidence of integration after a developmental shift (i.e., the transition after infancy). It is also interesting that the nonshared environment is influential across domains at most ages. It may be that transactions with the environment within which the child is experiencing development are particular, rather than general. The consistency of this pattern of findings with respect to nonshared environmental influences would seem to argue against such findings reflecting measurement error.

The final chapter of this part is the only chapter in this book that presents findings for twins beyond the age of 36 months. Schmitz et al. examined early predictors of problem behavior at age four as indicated by the Child Behavior Checklist (CBCL). The CBCL at age four can be considered the first standardized measure of adaptation that we have been able to analyze in our data set. As such, it provides a unique opportunity to look at antecedents of individual differences from the period of the child's second and third years. While the above-reviewed findings of a lack of correlation across domains seem to argue for specificity of behaviors rather than for interconnectedness, there may be antecedents across domains that could reflect heterotypic continuities in development.

The first set of findings concern earlier ratings of the child's temperament. Results are strikingly consistent with respect to parental ratings. Both mothers' and fathers' assessments of emotionally at all four ages of our study correlate with internalizing, externalizing, and total scale values of the CBCL at age four. Moreover, both mothers' and fathers' assessments of shyness at earlier ages correlate with the internalizing scale of the CBCL at age four. In contrast, observationally rated temperament yielded only occasional and inconsistent correlations with the CBCL outcomes at four years. With respect to emotion, earlier maternal ratings of sadness, anger, and enjoyment yielded low but significant correlations with the CBCL.

Observations of empathy at 24 months correlated with CBCL reports at age four for girls but not for boys. Earlier language and general cognitive ability were predictive at low levels of correlation with the CBCL at age four, a finding that again held for girls but not for boys. The perceived family environment was assessed by two measures and yielded an interesting pattern that differed for mothers and fathers. Mothers who experienced their families as less cohesive and having more conflict and on the Family Environment Scale tended to rate more problem behaviors in their child at age four on the CBCL. Fathers also had predictive correlations from the Family Environment Scale, but the pattern was different and suggested they paid attention to different features of family life—namely, activity and family organization rather than cohesion and conflict.

Although chapter 24 deals mainly with phenotypic analyses, twin analyses and quantitative modeling generate complex, but interesting results. These include findings with respect to associations between early temperament and the CBCL at age four and with respect to low parental agreement about child behavior. The conclusion of the chapter again brings emphasis to a theme of this part: findings differ according to who assessed the child. More research on situational and relationship specificity is needed.

18

Mothers' Perceptions of Differential Treatment of Infant Twins

Shirley McGuire
Anne-Catherine Roch-Levecq

Families with twins are an interesting context in which to study parenting. Mothers and fathers are faced with taking care of two demanding infants rather than one: two mouths to feed, two bottoms to change, two cries to answer. The children are immediately born into a sibling relationship: sharing toys, sharing rooms, sharing parental attention. A study of 200 parents of 2-year-old twins found that these families have to cope with a host of financial, health, and psychological problems (Robin et al., 1991). Yet, compared to research on the social development of singleton infants, studies of young twins are rare.

There have been some observational studies of parent–twin attachment and twin–twin social interaction (Brooks & Lewis, 1974; Clark & Dickman, 1984; Goldberg et al., 1986; Goshen-Gottstein, 1981; Vandell et al., 1988), including one study based on the the MacArthur Longitudinal Twin Study sample (Robinson & Little, 1994). This chapter examines mothers' perceptions of parenting their identical or fraternal twins. Thomas and Chess (1977) developed their infant temperament categories based on parental interviews concerning their infant's behavior. They argued that parents' perceptions of the children's behaviors were immensely important for understanding the early parent–child relationship, even when the behaviors were biologically based. Parents may react to temperamental differences and thus contribute to the stability of individual differences in children's behavior across development.

One could argue that parents would not admit that they treat their children differently. However, in a study of maternal reports of differential treatment, 32% of the mothers admitted to large discrepancies in their affectionate behavior and 29% admitted to large discrepancies in their controlling behavior toward their two nontwin children (Dunn et al., 1990). The older siblings were in middle childhood (average age was 7 years old), and the younger siblings were in early childhood (average age was 4 years old). These findings were replicated in a longitudinal follow-up conducted 3 years later (McGuire et al., 1994). Since the study consisted of full and adoptive siblings, it may be that these differences are due to age and gender differences. Mothers may not treat twins that differently. However, adolescent and adult twins report significant levels of parental differential treatment (Baker & Daniels, 1990; Pike et al., 1994). In fact, dizygotic twins reported more differential experiences than monozygotic twins, which suggests that parental behavior may be influenced by genetic differences among the children. Still, mothers of infant twins may not perceive or report differential parenting. Mothers may have few unique experiences with each child due to the amount of time the mothers and twins spend together in triads rather than in dyads. Consequently, parents of twins may not develop a different relationship with each child until after infancy.

There are a few studies of mothers' perceptions of parenting infant and toddler twins. Most of the studies used small sample sizes, but the findings are interesting. Studies conducted by Robin and colleagues (Robin et al., 1988, 1992, 1993) in France have focused on the extent to which mothers treat the twins as a unit, differentiate between the two, or develop an intermediate parenting strategy. They found that observed maternal behavior, but not attitudes about parenting, differed by zygosity. Mothers treated monozygotic (MZ) twins more alike than mothers of dizygotic (DZ) twins, but the two groups of mothers did not differ in their reports of parenting. In addition, Lytton (1977) found that parents treat twins based on their actual, not perceived, zygosity. For example, parents who thought their twins were fraternal, but blood tests showed that they are identical, treated their children like identical twins. That is, the parents treated the twins as similarly as parents who knew their twins were identical. These findings do not support the idea that monozygotic twins are treated more similarly than dizygotic twins because monozygotic twins are labeled "identical." Rather, genetic similarities and differences play a role in children's social interactions.

Another observational study across childhood found that parents treated MZ twins and DZ twins the same, while reporting that identical twins behaved more similarly than fraternal twins (Cohen et al., 1977). Thus, the data during infancy and childhood are mixed. Studies during middle childhood and adolescence, on the other hand, typically find that identical twins are treated more similarly than fraternal twins (see Plomin, 1994 for a review).

The few data that are available on the stability of differential treatment by mothers concerns parent–child relationships in middle childhood and adolescence (Brody et al., 1992; McGuire et al., 1994). Thus, an additional goal of this chapter is to provide the first longitudinal data on mothers' reports of differential treatment of infant twins. The data on twin similarity for intelligence shows that DZ twins become less similar across development, while MZ twin correlations remain high (see Plomin, 1986 for review). This suggests that

DZ twins may experience less similar environments over time compared to MZ twins. However, it is not clear whether different experiences with parents appear in infancy or began later during childhood.

This chapter addresses three research questions: (1) To what extent do mothers report treating their infant twins similarly? (2) How stable is differential treatment across infancy? and (3) Does stability in similar treatment differ by zygosity?

METHODS

Participants

The participants included 200 mothers of same-gender twin pairs who participated in measurements at four twin ages; 14, 20, 24, and 36 months (see chapter 3 for details about the sample). This chapter includes data for 176 mothers who completed the differential treatment measure at all four time points. Thus, there were complete data for 97 of the mothers with identical twins (i.e., MZ) and 79 of the mothers with fraternal twins (i.e., DZ).

Measures

As part of the MacArthur Longitudinal Twin Study assessment procedures, mothers were interviewed about each child's health and temperament, sibling interaction, and parenting (see chapter 3 for details about the measures). Only the questions about differential treatment were used in these analyses. Specifically, mothers were interviewed in a semistructured format about the amount of discipline, affection, and attention they displayed toward one twin compared to the other (see Appendix for questions). Mothers were encouraged to discuss their experiences at length. Then, testers marked the degree of differential treatment described using a 5-point scale: 1 = twin A much more than twin B; 3 = both about the same; 5 = twin B much more than twin A.

Data Analyses

In most cases, researchers have examined relative differences in parental treatment of their children (e.g., Dunn et al., 1990; McHale & Pawletko, 1992). They use this strategy when the direction of the treatment matters; that is, when they want to know whether particular children are treated better than their siblings. This type of comparison is meaningful when examining associations between differential treatment and individual differences in children's adjustment. Children who receive more favorable treatment often feel better about themselves. This chapter is concerned with zygosity differences in parenting. That is, to what extent are MZ and DZ twins treated differently regardless of the direction of treatment? Relative difference scores are difficult to interpret when comparing the treatment of identical and fraternal twins. Thus, absolute difference scores were used in our analyses. Specifically, differential treatment scores were collapsed into three categories: (1) treated the twins a lot different; (2) treated the twins a little different; and (3) treated the twins the same.

RESULTS

The first research questions was, To what extent do mothers report treating their infant twins similarly? Table 18.1 shows the percentage of mothers who reported treating children "a lot different," "a little different," and "the same" across all four time periods and by zygosity. At each time point, 2 (zygosity) × 3 (treatment) chi-square analyses were conducted to test whether differential treatment differed by zygosity. The chi-square analyses were significant at two of the four time points for differential affection, three of the four time points for differential attention, and at all four time points for differential discipline. Mothers of monozygotic twins reported treating their twins more similarly than mothers of dizygotic twins. For differential attention and discipline, it

Table 18.1. Percentage of mothers reporting each level of differential treatment at 14, 20, 24, and 36 months and by twins' zygosity

Measure/age	MZ			DZ			χ^2_2
	Same	A lot	A little	Same	A lot	A little	
Affection							
14 Months	49.5	45.4	5.1	43.0	34.2	22.8	12.09**
20 Months	41.2	46.4	12.4	41.8	45.6	12.7	0.02
24 Months	58.8	35.1	6.2	39.2	36.7	24.1	13.13**
36 Months	41.2	43.3	15.5	40.5	31.6	27.8	4.73
Attention							
14 Months	74.2	22.7	3.1	65.8	30.4	3.8	1.49
20 Months	71.1	24.7	4.1	51.9	36.7	11.4	7.76*
24 Months	71.1	25.8	3.1	50.6	32.9	32.9	12.27**
36 Months	70.1	25.8	4.1	53.2	29.1	29.1	10.05**
Discipline							
14 Months	59.8	30.9	9.3	45.6	31.6	22.8	6.83*
20 Months	57.7	27.8	14.4	36.7	31.6	31.6	10.02**
24 Months	56.7	32.0	11.3	35.4	34.2	30.4	12.17**
36 Months	53.6	36.1	10.3	31.6	30.4	38.0	19.89**

$**p < .01; *p < .05.$

Table 18.2. Stability coefficients for mothers' reports of differential treatment (absolute differences) across time

	14 Months	20 Months	24 Months	36 Months
Differential affection				
14 Months				
20 Months	.14			
24 Months	.30**	.27**		
36 Months	.22	.25*	.43**	
Differential attention				
14 Months				
20 Months	.24*			
24 Months	.11	.21		
36 Months	.16	.28**	.32**	
Differential discipline				
14 Months				
20 Months	.27**			
24 Months	.21	.32**		
36 Months	.30**	.33**	.28**	

$**p < .01; *p < .05.$

appears that mothers of dizygotic twins treat their twins less similarly across time.

The second question addressed was, How stable is differential treatment across infancy? Stability was assessed using contingency coefficients in the chi-square analyses, which are more sensitive to categorical data than Pearson product moment correlations (Bruning & Kintz, 1987). Twelve out of the 18 associations were significant, but were only in the low to moderate range (see table 18.2)

The last research question was, Does the stability of similar treatment or differential treatment differ by zygosity? This was examined three ways. First, we calculated contingency coefficients across time separately for monozygotic and dizygotic twins. The coefficients were not significantly different from one another. However, such an approach may not capture the consistency that does lie within the data. Thus, the second approach involved examining the percentage of mothers who reported treating the children "the same" across all four time periods. Table 18.3 shows the percentage of mothers who reported treating the twins "the same" for identical versus fraternal twins. Few mothers reported similar treatment across all points in time, which was also reflected in the low stability coefficients. For two of the areas, differential attention and discipline, the percentages did differ significantly by zygosity of the twins.

Table 18.3. Percentage of mothers reporting treating the twins the same across all four and three our of four times points, by zygosity

Measure	MZ	DZ
All four time points		
Differential affection	11.3	6.3
Differential attention	30.2	15.2**
Differential discipline	18.6	2.5**
Three out of four time points		
Differential affection	33.0	26.6
Differential attention	66.0	41.8**
Differential discipline	43.3	24.1**

**Proportions were significantly different from one another at $p < .01$.

Significance was tested using a z-test for independent proportions (Bruning & Kintz, 1987). Third, it could be that requiring mothers to report the same value across a 2-year period is too strict a criterion. Thus, the percentage of mothers who reported treating the twins "the same" for three out of the four time points was also examined. Again, the percentage of mothers reporting similar attention and discipline toward the two children differed significantly by zygosity. Mothers of monozygotic twins are more likely to consistently report treating the two twins the same across infancy compared to mothers of dizygotic twins.

DISCUSSION

Mothers' reports of differential attention, discipline, and affection (although less so) differed significantly by zygosity. While the majority of mothers of MZ twins reported treating the pair the same, only a third of the mothers of DZ twins reported treating the twins similarly. This suggests that even as early as 14 months, mothers are reacting to genetic differences among infant twins. This is in line with infancy temperament data that show that parents react to behavioral differences in children early in the parent–child relationship (e.g., Buss & Plomin, 1984; Thomas & Chess, 1977). The data are also consistent with the hypothesis that mothers of identical twins are more likely to treat the pair as a unit, while mothers of DZ twins are more likely to differentiate between the children (Robin et al., 1992).

Interestingly, continuity in reporting similar treatment toward the twins also differs by zygosity. Mothers of identical twins were more likely to continually report treating the two children the same compared to mothers of fraternal twins. However, this was true for differential attention and discipline, but not for differential affection.

Our results concerning genetic influences on parental affection versus parental attention and discipline differ from those typically found in the literature on genetics of the environment (Plomin, 1994). Most behavioral genetic studies involving older children and adults find either that parental warmth shows genetic influence but parental control does not (e.g., Plomin et al., 1988; Rowe, 1981, 1983) or that there is no difference in heritability for parental warmth and control (e.g., Baker & Daniels, 1990; Daniels & Plomin, 1985; Pike et al., 1994; Plomin et al., 1994; Robinson & Little, 1994). In fact, Lytton (1980) argued that parental acceptance is a reflection of the child's characteristics, while parental control is a result of the parent's planning and characteristics. Why are our data different?

It is possible that our study is simply a fluke? Certainly, our findings need to be replicated. However, these data may reflect an interesting phenomenon worth discussing: Children's characteristics may influence parental affection and control differently across development. It is possible that parental affection may be determined by child characteristics more during middle childhood than during infancy. Likewise, temperament researchers have shown that parental behaviors are linked to infants' early behavioral styles. A mother's controlling behavior may, in fact, be driven by her infant's behavior until she is able to adjust her parenting style to fit the child's disposition. For some parents such an adjustment may take years or may never occur. Thomas & Chess (1977) refer to this domain as the "goodness of fit" between the parent and child. As the parent gains more experience with the child, the parent learns what works and does not work in terms of parental discipline. Thus, parental discipline may be determined by child characteristics more during infancy than during middle childhood. It is clear that more data are needed on parental differential treatment at every age level.

Another interesting finding is that the stability coefficients in this study were lower than those found in a middle childhood sample of full and adoptive siblings (McGuire et al., 1994). There seems to be some instability in mothers' reports. This may be due to unreliability of parental reports or real change in the parents' and infants' behaviors. Infancy is a period of rapid development. Such low stability has been found in other studies of mothers' behavior toward children in infancy and toddlerhood using observational methods (Dunn et al., 1985, 1986). Parents may be adjusting their behavior as their children develop. More studies using multiple methods would be helpful here.

In summary, these data show that many parents report treating even infant twins differently. The fact that such treatment differed by zygosity suggests that genetically influenced infant behaviors, such as temperament, play a role in early parenting. However, there are still many unanswered questions.

APPENDIX

Differential Treatment Questions from Maternal Interview

Discipline

Do you find that one of them is easier to handle than the other? If there's something that you have to stop them doing, do you have to use different ways of controlling each of them? Does it mean that you have to discipline one of them more than the other?

Affection

Some parents find that they feel it's easier to be affectionate with one of their children than the other, that one child is more cuddly. Is this so with either of your twins?

Attention

When both the children are at home, do you find that you spend more time with one than with the other playing or talking? Or doing things outside the home with one more than the other?

1. Twin A much more than twin B
2. Twin A a bit more than twin B
3. Both about the same
4. Twin B a bit more than twin A
5. Twin B much more than twin A

REFERENCES

Baker, L. A., & Daniels, D. (1990). Nonshared environmental influences and personality differences in adult twins. *Journal of Personality and Social Psychology, 58,* 103–110.

Brody, G. H., Stoneman, Z., & McCoy, J. K. (1992). Associations of maternal and paternal direct and differential behavior with sibling relationships: Contemporaneous and longitudinal analyses. *Child Development, 63,* 82–92.

Brooks, J., & Lewis, M. (1974). Attachment behavior in thirteen-month-old, opposite-sex twins. *Child Development, 45,* 243–247.

Bruning, J. L., & Kintz, B. L. (1987). *Computational handbook of statistics* (3rd ed.). Glenview, IL: Scott, Foresman.

Buss, A., & Plomin, R. (1984). *Temperament: Early developing personality traits.* Hillsdale, NJ: Erlbaum.

Clark, P. M., & Dickman, Z. (1984). Features of interaction in infant twins. *Acta Geneticae Medicae et Gemellologiae: Twin Research, 33,* 165–171.

Cohen, D. J., Dibble, E., & Grawe, J. M. (1977). Parental style. *Archives of General Psychiatry, 34,* 445–451.

Daniels, D., & Plomin, R. (1985). Differential experiences of siblings in the same family. *Developmental Psychology, 21,* 747–760.

Dunn, J., Plomin, R., & Daniels, D. (1986). Consistency and change in mothers' behavior toward young siblings. *Child Development, 57,* 348–356.

Dunn, J., Plomin, R., & Nettles, M. (1985). Consistency and change in mothers' behavior toward infant siblings. *Developmental Psychology, 21,* 1188–1195.

Dunn, J., Stocker, C., & Plomin, R. (1990). Nonshared experiences within the family: Correlates of behavior problems in middle childhood. *Development and Psychopathology, 2,* 113–126.

Goldberg, S., Perrotta, M., Minde, K., & Corter, C. (1986). Maternal behavior and attachment in low-birth-weight twins and singletons. *Child Development, 57,* 34–46.

Goshen-Gottstein, E. R. (1981). Differential maternal socialization of opposite-sex twins, triplets, and quadruplets. *Child Development, 52,* 1255–1264.

Lytton, H. (1977). Do parents create, or respond to, differences in twins? *Developmental Psychology, 13,* 456–459.

Lytton, H. (1980). *Parent-child interaction: The socialization process observed in twin and singleton families.* New York: Plenum Press.

McGuire, S., Dunn, J., & Plomin, R. (1994). Maternal differential treatment of siblings and children's behavior problems: A longitudinal study. Unpublished manuscript.

McHale, S., & Pawletko, T. M. (1992). Differential treatment in two family contexts. *Child Development, 63,* 68–81.

Pike, A., Manke, B., Hetherington, E. M., Reiss, D., & Plomin, R. (1994) Differential experiences of adolescent siblings: Genetic and environmental influences. Unpublished manuscript.

Plomin, R. (1986). *Development, genetics, and psychology.* Hillsdale, NJ: Erlbaum.

Plomin, R. (1994). *Genetics and environment: The interplay between nature and nurture.* Thousand Oaks, CA: Sage.

Plomin, R., McClearn, G. E., Pedersen, N. L., Nesselroade, J. R., & Bergeman, C. S. (1988). Genetic influences on childhood family environment perceived retrospectively from the last half of the life span. *Development Psychology, 24,* 738–745.

Plomin, R., Reiss, D., Hetherington, E. M., & Howe, G. (1994). Nature and nurture: Genetic influence on measures of the family environment. *Development Psychology, 30,* 32–43.

Robin, M., Josse, D., Casati, I., Kheroua, H., et al. (1993). La gemellisation de l'environment ("twinning" of the physical environment of twins: Maternal attitudes and practices.). *Enfance, 47,* 393–406.

Robin, M., Josse, D., & Tourrette, C. (1988). Mother-infant interaction in early childhood. *Acta Geneticae Medicae et Gemellologiae: Twin Research, 37,* 151–159.

Robin, M., Josse, D., & Tourette, C. (1991). Forms of family reorganization following the birth of twins. *Acta Geneticae Medicae et Gemellologiae: Twin Research, 40,* 53–61.

Robin, M., Kheroua H., & Casati, I. (1992). Effects of early mother-twin relationships from birth to 3, on twin bonding. Seventh International Congress on Twin Studies (1992, Tokyo, Japan). *Acta Geneticae Medicae et Gemellologiae: Twin Research, 41,* 143–148.

Robinson, J., & Little, C. (1994). The emotional availability in mother-twin dyads: Effects on the organization of relationships. *Psychiatry: Interpersonal & Biological Processes, 57,* 222–231.

Rowe, D. C. (1981). Environmental and genetic influences on dimensions of perceived parenting: A twin study. *Developmental Psychology, 17,* 203–208.

Rowe, D. C. (1983). Biometrical genetic analysis of perceptions of family environment: A study of twin and singleton sibling kinships. *Child Development, 54,* 416–423.

Thomas, A., & Chess, S. (1977). *Temperament and development.* New York: Brunner/Mazel.

Vandell, D. L., Owen, M. T., Wilson, K. S., & Henderson, V. K. (1988). Social development in infant twins: Peer and mother-child relationships. *Child Development, 59,* 168–177.

19

Relationship Context as a Moderator of Sources of Individual Differences in Empathic Development

JoAnn L. Robinson
Carolyn Zahn-Waxler
Robert N. Emde

The expression of empathy is a complex constellation of emotions and behaviors that optimally convey a sense of caring and concern for another in distress with the goal of relieving that distress (Batson et al., 1981; Davis, 1983; Robinson et al., 1994). Empathic behaviors in children are generally considered to be subject to socialization or social forces that encourage or restrict their expression. Much research on empathic development has considered the importance of socializing influences, especially in early childhood (see Eisenberg & Mussen, 1989 for a review.) Some research has also considered the role of contextual factors influencing the degree of altruism or empathy expressed. This has included consideration of the role of degree of relatedness of the "victim" or beneficiary of prosocial acts. In general, children and adults are more likely to provide assistance to someone in distress if the individual is an acquaintance rather than a stranger (Dovidio, 1984; Krebs, 1975; Willis et al., 1977). Research has also shown that the degree of acquaintance is an important factor in the prediction of altruistic acts. For example, school-age children are more likely to share with or help a friend compared to a nonfriend (Bengtsson & Johnson, 1987). Few studies have considered these processes during the transition from infancy to early childhood, the developmental phases when empathic and altruistic behaviors are emerging in the child's behavioral repertoire.

The organization of behavioral responses within relations is strongly integrated into theory and research in the area of attachment relationships. Attachment theory posits that daily experiences of interaction with a caregiver provide the basis for the child's internal working model of a relationship, which, in turn, becomes the basis for the behavioral organization of the infant when separated from the caregiver and during reunions (Bretherton, 1985, 1987; Cicchetti et al., 1990). Empirical research has shown that the child's behavioral repertoire during reunions often differs when separations from mother and father are compared (Main & Weston, 1981; Marvin & Stewart, 1990). A general proposition we tested here is that empathic responses are also organized within relationships. Across family members, for example, empathic inclinations may vary considerably, with mother receiving greater empathy than father, or younger sibling receiving less empathy than an older sibling. Indeed, Lamb (1978) observed that toddlers were more likely to behave pro-socially with a parent than with a sibling. Similarly, the degree of empathic responding might vary across nonfamily and family members.

Socialization efforts within families are well known to be an important source of individual differences in empathy (Eisenberg & Mussen, 1989). Maternal socialization practices, in particular, have received considerable research attention. The mother's warmth and direct instruction about empathic and reparative behaviors are important predictors of empathy in very young children (Radke-Yarrow et al., 1983; Zahn-Waxler & Radke-Yarrow, 1990). While the mother's socialization efforts are often in consideration of other individuals in distress (such as a sibling), the mother's own distress may also be the focus of her socializing efforts. Zahn-Waxler et al. (1991) discussed the particular risks for children of depressed mothers, who may be drawn into a caretaking role in relation to the mother at a young age. Their empathic and reparative responses are excessive and may lead to the development of depression in later childhood. Children of depressed mothers may be subtly taught through emotional signals and verbal instruction that feeling sad for mother is welcome and necessary. In other families, children may be signaled that sadness and caretaking are not needed by the mother when she is distressed; she can take care of the situation herself or other adults may be asked to assume that role.

Through such subtle emotional communications between individuals, relationship-specific messages concerning the expression of affect and prosocial behavior become a part of the content of the child's early internal representations of expressing caring and concern. In the case of family members, we suggest that children develop specific scripts for how and when to express concern and care for them. However, when the person in distress is not well known to the child and relationship-specific rules have not been established over time, the child's "natural inclination" may be more apparent. By this we are suggesting that when encounters with unfamiliar individuals occur, genetic influences on behavior may more strongly affect the level of empathy the child expresses compared to environmental factors. Learned responses are less likely to apply in unfamiliar situations because of the absence of specific prior experiences and expectations. The child is, in a sense, on his or her own, especially if the unfamiliar individual is a researcher trained to provide minimal cues to the child about expected responses. In contrast, when empathic encounters involve individuals with an on-going relationship to the child, such as the child's mother, we anticipate that environmental influences may out-weigh genetic influences.

The second year of life is a period of rapid change in emotion regulation. Strong pressures for parents to assist the child in responding "safely" to a large range of situations require a rapid degree of teaching on their part to ensure generalization of children's behavior. One set of messages from parents en-courage inhibiting responses as the child becomes more mobile ("don't touch"), while another set of responses encourage action or proactive re-sponses ("pee in the toilet" or "tell me what you want"). These forms of so-cialization are in the service of the child's autonomous, self-regulating devel-opment and specifically of an internalization of rules and values that appear to accelerate during the second year with the onset of walking.

Other features of this socialization are more explicitly emotional and can be seen in the emergence of new emotional responses during the second year. Increases in the expression of empathy across the second year have been doc-umented in studies by Zahn-Waxler and colleagues (e.g., Zahn-Waxler et al., 1992a). Our previous research with the MacArthur Longitudinal Twin Study sample showed significant increases in components of the empathic response between 14 and 20 months. By 20 months, children generally expressed more concern, prosocial behavior, and hypothesis testing and less self-distress and indifference when multiple distress situations were presented (Zahn-Waxler et al., 1992b).

In another study that examined the development of prosocial and aggressive behaviors throughout the second year, no frequency differences were found for boys and girls in the first half of the second year. However, by 24 months girls exhibited fewer aggressive behaviors than boys (Cummings et al., 1986; Zahn-Waxler et al., 1992a). These findings, while highlighting gender-specific socialization of emotion, suggest that the end of the second year may be a time when environmental influences on behavior are in ascendency compared to earlier in the first and second years. Thus, we hypothesize that 24 months may be a time when environmental influences on behavior are more apparent than genetic influences.

In this chapter, we consider whether genetic and environmental sources of individual differences in empathic behavior during the period 14–36 months vary as a function of familiarity of the victim (i.e., mother versus an unfamiliar examiner). We examine how levels of empathic behavior may change as a function of victim as well as the relative coherence or intercorrelation of these behaviors during this time period.

METHODS

In this chapter we use the same sample scored for empathic responses as de-scribed in chapter 11. This sample included the first 250 twin pairs enrolled in the study. The number of monozygotic (MZ) pairs providing data across ages varied from 115 to 138, and the number of dizygotic (DZ) twin pairs varied from 95 to 108. Data were gathered in both home and laboratory visits. Mothers and examiners simulated distress once in each setting. In the home, the examiner's empathy simulation occurred first, after administration of the Bayley Scales of Infant Development (at 14, 20, and 24 months) or the Stanford-Binet (36 months). The examiner made sure that the child had an object to play with and then turned to her briefcase, pretending to find the next test probe. In closing her case, she feigned injury to her hand. This distress was vocalized in a moderate tone of voice for 30 s, at which time she began a recovery sequence, assuring the child that the injury no longer hurt. A hand-held stopwatch was used to control the duration of the phases of the simulation. The mother's empathy probe occurred 20–30 min later in the visit, following a transition period and a 7-min period of play with both twins simultaneously. In this simulation, mother and child played on the floor with a novel toy for a few moments, ensuring the child's interest in it. At the examiner's signal she got up off the floor and feigned injury to her knee. The same time-parameters were used to control the duration of each phase; however, the examiner signaled mother's termination of pain by inquiring whether she needed assistance. At 14, 20, and 24 months, the mother's simulation occurred first in the laboratory. Following a knock on the one-way mirror, mother feigned pinching her finger in a clipboard. The examiner's distress simulation occurred 15 min later and involved the examiner dropping a chair on her foot.

Five responses to the distress simulation are considered here: approach, hypothesis testing, concern, arousal, and prosocial acts. The approach dimension is a 4-point scale that categorizes the child's response as 4 = moving toward; 3 = remaining stationary; 2 = withdrawing or pulling back; or 1 = avoiding or turning away. Children engage in hypothesis-testing behaviors as they attempt to figure out what the victim is experiencing and what might have caused their distress. The frequency and complexity of these behaviors were ordered on a 5-point scale. Social referencing behaviors, where the child scans the victim's face and the injured area, and inquiring vocalizations are the most common forms of hypothesis testing in very young children. Expressions of empathic concern include, at the lower end of the scale, a sobering of facial features, while stronger expressions include furrowing of the brow and vocalizing concern with sympathetic intonations. Strength of concern was ordered on a 4-point scale. Arousal describes the child's level of tension during the distress simulation. Children who show very low arousal do not disrupt or slow their ongoing behaviors during the simulation and behave as if the distress is not recognized. Moderately aroused children do disrupt their activity and may have a period of stillness and intensity in their orientation toward the victim. Highly aroused children may have periods of freezing, where fingers or hands remain in an unusual posture for an extended period of time, or their activity might become excessive in its intensity, or they may cry and become quite distressed. Arousal level was ordered on a 5-point scale. Prosocial acts include acts directed toward the victim that are intended to relieve distress. These include comforting (hugging, patting, or rubbing), attempts to distract the victim from their distress, defensive behaviors such as pushing the chair that fell on the examiner's foot, and/or sharing a toy the child played with before the simulation. The percentage of episodes in which a prosocial act occurred was used as the index for this behavior.

A child's data were included in analyses if there was at least one examiner-directed or one mother-directed simulation completed. At all ages, the majority of children completed two simulations with each victim; 85% at 14 months, 83% at 20 months, 82% at 24 months, and 83% at 36 months.

RESULTS

Developmental Patterns of Responses

In our previous work with these data we have observed significant developmental changes from 14 to 20 months (Zahn-Waxler et al., 1992b). Table 19.1 presents the means and standard deviations for each behavioral dimension (hypothesis testing, concern, arousal, and prosocial acts) as a function of age and whether the behavior was directed to mother or victim. Multivariate analysis of variance models tested for age and sex differences within each behavioral dimension.

We found that responses directed toward both mother and examiner show an orderly increase in empathic tone across the second year (i.e., 14–24 months). Multivariate analyses testing for age and sex differences across behaviors within type of victim showed significant age ($F = 23.90$, $p < .001$) and sex effects ($F = 3.80$, $p < .005$). In general, girls had higher scores than boys. All age trends reflected increases in behaviors over time, except for the mother-directed responses on the arousal dimension, which showed an initial increase between 14 and 20 months of age and then a decrease between 20 and 24 months. In addition, mother-directed approach and concern showed a marked increase from 14 to 20 months but then stabilized and showed little change after 20 months. In general, mother-directed empathic behaviors were stronger or occurred more often at all ages compared to examiner-directed behaviors.

Stability of examiner-directed responses and mother-directed responses across the four age points was low. Correlations ranged from .00 to .30; the stronger correlations were found between adjacent ages rather than nonadjacent ages. For example, examiner-directed hypothesis testing showed negligible correlations between 14 months and all other ages ($r = .07$, .02, and .09 for 20, 24, and 36 months, respectively). But there was a significant but low correlation between 20 and 24 months ($r = .26$, $p < .001$) and between 24 and 36 months ($r = .23$, $p < .001$). The strongest stability was found for mother-directed concern, which had correlations ranging from .21 at 14 to 20 months, .30 from 20 to 24 months, and .10 from 24 to 36 months.

Table 19.1. Means and standard deviations of examiner-directed and mother-directed empathic responses within ages

Measure	Examiner, mean (SD)				Mother, mean (SD)			
	14 Months (n = 539)	20 Months (n = 487)	24 Months (n = 484)	36 Months (n = 469)	14 Months (n = 537)	20 Months (n = 488)	24 Months (n = 482)	36 Months (n = 464)
Approach	2.91	3.00	3.05	3.17	3.23	3.37	3.38	3.35
	(0.60)	(0.54)	(0.59)	(0.58)	(0.65)	(0.63)	(0.58)	(0.71)
Hypothesis testing	2.09	2.28	2.63	2.79	2.18	2.56	3.20	3.39
	(0.44)	(0.62)	(0.82)	(1.02)	(0.66)	(0.83)	(0.98)	(1.08)
Concern	2.29	2.47	2.51	2.57	2.16	2.48	2.45	2.50
	(0.53)	(0.50)	(0.47)	(0.50)	(0.61)	(0.58)	(0.55)	(0.59)
Arousal	1.86	1.75	1.67	1.63	2.02	1.86	1.96	1.91
	(0.43)	(0.45)	(0.38)	(0.45)	(0.44)	(0.48)	(0.42)	(0.41)
Prosocial acts	0.07	0.06	0.10	0.15	0.16	0.25	0.33	0.42
	(0.19)	(0.20)	(0.24)	(0.30)	(0.28)	(0.35)	(0.37)	(0.41)

Table 19.2. Correlation of mother-directed empathic responses with examiner-directed responses within ages

Measure	14 Months (n = 536)	20 Months (n = 485)	24 Months (n = 481)	36 Months (n = 458)
Approach	.10*	.10*	.02	.07
Hypothesis testing	.32**	.34**	.31**	.23**
Concern	.34**	.35**	.32**	.29**
Arousal	.32**	.49**	.24**	.27**
Prosocial acts	.09*	.19*	.20**	.21**

*p < .05; **p < .001.

Coherence of Responses across Victims

We correlated each behavioral dimension across victims (i.e., mother versus examiner) within ages to examine whether the coherence of children's behavioral responses differed over time. In general, we found a small increase in the correlation of mother-directed responses with examiner-directed responses from 14 to 20 months for four of the five behaviors (see table 19.2). However, approach showed little stability across victims at any age. At 36 months, the coherence between responses to examiner and mother was about the same magnitude as at 14 months, appearing slightly lower for some behaviors.

Heritability and Shared Environmental Influences

Twin correlations are presented in table 19.3, and heritability and common environment estimates were calculated for each behavior within age using the DeFries and Fulker (1985) regression method.

Examination of the pattern of twin correlations for examiner-directed responses suggests that there is greater resemblance among MZ than among DZ twins, especially at 36 months. The regression results confirm that there are little significant heritable or shared environment effects before 24 months of age for examiner-directed responses (see table 19.4). Indeed, all of the significant effects for examiner-directed behaviors were heritable influences. Noteworthy among these effects are the heritable effects on children's concern for examiner's distress at 14, 24, and 36 months. Hypothesis-testing behaviors directed at examiners at 24 and 36 months also had significant heritable influences. The level of heritability for hypothesis testing was particularly noteworthy ($h^2 = .44$).

Individual differences in mother-directed responses showed a different pattern of influences over time. A review of the pattern of twin correlations suggested that a predominance of DZ twin resemblance equaling or exceeding resemblance among MZ twins. At 14 months, a weak pattern of heritability for all five behaviors is suggested by the twin correlations but is born out in the

Table 19.3. MZ and DZ twin (intraclass) correlations

Measure	14 Months		20 Months		24 Months		36 Months	
	MZ	DZ	MZ	DZ	MZ	DZ	MZ	DZ
Examiner-directed responses								
Approach	.19*	.18	.07	.11	.07	.00	.16	.13
Hypothesis testing	.05	.13	.17*	−.10	.30**	.11	.50**	.04
Concern	.27**	.02	.30**	.26**	.33**	−.11	.29**	−.11
Arousal	.00	.02	.14	−.08	.10	.03	.20**	.16
Prosocial acts	.17*	−.05	.15	.12	−.09	.08	.20**	.13
Mother-directed responses								
Approach	.13	−.13	.25**	.04	.10	.36**	.06	.24*
Hypothesis testing	.20*	.12	.10	.36**	.35**	.27**	.29**	.25**
Concern	.19*	−.02	.09	.13	.15	.23*	.22*	.16
Arousal	.14	−.02	−.04	−.29**	.37**	.43**	.34**	.01
Prosocial acts	.37**	−.09	.09	.23*	.40**	.25*	.33**	.24*

*p < .05; **p < .01.

regression analyses only for prosocial acts (see table 19.4). Shared environmental influences were significant sources of individual differences for most mother-directed behaviors at 20 and 24 months of age. Concern in response to the mother's distress was significantly influenced by shared environment but not heritability at these two ages. Approach behaviors shifted from heritable influences at 20 months to shared environmental influences at 24 and 36 months, whereas hypothesis testing was associated with shared environment at 20 months. A final shift in the pattern of effects was observed for mother-directed behaviors at 36 months; low to moderate heritable and shared environment effects were seen across the five behaviors. Shared environment continued to exert a significant influence on approach and hypothesis testing; concern and arousal, however, had significant heritable effects.

DISCUSSION

The analyses presented in this chapter provide initial support consistent with our hypothesis that empathic responses are meaningfully differentiated by the relationship context in which they occur. Developmental trends showed increasing levels of all behaviors across time when the examiner was the victim; although trends for mother also generally increased, approach and concern levels remained fairly stable and high after 20 months of age. Arousal during mothers' simulations was the only behavior to decrease, which it did at 24 months compared to 14 months. This average drop-off is perhaps an intriguing

Table 19.4. Estimates of heritability (h^2) and common environmental (c^2) influences

Measure	Examiner-directed				Mother-directed			
	h^2	SE	c^2	SE	h^2	SE	c^2	SE
14 Months								
Approach	.02	.26	.17	.22	.09	.07	.00	.00
Hypothesis testing	.00	.00	.07	.06	.18	.26	.02	.22
Concern	.23**	.08	.00	.00	.13	.07	.00	.00
Arousal	.00	.00	.01	.06	.19	.08	.00	.00
Prosocial acts	.13	.07	.00	.00	.30**	.07	.00	.00
20 Months								
Approach	.00	.00	.09	.07	.22**	.08	.00	.00
Hypothesis testing	.10	.08	.00	.00	.00	.00	.22**	.06
Concern	.07	.25	.23	.21	.00	.00	.14*	.07
Arousal	.09	.08	.00	.00	.00	.00	.00	.00
Prosocial acts	.06	.26	.10	.21	.00	.00	.17*	.07
24 Months								
Approach	.06	.08	.00	.00	.00	.00	.21**	.06
Hypothesis testing	.29**	.08	.00	.00	.36	.25	.04	.21
Concern	.24**	.08	.00	.00	.00	.00	.21**	.06
Arousal	.10	.08	.00	.00	.00	.00	.37**	.06
Prosocial acts	.00	.00	.00	.00	.34	.25	.05	.21
36 Months								
Approach	.04	.27	.11	.21	.00	.00	.15*	.07
Hypothesis testing	.44**	.08	.00	.00	.00	.00	.30**	.06
Concern	.21**	.08	.00	.00	.22**	.08	.00	.00
Arousal	.07	.27	.13	.22	.25**	.08	.00	.00
Prosocial acts	.14	.27	.05	.22	.25	.26	.13	.21

*$p < .05$; **$p < .01$.

signal that children have become more emotionally defended in the presence of their mother's distress. This normative change may be a healthy sign that children, in general, may experience diminished arousal as they become more aware of mother as other and may rely more on language to resolve their uncertainty before becoming highly attuned with mother's distress.

Although there was a general developmental trend of increasing empathic responses, empathic concern and arousal were initially higher in relation to examiners at 14 months compared to mother-directed responses. Although at later ages mother-directed concern was roughly equivalent to that expressed toward examiners, arousal remained higher in relation to examiners than mothers throughout. Approach, hypothesis testing, and prosocial acts, however, were consistently higher in relation to the mother. This point is salient because attempts to assist the mother occurred at much higher rates than with the examiner, two to three times as often at every age. The motivation to help, as has been seen in many studies, is affected by the helpfulness of models, particularly when the model is familiar and loved and is in need.

Individual differences were preserved across victims at fairly stable levels from 14 to 36 months, as indicated by the intraclass correlations of examiner- and mother-directed behaviors. The slight drop in coherence of the behaviors at 36 months perhaps suggests a reorganization and differentiation of empathic response as a function of relationship to the victim. A further point suggesting the specificity of children's responses is the low correlation of the tendency to approach across mother and examiner at all four ages. It is interesting that the children most likely to approach examiners were not those most likely to approach their mother, although at every age children were more likely to approach their mother than examiner. One might speculate that the tendency to give comfort or aid to the mother was more common among those children that had a close mother-child relationship.

The interesting story in these data is the pattern of heritability and common environmental influence that emerged for behaviors directed at examiners versus mothers. Examiner-directed behaviors most often were associated with heritable influences, especially at 24 and 36 months, whereas mother-directed behaviors showed a mix of effects with shared environmental influences more evident at 20 and 24 months. Behaviors observed at 24 months showed the most contrast in sources of individual differences in relation to examiners and mothers, with heritability evident for levels of empathic concern (and hypothesis testing) directed toward examiners and shared environment for concern (and approach and arousal) directed toward mothers. This provides support for our hypothesis that relationships shape empathic involvement and may moderate the influence of genes and environment on behavior. The end of the second year is an important time in relation to emotion socialization, and the trend toward shared environmental sources of influence on the twins' behaviors at 20 and 24 months suggests a shaping of their responses by the mother.

The changing pattern of effects for mother-directed behaviors between 24 and 36 months from shared environment to a mix of heritable and shared environmental effects suggests that a reorganization of behavior may occur during this time. This is also supported by the emergence of heritability as an important influence on individual differences in hypothesis-testing behaviors directed toward the examiner at 24 and 36 months.

This investigation is limited by its reliance on a maximum of two observation opportunities for each construct, leading to lower reliability of estimates and less power to detect effects. In addition, the sample size constrains our ability to examine intriguing questions about gender specificity. Are girls more influenced by maternal socialization efforts in the second year? Are boys' empathic responses more influenced by heritability? An even larger sample of twins, using structural equation modeling, would be needed to adequately address these issues and to provide further tests of hypotheses suggested by the results presented here.

REFERENCES

Batson, C. D., Duncan, B. D., Ackerman, P., Buckley, T., & Birch, K. (1981). Is empathic emotion a source of altruistic motivation? *Journal of Personality and Social Psychology, 40,* 290–302.

Bengtsson, H., & Johnson, L. (1987). Cognitions related to empathy in 5 to 11 year-old children. *Child Development, 58,* 1001–1012.

Bretherton, I. (1985). Attachment theory: Retrospect and prospect. In I. Bretherton & E. Waters (Eds.), *Growing points of attachment theory and research: Monographs of the Society for Research in Child Development, 50* (1-2, Serial No. 209), 3–35.

Bretherton, I. (1987). New perspectives on attachment relations: Security, communication and internal working models. In J. Osofsky (Ed.), *Handbook of infant psychology* (2nd ed., pp. 1061–1100). New York: Wiley.

Cicchetti, D., Cummings, E. M., Greenberg, M. T., & Marvin, R. S. (1990). An organizational perspective on attachment beyond infancy: Implications for theory, measurement, and research. In M. T. Greenberg, D. Cicchetti, & E. M. Cummings (Eds.), *Attachment in the preschool years: Theory, research, and intervention* (pp. 3–49). Chicago: University of Chicago Press.

Cummings, E. M., Hollenbeck, B., Iannotti, R., Radke-Yarrow, M., & Zahn-Waxler, C. (1986). Early organization of altruism and aggression: Development patterns and individual differences. In C. Zahn-Waxler, E. M. Cummings, & R. Iannotti (Eds.), *Altruism and aggression: Biological and social origins* (pp. 165–188). New York: Cambridge University Press.

Davis, M. H. (1983). The effects of dispositional empathy on emotional reactions and helping: A multidimensional approach. *Journal of Personality, 51,* 167–184.

DeFries, J. C., & Fulker, D. W. (1985). Multiple regression analysis of twin data. *Behavior Genetics, 15,* 467–473.

Dovidio, J. F. (1984). Helping behavior and altruism: An empirical and conceptual overview. In L. Berkowitz (Ed.), *Advances in experimental social psychology* (Vol. 17, pp. 361–427). New York: Academic Press.

Eisenberg, N., & Mussen, P. H. (1989). *The roots of prosocial behavior in children.* Cambridge: Cambridge University Press.

Krebs, D. (1975). Empathy and altruism. *Journal of Personality and Social Psychology, 32,* 1134–1146.

Lamb, M. E. (1978). Interactions between eighteen-month-olds and their preschool-aged siblings. *Child Development, 49,* 51–59.

Main, M., & Weston, D. (1981). The quality of the toddler's relationship to mother and father: Related to conflict behavior and readiness to establish new relationships. *Child Development, 52,* 932–940.

Marvin, R. S., & Stewart, R. B. (1990). A family systems framework for the study of attachment. In M. T. Greenberg, D. Cicchetti, & E. M. Cummings (Eds.), *Attachment in the preschool years: Theory, research, and intervention.* Chicago: University of Chicago Press.

Radke-Yarrow, M., Zahn-Waxler, C., & Chapman, M. (1983). Children's prosocial dispositions and behavior. In P. H. Mussen & E. M. Hetherington (Eds.),
*Handbook of child psychology: Socialization, personality and social devel-
opment* (Vol. 4, pp. 469–545). New York: Wiley.

Robinson, J. L., Zahn-Waxler, C., Emde, R. N. (1994). Patterns of development
in early empathic behavior: Environmental and child constitutional influences. *Social Development, 3,* 125–145.

Willis, J. B., Feldman, N. S., & Ruble, D. N. (1977). Children's generosity as
influenced by deservedness of reward an type of recipient. *Journal of Educational Psychology, 69,* 33–35.

Zahn-Waxler, C., Cole, P., & Barrett, K. (1991). Guilt and empathy: Sex differences and implications for the development of depression. In K. Dodge &
J. Garber (Eds.), *Emotion regulation and dysregulation.* New York: Cambridge University Press.

Zahn-Waxler, C., & Radke-Yarrow, M. (1990). The origins of empathic concern.
Motivation and Emotion, 14, 107–130.

Zahn-Waxler, C., Radke-Yarrow, Wagner, E., & Chapman, M. (1992a). Development of concern for others. *Developmental Psychology, 28,* 126–136.

Zahn-Waxler, C., Robinson, J., & Emde, R. N. (1992b). The development of empathy in twins. *Developmental Psychology, 28,* 1038–1047.

20

The Development of Observed Shyness from 14 to 20 Months

Shyness in Context

Stacey S. Cherny
Kimberly J. Saudino
David W. Fulker
Robert Plomin
Robin P. Corley
John C. DeFries

There are many different temperament theories; however, most agree that temperament refers to biologically based, early-appearing behavioral dispositions that display consistency across time and situations (Goldsmith et al., 1987). The issue of consistency across time has been addressed in earlier chapters presenting longitudinal analyses of observer-rated and parent-rated temperament. We now turn to the issue of cross-situational consistency.

The present chapter focuses on shyness, employing videotaped observations of twins' behavior in both the laboratory and the home at 14 and 20 months of age. Multiple indicators of shyness in the laboratory and home at the two ages are available, forming laboratory and home composite measures of shyness at each age. Shyness is especially amenable to brief observations in the laboratory and home because shyness refers to initial social responding to unfamiliar persons. This is the natural situation when testers enter the home or when a child enters the laboratory.

Shyness is also interesting in terms of behavioral genetics for two reasons. First, it has been suggested that shyness is among the most heritable dimensions of temperament throughout the life span (Plomin and Daniels, 1986). Second, shyness is only modestly related to extraversion and neuroticism, the two dimensions of personality that have attracted most behavioral genetic research. Three previous infant twin studies suggest genetic influence for objective observations related to shyness. Using observations of shyness in the home, Plomin and Rowe (1979) found substantial genetic influence on social responding to a stranger. Goldsmith et al. (1999) report similar findings for twins' reactions to the approach of a stranger in the laboratory. An earlier report from the MacArthur Longitudinal Twin Study in the second year of life concerned behavioral inhibition and also provided relevant data. Behavioral inhibition was assessed by laboratory observations that included reluctance to approach an unfamiliar person or object and remaining close to the parent during such encounters (Robinson et al., 1992). Heritabilities for a composite index of behavioral inhibition were .53, .42, and .51 at 14, 20, and 24 months of age, respectively, with no evidence for shared environmental variance.

Behavioral observations of shyness in both the home and the laboratory permit novel analyses of the genetic underpinnings of the situational specificity of shyness. For example, does the magnitude of genetic influence on shyness differ across context or setting? A second, and perhaps less obvious, question involves the extent to which genetic effects on observed shyness in the home overlap with genetic effects on observed shyness in the laboratory. A trans-situational view of temperament would predict that what is common between the two contexts is heritable, even though the phenotypic correlation between shyness in the home and the laboratory may be modest. In contrast, a contextual view of temperament would be compatible with finding independent genetic influences on shyness in the home and the laboratory. It is also possible that some combination of these two opposing positions might apply to the temperament dimension of shyness.

We applied a cross-lagged panel design genetic model of observed shyness in the home and laboratory, allowing examination of genetic change and continuity from 14 to 20 months in each of the two shyness situations and their genetic and environmental relationships to each other. To our knowledge, the MacArthur Longitudinal Twin Study is the first longitudinal genetic investigation of shyness in infancy. We therefore used the cross-lagged panel design to address the aforementioned contextual issues relating to differential heritability and shared genetic effects from a developmental perspective. In other words, we sought to explore genetic and environmental contributions to the stability of shyness across age and developmental associations between shyness in one context at 14 months and shyness in another context at 20 months.

SUBJECTS AND MEASURES

Subjects were identical (monozygotic, MZ) and same-sex fraternal (dizygotic, DZ) twin pairs participating in the ongoing MacArthur Longitudinal Twin Study (Emde et al., 1992; Plomin et al., 1990) at the University of Colorado, Boulder. Longitudinal data from more than 350 MZ and DZ twin pairs tested at 14 and 20 months in the laboratory and the home are available on what is now the complete sample for these ages.

At both 14 and 20 months, videotaped records of the testers' entry into the twins' homes and the twins' entry into the laboratory formed the basis of the shyness ratings. At the home, two testers arrived at the front door; one greeted the mother and children while the other operated a videocamera. This entry procedure involved five 1-min intervals. In the first 2 min, the greeter offered the mother the consent form to read and conversed quietly with her. In the third minute, the greeter knelt on the floor and removed two high-interest toys from a bag and placed them on the floor in front of her, briefly demonstrating their use. In the fourth minute, she demonstrated the toys further and invited the children, by name, to play with them. In the fifth minute, if the children had not taken the toys, the greeter handed one to each child. Two identifying vests were removed from her bag and one put on each twin during this final minute of the procedure.

The laboratory visit typically occurred 1–2 weeks after the home visit. The children entered the reception area with their mother while a stationary videocamera recorded their activity. After removing any outerwear, the twins were free to explore a variety of high-interest toys on a low table while the examiner discussed the consent from and laboratory procedures with the mother. In the fourth minute, the examiner addressed the children by name, and in the fifth minute identifying vests were placed on each twin.

The videotaped episodes were subsequently rated for the occurrence of a number of discrete behaviors, the latency to approach the tester, and Likert-style scores of the child's initial response to the tester's addressing him or her, hesitation in responding to the tester, and a global evaluation of the child's shyness. Latency to approach, initial response to being addressed, and hesitation in responding were rated only once, whereas the discrete behaviors and the global ratings were done for each minute segment. A minimum of two segments had to be rateable for an average score to be given; the discrete behaviors and global shyness score were prorated for missing segments in creating a total score for the 5-min episodes. Six discrete behaviors were common to both contexts: (1) whether the child played with the tester-supplied toys, (2) whether the child was proximate to his or her mother, (3) whether the child clinged to his or her mother, (4) whether the child engaged in self-soothing behavior, (5) whether the child vocalized, and (6) whether the child cried. Three additional discrete behaviors were rated in the home situation: (7) whether the child was proximate to one of the testers, (8) whether the child touched his or her mother, and (9) whether the child played with his or her own familiar toys.

The latency in seconds, the child's initial response, the hesitation score, and the averaged global ratings and discrete behavior averages were then z-scored and subjected to a principal components factor analysis with the first component retained as the primary measure of shyness for that age and context. In each case, the averaged global rating of shyness was the measure that correlated highest with the first principal component (.90 for 14-month home, .92 for 14-month laboratory, .89 for 20-month home, and .92 for 20-month laboratory). The second-highest loading item was the average score for playing with the tester-supplied toys for three episodes; for the 20-month home visit this item was the third highest-loading, with the second-highest loading item the latency to approach the tester. The first principal components explained 33%,

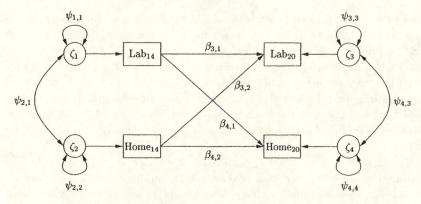

Figure 20.1. Cross-lagged panel model of shyness assessed in the laboratory and home at 14 and 20 months. The ζ are latent factors influencing observed shyness. Other symbols are described in the text. Paths with no coefficient are fixed at unity.

32%, 34% and 34% of the variance among the measures for laboratory and home assessments at 14 months and laboratory and home episodes at 20 months, respectively.

MULTIVARIATE MODEL

The cross-lagged panel model was fitted to these data. This model is often used with data on more than one measure or construct measured at multiple points in time. This model is shown in figure 20.1. The model is full-rank, completely explaining the observed covariance structure, as would, for example, the Cholesky decomposition introduced in chapter 2. However, this model allows us to determine the extent to which each situational measure of shyness at 14 months influences each measure of shyness and 20 months. For instance, we can determine whether the home or laboratory measure of shyness is a better predictor of home and laboratory shyness at 20 months.

The path diagram of the cross-lagged panel model, shown in figure 20.1, implies specific mathematical expectations for the covariance structure among the laboratory and home measures of shyness at 14 and 20 months. These expectations, given in Appendix Note 1, may either be derived from the path diagram, using path tracing rules (Li, 1975), or using matrix algebra and employing the LISREL model (Jöreskog and Sörbom, 1989).

GENETIC EXTENSION OF THE CROSS-LAGGED PANEL MODEL

With genetically informative data, such as those available on MZ and DZ twin pairs in the MacArthur Longitudinal Twin Study, we can examine the relationship among laboratory and home shyness at 14 and 20 months at both the environmental and the genetic levels in order to test if similar or different developmental process are operating at these different levels. Three sources of variation may be observed at each time point. The expectation for the correlation between MZ twin pairs is h^2, the proportion of variation due to additive differences in genotype among individuals, plus c^2, the proportion of variance due to environmental influences shared by members of a twin pair. For DZ twin pairs, the expected correlation is c^2 plus one half the genetic variance, h^2. A third component, e^2, is the proportion of variance due to environmental influences specific to individuals, and is equal to $1—h^2—c^2$. The straightforward extension of the cross-lagged panel model to genetically informative twin data is described in Appendix Note 2.

In summary, there are three versions of each of the parameters of the cross-legged panel model, one for each of the genetic, shared environmental, and nonshared environmental components. The model allows, for instance, the causal influence of laboratory shyness at 14 months on shyness measured in the home at 20 months to be a result of genetic or environmental influences, or any combination of those three sources of covariation.

FITTING THE MODELS

To test the adequacy of our models and to estimate the importance, or magnitude, of each type of influence, we compared the observed patterns of data, both among pairs of twins and across ages and measures, with those expected from the model. The parameters of the model can then be estimated by minimizing the discrepancy between the expected and the observed covariance structure, using optimization routines on the computer. Details of our model-fitting procedures are given in Appendix Note 3.

TWIN CORRELATIONS

Before doing any developmental modeling, examining the twin correlations is a useful first step. These correlations, at each age, are presented in table 20.1, along with the number of twin pairs on which these correlations were computed. It is clear that in all cases, the MZ correlations are higher than those of the DZs, implying genetic influence, although the magnitude of this difference is quite small for shyness assessed in the home at 20 months. Of course, simply examining twin correlations does not yield an optimal analysis of multivariate longitudinal data, but it is just a starting point. Of equal or greater interest are the cross-sibling, cross-measure correlations, and developmental models of them, such as the one described above, which will allow further dissection of the relationship between shyness assessed in the two contexts, laboratory and home, and at 14 and 20 months.

Table 20.1. Twin correlations for laboratory and home shyness at 14 and 20 months

Measure	MZ		DZ	
	r	N	r	N
14 Months				
Laboratory	.69	92	.48	79
Home	.63	192	.37	169
20 Months				
Laboratory	.77	83	.61	68
Home	.61	178	.58	149

$p < .0001$ for all correlations.

MODEL FITTING

Estimates of h^2, c^2 and e^2, obtained from fitting the full multivariate genetic model to the twin data, are presented in table 20.2. Heritability at 14 months was substantial, while the shared environment appeared to be the more salient component at 20 months. The proportion of variance attributable to the non-shared environment was relatively constant across all ages.

It should be noted that the genetic and shared environmental variance components estimated from fitting the full model to these data are somewhat different from what univariate analyses of the twin correlations at each age might yield. These differences arise because the multivariate models take into account the cross-twin, cross-measure covariance structure, which can impact the within-time parameter estimates. Use of the cross-twin, cross-measure information is an inherent advantage of a multivariate analysis. When samples sizes are small, the standard errors of the twin covariances are relatively large. However, the covariances between one twin on one measure and the other member of the pair and a different measure also contribute information to the within-measure twin covariances. This information would not be used in a

Table 20.2. Estimates of h^2, c^2, and e^2 for laboratory and home shyness at 14 and 20 months, obtained from fitting the full multivariate model

Variance component	14 Months		20 Months	
	Laboratory	Home	Laboratory	Home
h^2	.44	.51	.19	.09
c^2	.23	.12	.58	.53
e^2	.34	.37	.23	.39

Table 20.3. Tests of developmental patterns

Model	Form	$-2LL$[a]	NPAR[b]	χ^2	df	p
1.	Full Model	5451.392	46			
2.	Model 1, drop nonshared environmental $\beta_{3,2}$ and $\beta_{4,1}$	5451.480	44	0.088	2	$> .95$
3.	Model 2, drop nonshared environmental $\psi_{4,3}$	5465.520	43	14.040	1	$< .001$
4.	Model 2, drop nonshared environmental $\psi_{2,1}$	5455.008	43	3.528	1	$> .05$
5.	Model 4, drop nonshared environmental $\beta_{3,1}$ and $\beta_{4,2}$	5457.640	41	2.632	2	$> .25$
6.	Model 5, drop shared environmental $\beta_{3,2}$ and $\beta_{4,1}$	5457.716	39	0.076	2	$> .95$
7.	Model 6, drop shared environmental $\psi_{4,3}$	5459.516	38	1.800	1	$> .15$
8.	Model 7, drop shared environmental $\psi_{2,1}$	5463.283	37	3.767	1	$> .05$
9.	Model 8, drop shared environmental $\beta_{3,1}$ and $\beta_{4,2}$	5465.384	35	2.101	2	$> .30$
10.	Model 9, drop shared environmental $\psi_{1,1}$ and $\psi_{2,2}$	5466.418	33	1.034	2	$> .55$
11.	Model 10, drop shared environmental $\psi_{3,3}$ and $\psi_{4,4}$	5486.309	31	19.891	2	$< .001$
12.	Model 10, drop genetic $\beta_{4,1}$	5473.977	32	7.559	1	$< .01$
13.	Model 10, drop genetic $\beta_{3,2}$	5466.750	32	0.332	1	$> .55$
14.	Model 13, drop genetic $\psi_{4,3}$	5470.046	31	3.296	1	$> .05$
15.	Model 14, drop genetic $\psi_{2,1}$	5495.315	30	25.269	1	$< .001$
16.	Model 14, drop genetic $\beta_{3,1}$	5501.794	30	31.748	1	$< .001$
17.	Model 14, drop genetic $\beta_{4,2}$	5480.419	30	10.373	1	$< .002$
18.	Model 14, drop genetic $\psi_{4,4}$	5470.046	30	0.000	1	$> .999$
19.	Model 18, drop genetic $\psi_{3,3}$	5472.060	29	2.014	1	$> .15$

[a] $-2 \times$ log-likelihood of the data.
[b] Number of free parameters.

univariate analysis of each situation at each age; thus, resulting variance component estimates would have larger standard errors and might differ from those obtained from the multivariate analysis.

The next step was to systematically test components of the full model to determine which aspects of the model are essential for explaining these data and which are not. Tests of the nonshared environmental parameters were conducted first, as shown in table 20.3. We began by testing whether there was transmission over time across measures. That is, we tested whether nonshared environmental influences on laboratory-assessed shyness at 14 months had a direct effect on shyness observed in the home at 20 months and whether home shyness at 14 months had a direct effect on laboratory shyness at 20 months. The likelihood-ratio χ^2 test indicated that such influences are not present in these data (model 2). Next, we tested whether the covariance between nonshared environmental influences specific to laboratory and home shyness at 20 months was essential and found that there was a significant decrement in model fit if this component were omitted (model 3). We then tested the non-shared environmental covariance among the two measures of shyness at 14 months and found that such a covariance was not present in these data (model 4). The final nonshared environmental test was of the direct 14–20-month transmission of both laboratory and home shyness, and we conclude that such a process is not manifested in this data set (model 5). Model 5 was, therefore, chosen as the base model for the first test in the next series, that of the shared environmental processes. At the nonshared environmental level, there are variances specific to each measure of shyness at each age, which necessarily must be present, because these variances also contain measurement error, and, additionally, the nonshared environmental specific variances at 20 months are correlated. No other nonshared environmental correlations were present.

The shared environmental tests followed the same pattern as those performed on the nonshared environment, with the addition of tests of the measure-specific variances. The test of cross-measure age-to-age transmission indicated that at the shared environmental level, such a process is not a significant contributor to phenotypic covariance (model 6). The 20- and 14-month covariances were also found not necessary to explain the observed data (models 7 and 8). Age-to-age transmission of both the laboratory- and home-assessed measures of shyness was also not detected (model 9). Furthermore, no appreciable shared environmental variance was found at 14 months (model 10). However, the variances at 20 months were necessary to explain these data (model 11). We therefore chose model 10, one with only uncorrelated shared environmental variances present only at 20 months, as the base model for entry into tests of the last major component of covariation, the genetic structure.

Genetic transmission from the 14-month laboratory assessment of shyness to shyness observed in the home at 20 months was necessary to explain these data (model 12). The other cross-measure transmission parameter was not necessary, however (model 13). The 20-month covariance was also not needed (model 14). However, the 14-month covariance was necessary (model 15). Transmission of both laboratory (model 16) and home (model 17) shyness at 14 months to the respective measures at 20 months was necessary to explain the observed covariances structure. Finally, neither home (model 18) nor laboratory (model 19) new genetic variance at 20 months was found in these data. (We tested the home 20-month specific variance before the laboratory variance because the home variance was estimated at zero and so dropping it would not change the model fit.) That is, the genetic influences at 14 months were sufficient to account for the genetic variation observed at 20 months.

The final reduced model of shyness in the laboratory and home at 14 and 20 months is presented in figure 20.2. The observed variables are standardized to unit variance, using an extension of the procedures of LISREL (Jöreskog and Sörbom, 1989). This allows for the genetic factors and shared and nonshared environmental factors to have variances which, when summed, equal unity.

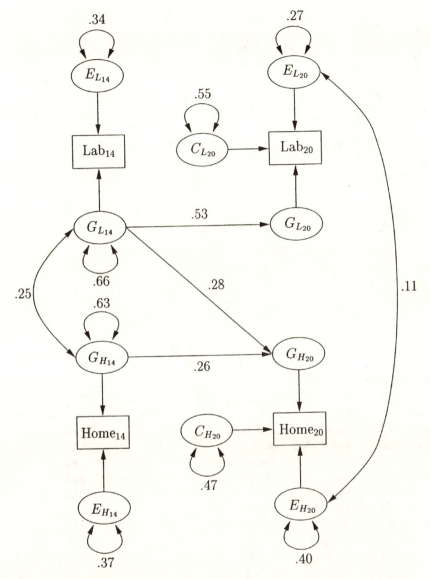

Figure 20.2. Final reduced model of shyness assessed in the laboratory and home at 14 and 20 months. Observed measures are shown in boxes; latent measures are shown in ovals. The *G*s in the ovals denote latent genetic factors, the *C*s denote latent shared environmental factors, and the *E*s denote latent nonshared environmental factors. Paths without coefficients are fixed at unity.

That is, their variances are also proportions of phenotypic variance. To recap, the nonshared environmental influences are, for the most part, measure specific, with only a single correlation between 20-month nonshared environmental influences. Shared environmental influences are only present at 20 months and do not contribute to the correlation between laboratory and home shyness. Finally, genetic influences on laboratory shyness at 14 months have a direct effect on shyness observed in both contexts at 20 months. Genetic influences on home shyness at 14 months only have a direct impact on home shyness assessed at the second time point. Laboratory and home shyness at 14 months are genetically correlated, and the genetic influences presented at the earlier age completely explain the genetic variance observed at 20 months.

DISCUSSION

The results from this first longitudinal twin study of objectively assessed shyness in infancy largely support the trans-situational view of genetic underpinnings of shyness. Genetic factors are primarily responsible both for stability from 14 to 20 months and for the phenotypic correlations between laboratory and home shyness at both ages. Moreover, the high genetic correlations between ages and situations suggest that there is substantial overlap in the genetic influences affecting the construct of shyness at both ages and in both situations. In other words, although shyness is not highly stable from 14 to 20 months and although laboratory and home measures of shyness are not highly correlated phenotypically, the same genetic effects are involved at both ages and in both situations. Consistent with a trans-situational view of temperament, what is stable across ages and situations is genetically mediated. However, the trans-situational view may not tell the whole story. At both ages, there is modest genetic influence on shyness that is unique to both the home and laboratory situations. Even though the covariance between ages and situations can be explained by genetic factors, there are significant genetic effects on shyness in the home that are independent of shyness assessed in the laboratory and vice versa. These findings hint that there may be some small contextual or situationally specific genetic influences on shyness in the home.

Developmental change from 14 to 20 months is mediated by the shared and nonshared environment. That is, the differences that arise across age can be attributed to environmental factors. In addition, at 14 months the differences between home and laboratory shyness were due to both genetic and nonshared environmental influences, whereas at 20 months of age, shared and nonshared environmental influences contribute to situational differences in shyness.

It is interesting to note that, unlike some parental rating measures, these objective measures of shyness in the laboratory and home show evidence for shared environmental influence in addition to genetic influence, at least at age

20 months. However, the nature of these shared environmental influences is puzzling because they are only detectable at 20 months and not at 14 months. The shared environmental influences might be due to observing both twins together in the laboratory and home. Observing both members of a twin pair together could create occasion-specific and situation-specific factors that are experienced similarly by both members of a twin pair but do not lead to co-variance across situations or occasions. The presence of such influences only at the later time point might be due to the emergence of self-perception by age 20 months, along with an appreciation of rules and standards (Kagan, 1981), all of which could result in self-evaluative anxiety as a new aspect of shyness as the child matures.

In conclusion, most temperament theories would predict that shyness, as a temperamental dimension, would be heritable and would display consistency across time and situations. The present analyses suggest that these are not independent predictions. That is, the stability of shyness across age and the associations between shyness in the laboratory and shyness at home arise because of shared genetic influences.

APPENDIX

NOTE 1

Using the LISREL approach, the cross-lagged panel model can be parameterized in two matrices, $\mathbf{B}$ and $\mathbf{\Psi}$. As shown on the path diagram, the $\mathbf{B}$ matrices contain the transmission parameters from the 14- to the 20-months measures, and the $\mathbf{\Psi}$ matrices contain the measure- and time-specific variances and co-variances. The two matrices are parameterized as follows:

$$\mathbf{B} = \begin{bmatrix} 0 & 0 & 0 & 0 \\ 0 & 0 & 0 & 0 \\ \beta_{3,1} & \beta_{3,2} & 0 & 0 \\ \beta_{4,1} & \beta_{4,2} & 0 & 0 \end{bmatrix} \tag{1}$$

and the symmetric matrix,

$$\mathbf{\Psi} = \begin{bmatrix} \psi_{1,1} & & & \\ \psi_{2,1} & \psi_{2,2} & & \\ 0 & 0 & \psi_{3,3} & \\ 0 & 0 & \psi_{4,3} & \psi_{4,4} \end{bmatrix}. \tag{2}$$

The free parameters are indicated by the appropriate Greek letter, and all other elements of the above matrices are fixed. The diagonal elements in ψ were constrained to be positive because negative variances would be nonsensical. The expectation for the phenotypic covariance structure, $\mathbf{\Sigma}$ is:

$$\mathbf{\Sigma} = (\mathbf{I} - \mathbf{B})^{-1}\mathbf{\Psi}(\mathbf{I} - \mathbf{B}')^{-1}. \tag{3}$$

Note 2

The extension of the cross-lagged panel model to genetically informative twin data involves increasing the number of parameter matrices and parameters threefold from the phenotypic model described above. Each parameter matrix has a genetic, shared environmental, and nonshared environmental analogue, with three corresponding expected covariance matrices for each of these three sources of individual differences. The MZ and DZ expected covariance matrices take a special form whereby they are partitioned into four equal quadrants. The top-left and bottom-right quadrants contain the within-pair variances and covariances, or the phenotypic variances and covariances. The other two quadrants contain the cross-twin variances and covariances. These are expected to differ between MZ and DZ twin pairs. The expected covariance matrices takes the following form:

$$\Sigma = \begin{bmatrix} \Sigma_G + \Sigma_C + \Sigma_E & r\Sigma_G + \Sigma_C \\ r\Sigma_G + \Sigma_C & \Sigma_G + \Sigma_C + \Sigma_E \end{bmatrix}, \tag{4}$$

where r is the genetic correlation, $r = 1$ for MZ pairs and .5 for DZ pairs, and the Σ_G, Σ_C, and Σ_E are the expected genetic, shared environmental, and nonshared environmental covariance matrices obtained from equation 3 computed using the component parameter matrices for each of these three sources of individual differences. When suitably standardized, these expected covariance matrices are analogous to the univariate h^2, c^2, and e^2 quantities discussed above.

Note 3

As in any longitudinal study, we do not have data on every single subject at each time point. In addition, we have more missing data for the laboratory than for the home measurements. To make optimal use of the data we have, we must fit the model directly to the raw data rather than to observed covariance matrices. A maximum-likelihood (ML) pedigree approach (Lange et al., 1976) was used to analyze the data in an optimal manner, allowing one to accommodate any pattern of missing data. We maximized the following ML pedigree log-likelihood (LL) function, using the freely available and highly flexible Mx modeling package (Neale, 1994):

$$LL = \sum_{i=1}^{N} [-\frac{1}{2} \ln |\Sigma_i| - \frac{1}{2}(x_i - \mu_i)' \Sigma_i^{-1}(x_i - \mu_i) + constant], \tag{5}$$

where χ_i = vector of scores for twin pair i; Σ_i = appropriate MZ or DZ expected covariance matrix; N = total number of twin pairs; μ_i = vector of expected MZ or DZ means; and where

$$2(LL_1 - LL_2) = \chi^2 \tag{6}$$

for testing the difference between two alternative models. The vector of means is modeled as a set of free papameters, allowing estimation of different means in MZs and DZs and for twin 1 and twin 2. Mean structure is not of primary interest in the present chapter, and so no theoretical model was imposed. However, use of the present fit function allows for such a possibility if modeling mean structure is of interest. Use of this fit function, as opposed to the more common ML function used by such programs as LISREL (Jöreskog and Sörbom, 1989) and EQS (Bentler, 1989) allows all the data to be analyzed. This method will recover population parameters provided that the missing data are "missing at random" (MAR), in the terminology of Little and Rubin (1987). That is, the data that are missing may be either a random subset of those that are observable (or "missing completely at random"; MCAR) or are missing based on scores which are observed at random (MAR). Use of the more common fit function for complete data would necessitate elimination of those sibling pairs from the analysis who were not measured at both time points and in both situations, which would mean losing information unnecessarily. Furthermore, had this been done, we would need to make the more stringent assumption that those data that were not obtained are MCAR and not merely MAR. In the case where there are no missing data, this pedigree function yields the same results as the ML function for covariance matrices.

REFERENCES

Bentler, P. M. (1989). *EQS structural equations program manual.* Los Angeles: BMDP Statistical Software.

Emde, R. N., Plomin, R., Robinson, J., Reznick, J. S., Campos, J., Corley, R., DeFries, J. C., Fulker, D. W., Kagan, J., & Zahn-Waxler, C. (1992). Temperament, emotion, and cognition at 14 months: The MacArthur Longitudinal Twin Study. *Child Development, 63*, 1437–1455.

Goldsmith, H. H., Buss, A. H., Plomin, R., Rothbart, M. K., Thomas, A., Chess, S., Hinde, R. A., & McCall, R. B. (1987). Roundtable: What is temperament? Four approaches. *Child Development, 58*, 505–529.

Goldsmith, H. H., Lemery, K. S., Buss, K. A., Campos, J. (1999). Genetic analyses of focal aspects of infant temperament. *Developmental Psychology*, 35, 972–985.

Jöreskog, K. G., & Sörbom, D. (1989). *LISREL 7: A guide to the program and applications* (2nd ed). Chicago SPSS, Inc.

Kagan, J. (1981). *The second year: The emergence of self-awareness.* Cambridge, MA: Harvard University Press.

Lange, K., Westlake, J., & Spence, M. A. (1976). Extensions to pedigree analysis: III. Variance components by the scoring method. *Annals of Human Genetics*, 39, 485–491.

Li, C. C. (1975). *Path analysis: A primer.* Pacific Grove, CA: Boxwood Press.

Little, R. J. A., & Rubin, D. B. (1987). *Statistical analysis with missing data.* New York: Wiley.

Neale, M. C. (1994). *Mx: Statistical modeling* (3rd ed.). Richmond, VA: Department of Psychiatry, Medical College of Virginia.

Plomin, R., Campos, J., Corley, R., Emde, R. N., Fulker, D. W., Kagan, J., Reznick, J. S., Robinson, J., Zahn-Waxler, C., & DeFries, J. C. (1990). Individual differences during the second year of life: The MacArthur Longitudinal

Twin Study. In J. Columbo & J. Fagan (Eds.), *Individual differences in infancy: Reliability, stability and predictability* (p. 431–455). Hilldale, NJ: Erlbaum.

Plomin, R, & Daniels, D. (1986). Genetics and shyness. In W. H. Jones, J. M. Cheek, & S. R. Briggs (Eds.), *Shyness* (p. 63–80). New York: Plenum Press.

Plomin, R., & Rowe, D. C. (1979). Genetic and environmental etiology of social behavior in infancy. *Developmental Psychology, 15*, 62–72.

Robinson, J. L., Kagan, J., Reznick, J. S., & Corley, R. (1992). The heritability of inhibited and uninhibited behavior: A twin study. *Developmental Psychology, 28*, 1030–1037.

21

Perinatal Effects on General Cognitive Ability, Weight, Height, and Temperament

Charles Martin
Scott L. Hershberger

The first 3 years of life are a time of rapid growth and development of children. How children grow and develop may be linked to factors operating at the time of delivery. The purpose of this chapter is to examine perinatal effects on the development of general cognitive ability, height, weight, and temperament at 14, 20, 24, and 36 months of age.

Undoubtedly the most examined perinatal variable, with respect to later development, has been birth weight. Significant deficits in general cognitive ability in infancy and childhood are found frequently between low and high birth-weight infants (Broman et al., 1975; Klein et al., 1989; McCormick, 1989; Scott, 1987). Low birth weight is commonly defined as <2500 g, although some use <1500 g as the criterion. Deleterious effects of a low birth-weight status have also been shown for more specific cognitive abilities, including visual memory (Rose et al., 1991) and visual-motor coordination (Levy-Shiff et al., 1994). An interesting exception to this occurs for speech and language development, for which no association is found with birth weight once cognitive ability and perinatal complications are controlled (Aram et al., 1991; Hirata et al., 1983). Although the effects of birth weight are most pronounced when low and normal birth-weight groups are compared, a relation between birth weight and cognitive performance in infants and children exists throughout the entire birth-weight range (Churchill, 1965; Willerman & Churchill, 1967).

Low birth weight has also been related to various noncognitive variables: low birth-weight children are more depressed, have less social acceptance, and poorer social skills in grade school (Hoy et al., 1992). Certainly, physical problems apparent at birth (e.g., intraventricular hemorrhage, hydrocephalus) are related to later cognitive deficits (Hack et al., 1991). Autism has also been related to a number of perinatal events, including malposition, general anesthesia, and amniotic fluid problems (Finegan & Quarrington, 1979) and Cesarean section and prolonged labor (Deykin & McMahon, 1980).

There is little dispute that a birth weight <2500 g is detrimental to later cognitive ability. Nonetheless, a problem commonly shared by studies examining the effects of low birth weight is the coexistence of other perinatal difficulties. In many studies, the extent to which these other factors (not low birth weight) may be the primary causes of later deficits in cognitive ability is unclear. For example, early gestational age (prematurity) is highly correlated with both low birth weight and deficits in cognitive ability (Brooks-Gunn et al., 1994; Saigal et al., 1991). There are many nonperinatal factors in the birth weight studies that also remain uncontrolled. Genotype, intrauterine environment, maternal age, initial and prenatal care are all factors influencing later cognitive ability.

For a more accurate picture of how birth weight influences cognition and other variables, these factors must be controlled. Monozygotic twins, because their of identical genes, are a valuable tool for genetic research. Twin studies provide the control that is lacking in many studies of the birth weight–cognitive ability relation. The use of twins for the analysis of this relation is essential because factors such as gestational age, intrauterine environment, and prenatal care are equated. Several cross-sectional twin studies have found a relation between birth weight and general cognitive ability in children throughout the entire range of birth weight, although the relation is more pronounced for low birth-weight infants (Babson et al., 1964; Churchill, 1965; Scarr, 1969; Willerman & Churchill, 1967).

Two longitudinal twin studies have provided valuable information concerning the relation between birth weight and cognitive ability. The first of these studies is the Louisville Twin Study (LTS; Wilson, 1983). In the LTS, approximately 450 pairs of monozygotic and dizygotic twins were tested with the Bayley Scales of Infant Development (Bayley, 1969) at 6, 12, 18, and 24 months, the Stanford-Binet (Terman & Merrill, 1973) at 3 years, and the Wechsler Preschool and Primary Scale of Intelligence (Wechsler, 1967) at 6 years. Across the 6 years, low birth-weight (< 1750 g) infants scored significantly lower on these tests of cognitive ability compared to normal birth-weight infants. However, low birth-weight infants from upper socioeconomic statuses ultimately scored comparably to the normal birth-weight infants; low birth-weight infants from lower socioeconomic statuses did not. In addition, in those pairs of monozygotic twins in which the weight of one twin significantly exceeded the weight of the other twin, by 6 years the cognitive performance of the cotwins was equivalent (Wilson, 1985).

A second longitudinal study is the Colorado Adoption Project (CAP), in which matched adopted and nonadopted infants and their families were evaluated at 12 and 24 months. No significant differences were found between low and normal birth-weight infants at either 12 or 24 months of age on the Bayley

Mental Development Index (MDI); however, differences were found at 12 months on the verbal skill scale and at 24 months on the spatial skill scale of the MDI (Plomin & DeFries, 1985). Gestational age was correlated with the MDI at 12 months but not at 24 months; the means–end and imitation scales of the MDI were also correlated with gestational age at 12 months; the spatial skill scale was correlated with gestational age at 24 months. No more significant associations than expected by chance were found between birth weight/gestational age and temperament/behavioral problems at 12 and 24 months.

Although the information from these two longitudinal studies is of great value, it should be emphasized that cognitive ability and temperament were examined in infants and children who represented the entire continuum of perinatal difficulties; for example, approximately 3% of the participants in the CAP had gestational ages <36 weeks, a percentage which is representative of the population. In the LTS, 14% of the participants had birth weights <1750 g; <1% had gestational ages <36 weeks. Even though a relation appears to exist between some perinatal events (e.g., birth weight, gestational age) and later development, the relation is much stronger for infants of low birth weight and early gestational age. Low birth weight and early gestational age are also frequent concomitants of other deleterious perinatal events such as malposition, making causal inferences between the individual perinatal events and later development difficult.

In the MacArthur Longitudinal Twin Study, twins were preferentially selected for normal range birth weight and health. Twins with very low birth weights (<1000g) and with very poor health (requiring >2 weeks hospitalization) were excluded from the study. Thus, unlike previous studies, the MacArthur Longitudinal Twin Study permits the longitudinal investigation of the relation between perinatal events and cognitive ability and temperament in a sample of infants whose perinatal environment was relatively normal with respect to birth weight and gestational age. Associations found between other perinatal events and cognitive ability and temperament, therefore, cannot be attributed to an abnormal gestation or low birth weight.

Measures

The perinatal measures used in the study included (1) Length of labor: How long (in hours) was labor, that is, from the time of the first contractions to delivery? (2) Gestational age: What was their (the twins') gestational age at birth? (3) Birth weight: What did the babies weigh at birth? (4) Abnormal conditions: Were there any abnormal conditions in the baby—anoxia, high bilirubin, tremors, convulsions, anatomical abnormalities? (5) Length of infants' hospital stay: Were either of the babies hospitalized beyond your stay at the hospital? How long? Were any (6) analgesics, (7) tranquilizers, (8) anaesthesia, or (9) general anaesthesia given during delivery? (10) Delivery type: Were the twins delivered vaginally or by Cesarian? (11) Womb position: Were the infants in a normal position or breech? and (12) Problems breathing: Were there any problems with the babies starting to breathe after delivery?

Measures of physical development at 14, 20, 24, and 36 months of age were weight (grams) and height (centimeters).

Temperament of the twins was assessed at 14, 20, 24, and 36 months using the six scales of the Colorado Childhood Temperament Inventory (CCTI; Rowe & Plomin, 1977)—emotionality, activity, persistence, soothability, shyness, and sociability; and the nine scales of the Toddler Temperament Survey (TTS; Fullard et al., 1984)—activity, rhythmicity, approach-withdrawal, adaptability, intensity, mood, persistence, distractibility, and threshold. Mothers were asked these questions at the 14-month interview.

General cognitive ability was measured using the Bayley MDI derived from the Bayley Scales of Infant Development (Bayley, 1969) at 14, 20, and 24 months and the Stanford-Binet IQ at 36 months (Terman & Merrill, 1973).

Height and weight at 14, 20, 24, and 36 months were measured in a laboratory setting as indicated in chapter 3.

RESULTS

Phenotypic analyses

As shown in table 21.1, sex differences were found for several of the perinatal variables. Specifically, males had a lower birth weight than females, males experienced abnormal perinatal conditions with a greater frequency than did females, males had longer hospital stays than females, local anesthesia was given more frequently during births of females, and general anesthesia was given more frequently during births of males. No other sex differences were found for the perinatal variables.

Associations between perinatal variables and general cognitive ability, temperament, height, and weight were examined at 14, 20, 24, and 36 months. Each of these variables measured at 14, 20, 24, and 36 months were regressed onto the set of 12 perinatal variables in a multiple regression. An examination of the regression equations revealed that birth weight was the only perinatal variable of the 12 to be significantly associated with later weight, height, and general cognitive ability (see table 21.2). Moreover, there was generally no association between the perinatal variables and temperament. Due to this lack of association, the temperament variables were not used in subsequent phenotypic analyses.

To examine the relation between birth weight and later weight, height, and general cognitive ability in greater detail, their relation was examined at each age separately. The infants were first classified into four groups based on their birth weight. The first group consisted of infants with birth weights of 2800 g or more. The second group of infants had birth weights ranging from 2500 to

Table 21.1. Perinatal variable (means ± SD)

Variable	Total	n	M	n	F	n	t
Length of labor (hr)	7.80 ± 9.10	331	7.58 ± 8.81	168	8.02 ± 9.41	163	0.46
Gestational age (weeks)	37.31 ± 2.32	381	37.19 ± 2.37	189	37.43 ± 2.27	192	1.04
Birth weight (g)	2563.23 ± 467.98	766	2521.01 ± 442.02	382	2605.68 ± 489.62	384	−2.51**
Problems breathing (%)	0.28 ± 0.45	768	0.31 ± .46	382	0.25 ± .43	386	−1.94
Abnormal conditions (e.g., anoxia)	0.38 ± 0.48	768	0.42 ± 0.49	382	0.34 ± 0.47	368	−2.28*
Length of infants' hospital stays (days)	3.11 ± 7.80	762	3.71 ± 9.08	378	2.52 ± 6.24	384	−2.12*
Analgesics given (%)	0.08 ± 0.27	758	0.06 ± 2.4	378	0.09 ± 0.29	380	1.62
Tranquilizers given (%)	0.01 ± 0.09	758	0.01 ± 0.07	378	0.01 ± 0.10	380	0.81
Local anesthesia given (%)	0.61 ± 0.49	758	0.58 ± 0.49	378	0.65 ± 0.48	380	2.07*
General anesthesia given (%)	0.09 ± 0.29	758	0.12 ± 378	378	0.07 ± 0.25	380	−2.51**
Delivery type (vaginal)	0.47 ± 0.50	768	0.45 ± 0.50	388	0.49 ± 0.50	386	1.38
Womb position (normal)	0.68 ± 0.47	752	0.65 ± 0.48	374	0.71 ± 0.45	378	1.67

$*p < .05; **p < .01.$

Table 21.2. Regression coefficients using birth weight (g) to predict selected variables at 14, 20, 24, and 36 months of age

Variable	14 Months	20 Months	24 Months	36 Months
Weight (g)	33.17**	32.89**	33.74**	44.23**
Height (cm)	.47**	.46**	.59**	.44**
General cognitive ability	.16*	.12*	.16*	0.04

$*p < .05; **p < .01; ***p < .001.$

Table 21.3. Correlation of birth weight with general cognitive ability (mean) at 14, 20, 24, and 36 months of age

Birth weight	Total n	14 Months			20 Months			24 Months			36 Months		
		GCA	n	r	GCA	n	r	GCA	n	r	GCA	n	r
2800 + g	249	107.6	124	.07	108.37	115	.06	111.26	113	.15	104.94	103	.01
2500–2800 g	173	105.19	80	.20	105.96	67	.11	105.39	66	.10	100.75	68	.01
2200–2500 g	177	104.22	87	.06	104.17	78	.30**	106.49	84	.09	103.88	74	.20
<2200 g	167	99.75	83	.22*	99.76	71	.22	105.24	72	.24*	103.08	61	.13
F		5.89***			3.89*			2.30			.82		
df		(3,370)			(3,329)			(3,331)			(3,302)		
Total sample r		.24			.21			.17			.05		

*$p < .05$; **$p < .01$; ***$p < .001$.

2800 g. The third group was composed of infants with birth weights ranging from 2200 to 2500 g. The fourth group consisted of infants with birth weights <2200 g. Within each of the four birth-weight groups, the correlation between birth weight and later weight, height, and general cognitive ability was computed at each age.

A significant difference was present among the four birth-weight groups with respect to later weight at each of these ages. This was shown both by the significant F-ratio computed at each age, as well as the significant, overall correlation between birth weight and weight at 14, 20, 24, and 36 months. Of greatest interest was the source of this significant, overall correlation: It appeared to be induced only in the 2800 + g birth-weight group. Birth weight was significantly correlated with later weight at each age only for those infants in the 2800 + group. However, there appeared to be no pattern to the magnitude of the correlations: the correlation increased from .23 at 14 months to .40 at 20 months, then decreased to .33 at 24 and to .27 at 36 months. Further, birth weight within the <2200 g group was significantly correlated with height only at 14 months. There were no other significant correlations between birth weight and height conditional on birth-weight group.

The correlational pattern is less clear for general cognitive ability, as shown in table 21.3. Birth weight was correlated with general cognitive ability at 14 months within the <2200 g group; it was also correlated with general cognitive ability at 20 and 24 months within the 2200–2500 g group. The overall correlation between birth weight and general cognitive ability was also significant at 14, 20, and 24 months. Neither the group conditional or unconditional correlations were significant at 36 months. Further, the significant F-ratio at 14, 20, and 24 months implies that general cognitive ability differences exist among the four birth-weight groups at these ages but that this difference disappears by 36 months (when the F-ratio is nonsignificant).

To summarize, the only significant perinatal predictor of later weight, height, and general cognitive ability is birth weight. Not surprisingly, birth weight most strongly predicts later weight (especially for those infants weighing >2800 g at birth). Temperament at 14, 20, 24, and 36 months is not predictable from any of the perinatal variables.

Behavioral Genetic Analyses

It is possible that the significant associations between birth weight and later weight, height, and general cognitive ability are due to common genetic influences and/or common environmental influences. One indication that nongenetic or environmental influences mediate the relation between birth weight and these variables is a significant correlation between monozygotic intrapair differences in birth weight and intrapair differences in later weight, height, and general cognitive ability. Because monozygotic twins share all their genes, any phenotypic differences found between them must be due to environmental differences. Thus, a significant correlation between a birth weight difference score and a height difference score implies that whatever environmental effects influence birth weight to some extent also influence height. Conversely, a nonsignificant correlation between two difference scores computed within monozygotic twins implies that if the correlation between the (nondifference) scores of these variables is significant, common genetic effects mediate the relation. Thus, difference score correlations were computed within monozygotic twins between birth weight and later weight, height, and general cognitive ability at 14, 20, 24, and 36 months. The only difference score correlations that were significant were between birth weight and weight at 20 and 24 months of age. Therefore, with the exception of later weight, it may be concluded that environmental effects do not significantly mediate the relation between birth weight and height or general cognitive ability. Not surprisingly, in light of their previously reported nonsignificant correlation with birth weight, the difference score correlations for the temperament variables were not significant.

DISCUSSION

Perhaps the most significant result of this study is the significant relation that exists between birth weight and weight, height, and general cognitive ability at later time periods, within a sample that intentionally excluded infants whose birth weights and gestational ages did not fall within the normal range. This study replicates and extends others by confirming that even within the normal range of birth weight and gestational age, birth weight is still a predictor of later general cognitive ability. By 36 months, however, this relation disappears. Later weight is significantly predictable even at 36 months, although this relation is strongest for the heaviest birth-weight infants (>2800 g). Contrary to the height–weight association found in adults, birth weight and height are not associated throughout most of the period studied here, with the exception of 14 months.

It is somewhat surprising that birth weight was the only perinatal factor significantly related to variables measured during later infancy. The most likely explanation for this is the normalcy of the sample studied. Unusually low birth weights are a correlate of other perinatal difficulties; once infants are selected with birth weights in the normal range, the likelihood of deleterious perinatal events decreases. For example, as reported in table 21.1, less than half the infants (38%) experienced any sort of abnormal condition at birth. This is certainly not the case for samples of infants with unusually low birth weights, in which the majority experience some type of abnormal condition at birth. Other than birth weight, of all the perinatal factors examined in this study, low gestational age has been found to be a predictor of later cognitive difficulties. Yet, because this study preferentially included infants within a normal range of gestational age, no relation with later general cognitive ability was expected, and none was found. None of the perinatal variables was associated with later temperament. This is consistent with previous work that has found perinatal difficulty to be associated not with personality characteristics per se, but with inadequate social and interpersonal skills (e.g., Hoy et al., 1992).

REFERENCES

Aram, D. M., Hack, M., Hawkins, S., Weissman, B. M., & Borawski-Clark, E. (1991). Very-low-birthweight children and speech and language development. *Journal of Speech and Hearing Research, 34,* 1169–1179.

Babson, S., Kangas, J., Young, N., & Bramhill, J. (1964). Growth and development of twins of dissimilar size at birth. *Pediatrics, 30,* 327–333.

Bayley, N. (1969). *Bayley Scales of Infant Development.* New York: Psychological Corporation.

Broman, S. H., Nichols, P. L., & Kennedy, W. A. (1975). *Preschool IQ: Prenatal and early development correlates.* Hillsdale, NJ: Erlbaum.

Brooks-Gunn, J. et al. (1994). Early intervention in low-birth-weight premature infants. *Journal of the American Medical Association, 272,* 1257–1262.

Churchill, J. A. (1965). The relationship between intelligence and birth weight in twins. *Neurology, 15,* 341–347.

Deykin, E. Y., & MacMahon, B. (1980). Pregnancy, delivery, and neonatal complications among autistic children. *American Journal of Diseases of Children, 134,* 860–864.

Finegan, J., & Quarrington, B. (1979). Pre-, peri-, and neonatal factors and infantile autism. *Journal of Child Psychology and Psychiatry, 20,* 119–128.

Fullard, W., McDevitt, S. C., & Carey, W. B. (1984). Assessing temperament in 1 to 3-year-old children. *Journal of Pediatric Psychology, 9,* 205–217.

Hack, M., Breslau, N., Weissman, B., Aram, D., Klein, N., & Borawski, E. (1991). Effect of very low birth weight and subnormal head size on cognitive abilities at school age. *New England Journal of Medicine, 325,* 231–237.

Hirata, T., Epcar, J. T., Walsh, A., Mednick, J., Harris, M., McGinnis, M. S., Sehring, S., & Papedo, G. (1983). Survival and outcome of infants 501 to 750 gm: A six-year experience. *Journal of Pediatrics, 102,* 741–749.

Hoy, E. A., Sykes, D. H., Bill, J. M., Halliday, H. L., McClure, B. G., & Reid, M. M. (1992). The social competence of very-low-birthweight children: Teacher, peer, and self-perceptions. *Journal of Abnormal Child Psychology, 20,* 123–150.

Klein, N. K., Hack, M., & Breslau, N. (1989). Children who were very low birthweight: Development and academic achievement at nine years of age. *Development Behavioral Pediatrics, 10,* 32–37.

Levy-Shiff, R., Einat, G., Mogilner, M. B., Lerman, M., & Krikler, R. (1994). Biological and environmental correlates of developmental outcome of prematurely born infants in early adolescence. *Journal of Pediatric Psychology, 19,* 63–78.

McCormick, M. C. (1989). Long-term follow-up of infants discharged from neonatal intensive care units. *Journal of the American Medical Association, 261,* 2–31.

Plomin, R., & DeFries, J. C. (1985). *Origins of individual differences in infancy: The Colorado Adoption Project.* Orlando, FL: Academic Press.

Rose, S. A., Feldman, J. F., Wallace, I. F., & McCarton, C. (1991). Information processing at 1 year: Relation to birth status and development outcome during the first 5 years. *Developmental Psychology, 27,* 723–737.

Rowe, D. C., & Plomin, R. (1977). Temperament in early childhood. *Journal of Personality Assessment, 41,* 150–166.

Saigal, S., Rosenbaum, P., Szatmari, P., & Campbell, D. (1991). Learning disabilities and school problems in a regional cohort of extremely low birth weight (< 1000 g) children: A comparison with term controls. *Developmental and Behavioral Pediatrics, 12,* 294–300.

Scarr, S. (1969). Effects of birth weight on later intelligence. *Social Biology, 16,* 249–256.

Scott, D. T. (1987). Premature infants in later childhood: Some recent follow-up results. *Seminar in Perinatology, 11,* 191–199.

Terman, L. M., & Merrill, M. A. (1973). *Stanford-Binet Intelligence Scale: 1972 norms edition.* Boston: Houghton-Mifflin.

Wechsler, D. A. (1967). *Wechsler Preschool and Primary Scale of Intelligence.* New York: Psychological Corporation.

Willerman, L., & Churchill, J. A. (1967). Intelligence and birth weight in identical twins. *Child Development, 38,* 623–629.

Wilson, R. S. (1983). The Louisville twin study: Developmental synchronies in behavior. *Child Development, 54,* 298–316.

Wilson, R. S. (1985). Risk and resilience in early mental development. *Developmental Psychology, 21,* 795–805.

22

Height, Weight, and Body Mass Index

Marcie L. Chambers
John K. Hewitt
Stephanie Schmitz
Robin P. Corley
David W. Fulker

Being overweight or obese is a health problem of increasing concern in industrialized societies. The prevalence of being overweight in the United States appears to be rising both in adults (Kuczmarski et al., 1994) and in children (Gortmaker et al., 1987). Obesity in childhood is predictive of obesity in adulthood (Rolland-Cachera et al., 1987), and being overweight or obese is related to morbidity for a variety of diseases, including non-insulin-dependent diabetes, hypertension, and coronary heart disease (Hubert et al., 1983; Pi-Sunyer, 1991; Van Itallie, 1985;) as well as being associated with reduced longevity (Lew, 1985; Lew and Garfinkel, 1979).

In addition to physical health risks associated with being overweight, there are some psychosocial consequences for overweight children and adults. In their review, Wadden and Stunkard (1993) cite studies that show that even quite young children characterize drawings of an overweight or obese child in negative terms (e.g., "lazy") or as being a child they would prefer not to play with (Goodman et al., 1963; Maddox et al., 1968; Staffierie, 1967). Some obese individuals suffer from excessive self-disparagement in relation to their weight, and this problem is more likely to be severe in individuals who are obese from childhood (Stunkard and Burt, 1967; Stunkard and Mendelson, 1967). Thus, whether or not a child is heavy for his or her height may have implications for both physical and mental well being either as a child or later in life. One contribution a study such as ours can make is to provide a greater understanding of the causes, both genetic and environmental, of the early development of individual differences in height and weight and the health-related weight-for-height measure, the body mass index (BMI).

GENETIC STUDIES OF ADULT HEIGHT, WEIGHT, AND BODY MASS INDEX

The genetic contribution to height, weight, and BMI in adults has been the focus of numerous studies (Price et al., 1987; Sorensen et al., 1989; Stunkard et al., 1986a,b). Adoption and twin studies have played a key role in determining what proportion of the variance in height, weight, and BMI can be attributed to genetic influences and what proportion can be attributed to the environment.

In adoption studies, correlations for adoptive and nonadoptive siblings are often used to estimate heritability and the influence of the environment. Biological siblings share approximately 50% of their segregating genes, adoptive siblings share none. If genetic influences affect a trait, the correlation for biological siblings will exceed the correlation for adoptive siblings, and the adopted children will come to resemble their biological parents to some extent. Data from genetically unrelated individuals raised in the same environment provide a direct assessment of the contribution of the shared or common family environment (Plomin et al., 1988).

Adoption studies have indicated that genes play a significant role in the BMI of adults. Sorensen et al. (1989) conducted an adoption study in which male and female adoptees, separated from their biological parents at an early age, were compared to their biological full and half siblings. The authors reported that as the weight groups (thin, medium, overweight, and obese) of the adoptees increased, the BMI of their full siblings ($n = 115$) also increased significantly. Their half siblings ($n = 850$) showed a similar, but weaker relationship. These results indicate a genetic influence on human body fat over the entire range of thinness to obesity.

Price et al. (1987) conducted an adoption study and found that the BMI of biological mothers ($n = 123$) was significantly correlated with that of their daughters ($n = 172$, $r = .40$, $p < .001$). A trend was reported for the BMI of the biological fathers to correlate with their daughters, but it was not significant ($r = .18$, $p < .16$). This lack of significance may be due to the small sample size of fathers ($n = 62$) and possible errors in paternal weight, most of which were reported by the biological mothers. The correlations between the BMI of biological parents and their sons ($n = 185$) were positive, but not statistically significant. The correlations between adoptive parents and adoptees were low and nonsignificant.

Stunkard et al. (1986b) examined adoption data of 540 adult Danish adoptees and found a strong relationship between the weight class of adoptees and the BMI of their biological parents. The relationship was found both for mothers and their biological children ($p < .0001$) and for fathers and their biological children ($p < .02$). They did not find a significant relationship between the weight class of adoptees and the BMI of their adoptive parents. It was concluded that genes played an important role in determining fatness in adults but that family environment has no apparent effect.

In twin studies, genetic influence is assessed by comparing the resemblance of identical twins (monozygotic; MZ) to fraternal twins (dizygotic; DZ). If, for example, heredity is the sole familial factor that affects height, the correlation for height in MZ twins should be approximately twice that of the DZ twins. This is because MZ twins share 100% of their segregating genes, whereas DZ twins share, on average, 50%.

Stunkard et al. (1986a) conducted a twin study of adult obesity. This study assessed height, weight, and BMI for male MZ and DZ twins at two time points, first when the average age was 20 (MZ pairs, $n = 5884$, DZ pairs, $n = 7492$) and then again at a 25-year followup (MZ, $n = 1974$, DZ, $n = 2097$). They reported that fatness is under strong genetic control. Estimates of heritability for BMI were .77 at the initial assessment and increased to .84 25 years later. Height and weight showed similar results, with heritability for height estimated at .80 at both time points and heritability for weight estimated at .78 at the initial assessment and .81 at the 25-year follow-up.

CHILDHOOD STUDIES

Do the genetic and environmental influences that affect these traits remain constant over a lifetime, or do they differ from childhood to adulthood? Children have been examined, but to a lesser extent than adults, with several attempts being made to partition the phenotypic variance of height, weight, and BMI into genetic and environmental components (Biron et al., 1977; Cardon, 1994; Phillips & Matheny, 1990).

Biron et al. (1977) examined weight and weight/height (weight divided by height) in 374 Canadian families to assess whether the shared environment contributed to the familial resemblance of weight in children aged 1–21 years. In their sample, 535 children were adopted and 250 children were biological. The authors reported the mid-parent (parents' average score)–biological child r^2 to be .096 for weight and .066 for weight/height. The mid-parent–adopted child r^2 was 0.00. The sibling correlation in 80 homes with more than one biological child was .152 for weight and .135 for weight/height ($p < .001$). Where there was more than one adoptee in a home, the adoptee–adoptee correlations were zero for weight and for weight/height. They concluded that heredity explains most of the familial aggregation patterns of weight and weight/height in children.

Cardon (1994) analyzed sibling and parent–offspring resemblance in subjects from the Colorado Adoption Project (CAP). The purpose of this study was to characterize the pattern of influences for genetic and environmental effects on BMI in children from birth to age nine. The sibling analysis indicated that there is an overall trend of cumulative genetic influence throughout early and middle childhood. Parent–offspring heritability estimates ranged from .01 to .09 from birth to age two and then increased from age three to age nine to between .37 and .57.

Phillips and Matheny (1990) examined height in children from the Louisville Twin Study from birth to 18 years of age. Heritability of height from 6 years on was estimated at about .90. Shared environmental effects (unrelated to parental stature) were found only for birth length and height through 3 years. The results indicated a new genetic influence manifesting between birth and 3 years that continues through the growth period into maturity. Heritability for height increased from birth to 3 years from .38 to .81 and .54 to .82 in males and females, respectively. Sex differences were also examined, and the only difference was found at birth, with heritability accounting for 38% of the variance in females and 54% of the variance in males. Common environmental influences at birth accounted for 25% of the variance in females and 18% in males.

Frisancho et al. (1980) examined the growth in height of 1202 Peruvian Quechua and Mestizo children aged 6–9 years. Evaluations of ABO blood type, Rh systems, and skin reflectance measurements showed the Quechua children to be genetically different from the Mestizo children. During childhood, the two groups of children, matched for nutritional status, attained similar heights. After age 11 however, the Mestizos grew significantly taller. This finding suggests that environmental factors, such as nutrition, have a greater influence on height during childhood than during adolescence and, conversely, that the influence of genetic factors on height is greater during adolescence.

Thus, several studies, using different approaches, have suggested that the heritability of height, weight, or BMI increases during childhood. Phillips and Matheny (1990) suggested that there are new genetic influences on height between birth and 3 years, and Cardon (1994) also reported an increase in BMI heritability between age 2 and 3 years. In this chapter we examine the genetic and environmental influences during this early period of development more closely and explore the possibility that BMI may be related to temperament or psychosocial variables in early childhood.

MEASURES

Twins were assessed at 14, 20, 24, and 36 months of age in their homes and during laboratory visits. Height was measured while the child was in a prone position from 14 to 24 months. The child's mother held a tape measure at the top of the child's head while the tester measured to the feet. Height was measured while standing against a wall or door at age three. Weight was obtained in the laboratory at each time point using a baby scale. Body mass index was used as a measure of fatness because it approximates laboratory measurements of body fat (Price et al., 1990). BMI was calculated as weight (in kilograms) divided by height (in meters) squared.

MZ and DZ twins were compared to determine if the correlation for MZ twins exceeded that of DZ twins. Height, weight, and BMI correlations were calculated for each group at all four time points.

MODELS

Using a full-rank Cholesky decomposition model (Neale & Cardon, 1992; see chapter 2, this volume) phenotypic variance in height, weight, and BMI was partitioned into additive genetic (A), shared or common environment (C), and nonshared environmental (E) components (see Appendix for details of the mathematical model).

The full model and seven submodels were tested using a chi-square test statistic for each of the three variables (height, weight, BMI). The first model (ACE) is the full model where additive genetic, common environment, and individual environmental components load on each time point. The second model (AE) constrains the common environmental component to be zero, and the third model (CE) constrains the additive genetic component to be zero. These models test the significance of the overall common environmental and genetic influences.

The fourth and sixth models force the additive genetic and common environmental components to be represented by a single common factor (A_1 and C_1, respectively). The fifth, seventh, and eighth models force the additive genetic, common environment and individual environments to be age specific (A_d, C_d, E_d). The chi-square statistics for each of these submodels were compared to the full model to obtain the best-fitting model. In all cases the model chosen was also the best according to the Akaike Information Criterion ($\chi^2 - 2$ degrees of freedom).

RESULTS

Descriptive Statistics

The mean height, weight, and BMI, as well as the twin correlations, are presented in table 22.1. Height and weight correlations were considerably higher for MZ twins than for DZ twins. For BMI, MZ twins had higher correlations than DZ twins, the difference being the least at 14 months and the greatest at 36 months. The difference in the correlations between MZ and DZ twins for height, weight, and BMI indicates genetic influences for all three phenotypes.

Height

The full Cholesky model for height had a χ^2 of 54.25 (42 df, $p = .098$), which indicates a moderate fit. The results of the model comparisons for height data are shown in table 22.2. In this table, A represents the additive genetic effects, C represents the common environmental effects, and E represents the individual environmental effects. The subscript 1 indicates that a component (A, C, or E) is forced to be represented by a single common factor. For example, in model 4, the additive genetic component is forced to be represented by a single

Table 22.1. Descriptive statistics for height, weight, and body mass index (BMI)

Age	MZ				DZ			
	n	Mean	SD	r	n	Mean	SD	r
Height (inches)								
14 Months	197	30.06	1.49	.82	168	30.30	1.24	.42
20 Months	179	32.58	1.54	.81	152	32.86	1.36	.47
24 Months	167	34.04	1.52	.82	147	34.47	1.59	.60
36 Months	146	36.83	1.57	.94	130	37.29	1.45	.58
Weight (lbs)								
14 Months	157	21.87	2.57	.89	134	22.26	2.39	.50
20 Months	143	24.78	2.87	.90	124	25.15	2.53	.56
24 Months	147	26.44	3.18	.93	139	26.82	2.92	.55
36 Months	134	30.19	3.89	.91	124	31.06	3.19	.46
BMI (weight in kg/height in meters)^2								
14 Months	156	16.97	1.53	.69	131	17.11	1.74	.50
20 Months	143	16.43	1.61	.81	121	16.25	1.42	.63
24 Months	136	15.92	1.51	.83	117	15.85	1.51	.64
36 Months	117	15.62	1.34	.82	105	15.69	1.37	.41

All correlations (r) are statistically significant at p-.01 level.

Table 22.2. Maximum likelihood test statistics for height model comparisons

Model	χ^2	df	p	Models compared	χ^2 Diff	df	p
1. ACE	54.24	42	.098				
2. AE	85.98	52	.002	1 vs. 2	31.74	10	<.001
3. CE	196.49	52	.000	1 vs. 3	142.25	10	<.001
4. A_1CE	85.12	48	.001	1 vs. 4	30.88	6	<.001
5. A_dCE	132.64	48	.000	1 vs. 5	78.40	6	<.001
6. AC_1E	65.79	48	.045	1 vs. 6	11.55	6	>.05
7. AC_dE^a	61.30	48	.094	1 vs. 7	7.06	6	>.09
8. ACE_d	108.54	48	.000	1 vs. 8	54.30	6	<.001

A = additive genetic component; C = common environmental component; E = individual environmental component. Subscript 1 denotes a single common factor. Subscript d denotes age-specific factors. χ^2 Diff is the difference between the chi-squares of model 1 and the model with which it is being compared.
[a] Best-fitting model chosen.

Table 22.3. Genetic and environmental parameter estimates for height

Age	h^2					c^2					e^2				
	A_1	A_2	A_3	A_4	Total	C_1	C_2	C_3	C_4	Total	E_1	E_2	E_3	E_4	Total
14 Months	.78				.78	.00				.00	.22				.22
20 Months	.38	.23			.61		.18			.18	.00	.21			.21
24 Months	.36	.29	.00		.65			.17		.17	.02	.00	.16		.18
36 Months	.35	.31	.16	.00	.82				.12	.12	.00	.00	.00	.06	.06

h^2 = heritability; c^2 = common environment; e^2 = individual environment.

common factor. The d subscript indicates that a component is forced to be represented by age specific factors (d for a diagonal matrix). In model 5, the additive genetic component is forced to be diagonal. As indicated by the chi-square difference and p difference values of models 2 and 3 ($p < .001$ when common environment and genetic influences are constrained to zero), the additive genetic and the common environmental components are both needed for the best-fitting model. Constraining additive genetic and individual environmental effects to either a single common factor (model 4) or an age-specific model (models 5 and 8) leads to a worse-fitting model ($p < .001$). Genetic and individual environmental effects need to be represented by a full model. Common environment, however, is best represented by an age-specific model (model 7; $p > .30$). This indicates that, although new and residual genetic and individual environmental effects are important in determining height, common environmental effects are occasion specific.

After the best fitting model had been determined, parameters were estimated for each component, as shown in table 22.3. This analysis indicates that genetic influences account for the majority of the variation in height at each time point. Between 61% and 82% of the variation can be accounted for by genes. Most of the genetic influence is present at 14 months and has a residual effect on the other time points. There is new genetic influence coming in at 20 months that also affects the remainder of the time points. The common environment accounts for none to 18% of the variance in height. Individual environmental influences account for 22% of the variance at 14 months and decreases to 6% at 36 months.

Weight

The full Cholesky model for weight resulted in a χ^2 of 175.89 (df = 42, p = .000), which is highly significant and implies a very poor fit to the data. Table 22.4 shows the results of the model comparisons. Constraining the common environmental component (model 2) and the additive genetic component (model 3) to be zero resulted in worse-fitting models ($p < .05$ and $p < .001$, respectively). Forcing the additive genetic component to be a single factor

Table 22.4. Maximum likelihood test statistics for weight model comparisons

Model	χ^2	df	p	Models compared	χ^2 Diff	df	p Diff
1. *ACE*	175.89	42	.000				
2. *AE*	195.99	52	.000	1 vs. 2	20.10	10	<.05
3. *CE*	1760.09	52	.001	1 vs. 3	1584.20	10	<.001
4. A_1CE	256.79	48	.000	1 vs. 4	80.90	6	<.001
5. A_dCE	332.16	48	.000	1 vs. 5	156.27	6	<.001
6. AC_1E[a]	181.03	48	.000	1 vs. 6	5.14	6	>.50
7. AC_dE	195.81	48	.000	1 vs. 7	19.92	6	<.001
8. ACE_d	649.72	48	.000	1 vs. 8	473.83	6	<.001

A = additive genetic component; *C* = common environmental component; *E* = individual environmental component. Subscript 1 denotes a single common factor. Subscript *d* denotes age-specific factors. χ^2 Diff is the difference between the chi-square of model 1 and the model with which it is being compared.
[a] Best-fitting model chosen.

(model 4) or age specific (model 5) also resulted in worse-fitting models ($p <$ .001). Shared environment is best represented by a single common factor (model 6). This indicates that the variance in weight accounted for by the common environment is present at 20 months and continues to influence weight at the other two time points. Individual environment constrained to a time-specific factor (model 8) results in a worse-fitting model ($p < .001$). Using the best fitting model, parameters were estimated for each component. The proportion of variance that can be attributed to genetic and environmental components is shown in table 22.5. Genetic influence accounts for the majority of the variation in weight at all four time points (80% to 87%), with some new genetic influences at each age.

Common environment accounts for 0–10% of the variance, and individual environment accounts for 8–12%. No new common environmental influences come in after 20 months to account for the variance in weight. Individual environment accounts for 8–12% of the variance. Although environment is important, genes play a key role in determining weight in these children.

Table 22.5. Genetic and environmental parameter estimates for weight

Age	h^2					c^2					e^2				
	A_1	A_2	A_3	A_4	Total	C_1	C_2	C_3	C_4	Total	E_1	E_2	E_3	E_4	Total
14 Months	.87				.87	.00				.00	.12				.12
20 Months	.72	.06			.78	.10				.10	.05	.07			.12
24 Months	.69	.13	.05		.87	.04				.04	.05	.01	.02		.08
36 Months	.64	.02	.14	.00	.80	.10				.10	.04	.02	.00	.04	.10

h^2 = heritability; c^2 = common environment; e^2 = individual environment.

Table 22.6. Maximum likelihood test statistics for BMI model comparisons

Model	χ^2	df	p	Models compared	χ^2 Diff	df	p Diff
1. ACE	61.79	42	.025				
2. AE	80.43	52	.007	1 vs. 2	18.64	10	<.02
3. CE	96.93	52	.000	1 vs. 3	45.14	10	<.001
4. A_1CE	72.54	48	.013	1 vs. 4	10.75	6	<.10
5. A_dCE	77.48	48	.004	1 vs. 5	15.69	6	<.02
6. AC_1E [a]	65.85	48	.044	1 vs. 6	4.06	6	>.70
7. AC_dE	73.74	48	.010	1 vs. 7	11.95	6	<.10
8. ACE_d	94.23	48	.000	1 vs. 8	32.44	6	<.001

A = additive genetic component; C = common environmental component; E = individual environmental component. Subscript 1 denotes a single common factor. Subscript D denotes age-specific factors. χ^2 Diff is the difference between the chi-squares of model 1 and the model with which it is being compared.
[a] Best-fitting model chosen.

Body Mass Index

For BMI, the full Cholesky resulted in χ^2 of 61.79 (df = 42, p = .025), which is significant and implies an imperfect fit to the data. Table 22.6 shows the model comparison results. Constraining the common environment (model 2) and additive genetic effects (model 3) to zero results in a worse-fitting model. Constraining the additive genetic effects to a single common factor (model 4) results in a slightly better-fitting model. When the additive genetic effects were constrained to age-specific factors (models 5), the model again had a worse fit. The common environmental effects are best represented by a single common source (model 6).

The parameters were estimated for the best fitting model and are shown in table 22.7. Additive genetic effects account for 20–67% of the variance in BMI, the greatest being at 36 months. By 36 months, 67% of the variation in BMI can be accounted for by genes. Most of the genetic influence is present at 14 months and has a residual effect on all four time points. The common envi-

Table 22.7. Genetic and environmental parameter estimates for BMI

Age	h^2					c^2					e^2				
	A_1	A_2	A_3	A_4	Total	C_1	C_2	C_3	C_4	Total	E_1	E_2	E_3	E_4	Total
14 Months	.61				.61	.19				.19	.20				.20
20 Months	.10	.10			.20	.59				.59	0	.21			.21
24 Months	.06	.01	.25		.32	.48				.48	0	.01	.19		.20
36 Months	.23	.27	.17	0	.67	.15				.15	.02	0	.04	.12	.18

h^2 = heritability; c^2 = common environment; e^2 = individual environment.

ronment accounts for 15–59% of the variance, the least being at 36 months and the most at 20 months. The individual environment accounts for 18–20% of the variance.

In summary, between 61% and 82% of the variation in height can be accounted for by genetic influences. The proportion of variance in weight due to genetic factors is between 78% and 87%. For BMI, genetic influences account for 20% to 75% of the variance. Again, it is important to keep in mind that BMI is a measure of fatness independent of height.

Other Measures

Is there a relationship between anthropometric and psychosocial variables in early childhood? From time to time in the study of individual differences, hypotheses about a relationship between body size or type and temperament have been suggested. For example, Sheldon and Stevens (1970) hypothesized that three body types, endomorph, mesomorph, and ectomorph, could predict temperament. Sheldon and Stevens predicted that an endomorph would have a "viscerotonia" temperament. This type of person was also known as a "gut" person and was outgoing and sociable. A mesomorph had a "somatotonia" temperament and had a muscular body type. This type of person was aggressive and dominating. The ectomorph had a "cerebrotonia" temperament and was known as nervous. This type was quiet, restrained, and not necessarily intelligent.

Sheldon and Stevens (1970) did not use height or weight alone to predict temperament; instead, they used body type. An endomorph would have a high BMI. If Sheldon and Stevens's theory is correct, the correlation between BMI and shyness, the opposite of outgoing, would be negative. An ectomorph, having a delicate, slender body type, would "shrink away from sociality." In this case, the correlation between BMI and shyness is expected to be negative. In other words, a skinny individual would be shy. Using the MacArthur Longitudinal Twin Study data, we explored the possibility of relationships between height, weight, and/or body mass index and temperament along with a number of other psychosocial variables.

Temperament measures were obtained from the Colorado Child Temperament Inventory (Rowe & Plomin, 1977). In this study, parents rated their children on different aspects of temperament using a scale of one to five. Examples of the question asked are, Does the child make friends easily? Is the child friendly with strangers? The correlations between shyness and height, weight, and BMI were examined, but all were found to be small, and none proved to be statistically significant. Shyness and height were negatively correlated and ranged from −.04 to −.07. Weight and shyness were also negatively correlated at all four time points and ranged from −.06 to −.10. For BMI, the correlations ranged from .03 to −.10. We also examined the relationship between our anthropometric indices and a number of other psychosocial and cognitive measures that included approach-withdrawal, adaptability, intensity, mood, persistence, distractibility, soothability, sociability, and Bayley/Stanford-Binet measures of IQ. The results also revealed a picture of independence. In total, 240 correlations were examined, and only 17 of these were nominally significant at the .05 level. The significant values were low, ranging from .10 to .24. It therefore seems most reasonable to conclude there is little or no evidence for significant associations among these variables.

DISCUSSION

Height, weight, and BMI in children 14–36 months of age have been examined to determine what proportion of the variance in these variables can be accounted for by genetic effects and what proportion can be accounted for by the environment. The results indicate that genetic influences account for the majority of the variation in height and weight. Between 61% and 82% of the variation in height can be accounted for by genetic influences. During this period of rapid growth, new genetic variation is expressed between 14 and 36 months. These results are compatible with those reported by Phillips and Matheny (1990), who found heritability at age three to be .81. The common environment accounts for 0–18% of the variance in height and is best represented by age-specific factors that would include any correlated errors of measurement. The remaining variation is due to individual environment (6% to 22%), which includes random measurement error.

The findings for weight are similar to those for height. The proportion of variance due to genetic influence is between 78% and 87%. Although the magnitude of new genetic influence coming in after 14 months is small, the parameters estimating these effects in our model are significant. These findings are consistent with those of Biron et al. (1977), who reported that heredity explains most of the familial aggregation patterns of weight in children. The proportion of variance accounted for by common environment is minimal, 0–10%. At 14 months the common environment does not account for any of the variation in weight, while genetic effects account for 87%. The common environmental influence on weight is best represented by a single source, perhaps a genuine influence of the household nutritional environment, but one that has limited impact compared to genetic influences. Again, the remainder of the variance is due to individual environment and measurement error (8–12%).

The 22-month period studied is one of rapid growth, with a 35–40% increase in weight and a 25% increase in height. Even though there is considerable stability for individual differences in body weight (around the increasing mean), children are becoming taller and leaner on average (prior to the "adiposity rebound" a little later in childhood), and some children are becoming taller and leaner faster, or to a greater extent, than others. These patterns of change are reflected in individual differences in the BMI. The BMI is important because it is the most widely used summary index relating to body fatness and obesity. For BMI, the measure that may be most directly relevant to physical or psychological health, our data suggest several conclusions. By age 3 years, genetic effects account for two thirds of the individual differences among children. However, most of these genetic influences were not those manifested at the end of the first year of life (14 months in our study), even though the heritability was also high at that time. In other words, as predicted, there are substantial changes in the genetic influences on BMI between the first and third year of life. This kind of genotype-by-age interaction is consistent with the low adult–parent offspring resemblance observed during the offspring's first year of life (Cardon, 1994). Obese parents may not have obese babies, even though there is an increased chance that these babies have inherited a genetic vulnerability to develop obesity later.

Our analysis of BMI suggests that there may be substantial influences of the family environment on heavier versus leaner build around 2 years of age, although this was not as evident in the analysis of either weight or height per se. This may be an age when children are sensitive to household dietary or physical activity norms, in as much as this influences relative weight for height. If so, the impact of these influences must be relatively short lived. By 36 months, genetic influences predominate.

Our final observation is that, for the psychosocial measures we have examined, there is little or no evidence of any relationship to our anthropometric indices at this age. Although this does not preclude such relationships at later ages, it is clear that genetic influences on early physical development in the normal range are effectively uncorrelated with those on early childhood temperament, psychosocial, or cognitive development. This study has confirmed both the substantial influence of genetic factors on indices of body size in early childhood and their independence during this period, of normal psychosocial development.

APPENDIX

In the full model there is a separate Λ matrix for additive genetic (represented by (A), common or shared environment (represented by C), and nonshared environmental effects (represented by E). These components combine to specify the expected covariance matrix, Σ, as shown:

$$\Sigma = \Lambda_A \Lambda_A' + \Lambda_C \Lambda_C' + \Lambda_E \Lambda_E',$$

where Λ takes the form:

$$\Lambda = \begin{bmatrix} \lambda_{11} & 0 & 0 & 0 \\ \lambda_{21} & \lambda_{22} & 0 & 0 \\ \lambda_{31} & \lambda_{32} & \lambda_{33} & 0 \\ \lambda_{41} & \lambda_{42} & \lambda_{43} & \lambda_{44} \end{bmatrix}.$$

A maximum likelihood technique (Hayduk, 1987) is used to estimate the parameters in Σ from the observed covariance matrix (S) by minimizing the fit function:

$$F = \log |\Sigma| + tr(S\Sigma^{-1}) - \log |S| - (p + q).$$

The observed indicators are p (endogenous indicators) and q (exogenous indicators).

To test the significance of each component, submodels were evaluated by constraining the components to be tested to zero. For example, to test the significance of genetic effects, the additive genetic component was excluded from the model. The resulting χ^2 statistic of this model was then compared to the χ^2 of the full model to determine which model best represents the data (see table 22.2 model comparisons). The χ^2 of the first model is subtracted from the χ^2 of the second model to obtain a χ^2 difference. The degrees of freedom from the first model are also subtracted from the degrees of freedom from the second model to obtain a df difference. This new χ^2 and df indicate whether the additional constraints have significantly reduces the original model's ability to fit the data. If the new p value is $> .05$, the model fit is improved by the additional constraints (Hayduk, 1987).

The following are lambda matrices representing a single common factor and age specific factors.

Single common factor:

$$\Lambda = \begin{bmatrix} \lambda_{11} & 0 & 0 & 0 \\ \lambda_{21} & 0 & 0 & 0 \\ \lambda_{31} & 0 & 0 & 0 \\ \lambda_{41} & 0 & 0 & 0 \end{bmatrix}$$

Age-specific factor:

$$\Lambda = \begin{bmatrix} \lambda_{11} & 0 & 0 & 0 \\ 0 & \lambda_{22} & 0 & 0 \\ 0 & 0 & \lambda_{33} & 0 \\ 0 & 0 & 0 & \lambda_{44} \end{bmatrix}$$

REFERENCES

Biron, P., Mongeau, J. G., & Bertrand, D. (1977). Familial resemblance of body weight and weight/height in 374 homes with adopted children. *Journal of Pediatrics, 91,* 555–558.

Cardon, L. R. (1994). Height, weight, and obesity. In J. C. DeFries, R. Plomin, & D. W. Fulker (Eds.), *Nature and nurture during middle childhood* (pp. 165–172). London: Blackwell.

Frisancho, A. R., Guire, D., Babler, W., Borkan, G., & Way, A. (1980). Nutritional influence on childhood development and genetic control of adolescent growth of Quechuas and Mestizos from the Peruvian lowlands. *American Journal of Physical Anthropology, 52,* 367–375.

Goodman, N., Dornbusch, S. M., Richardson, S. A., & Hastorf, A. H. (1963). Variant reactions to physical disabilities. *American Sociological Review, 28,* 429–435.

Gortmaker, S. L., Dietz, W. H., Sobol, A. M., & Wehler, C. A. (1987). Increasing pediatric obesity in the United States. *Journal of Diseases of Childhood, 141,* 535–540.

Hayduk, L. A. (1987). *Structural equation modeling with LISREL.* Baltimore, MD: Johns Hopkins University Press.

Hubert, H. B., Feinleib, M., McNamara, P. M., & Castelli, W. P. (1983). Obesity as an independent risk factor for cardiovascular disease: A 26-year follow-up of participants in the Framingham Heart Study. *Circulation, 67,* 968–977.

Kuczmarski, R. J., Flegal, K. M., Campbell, M. H. S., & Johnson, C. L. (1994). Increasing prevalence of overweight among US adults. *Journal of the American Medical Association, 272,* 205–211.

Lew, E. A. (1985). Mortality and weight: Insured lives and the American Cancer Society studies. *Annals of Internal Medicine, 103,* 1024–1029.

Lew, A., & Garfinkel, L. (1979). Variations in mortality by weight among 750,000 men and women. *Journal of Chronic Diseases, 32,* 563–576.

Maddox, G. L., Back, K., & Liederman, V. (1968). Overweight as social deviance and disability. *Journal of Health and Social Behavior, 9,* 287–298.

Neale, M. C., & Cardon, L. R. (1992). *Methodology for genetic studies of twins and families.* Boston: Kluwer.

Phillips, D., & Matheny, A. P. (1990). Quantitative genetic analysis of longitudinal trends in height: Preliminary results from the Louisville twin study. *Acta Genet Medical Gemellol, 39,* 143–163.

Pi-Sunyer, F. X. (1991). Health implications of obesity. *American Journal of Clinical Nutrition, 53,* 1595S–1603S.

Plomin, R., DeFries, J. C., & Fulker, D. W. (1988). *Nature and nuture during infancy and early childhood.* New York: Cambridge University Press.

Price, R. A., Cadoret, R. J., Stunkard, A. J., & Troughton, E. (1987). Genetic contributions to human fatness: An adoption study. *American Journal of Psychiatry, 144,* 1003–1008.

Price, R. A., Ness, R., & Laskarzewki, P. (1990). Common major gene inheritance of extreme overweight. *Human Biology, 14,* 747–765.

Rolland-Cachera, M. F., Deheeger, M., & Guilloud-Bataille, M. (1987). Tracking the development of obesity from one month of age to adulthood. *Annals of Human Biology, 14,* 219–229.

Rowe, D. C., & Plomin, R. (1977). Temperament in early childhood. *Journal of Personality Assessment, 41,* 150–156.

Sheldon, W., & Stevens, S. S. (1970). *The varieties of temperament.* New York: Hafner.

Sorensen, T. I., Price, R. A., Stunkard, A. J., & Schulsinger, F. (1989). Genetics of obesity in adult adoptees and their biological siblings. *British Medical Journal, 298,* 87–90.

Staffierie, J. R. (1967). A study of social stereotype of body image in children. *Journal of Personality and Social Psychology, 7,* 101–104.

Stunkard, A. J., & Burt, V. (1967). Obesity and the body image: II. Age at onset of disturbances in the body image. *American Journal of Psychiatry, 123,* 1443–1447.

Stunkard, A. J., & Mendelson, M. (1967). Obesity and the body image: 1. Characteristics of disturbances in the body image of some persons. *American Journal of Psychiatry, 123,* 1296–1230.

Stunkard, A. J., Foch, T. T., & Hrubec, Z. (1986a). A twin study of human obesity. *Journal of the American Medical Association, 256,* 51–54.

Stunkard, A. J., Sorensen, T. I., Hanis, C., Teasdale, T. W., Chakraborty, R., Schull, W. J., & Schulsinger, F. (1986b). An adoption study of human obesity. *New England Journal of Medicine, 314,* 193–198.

Van Itallie, T. B. (1985). Health implications of overweight and obesity in the United States. *Annals of Internal Medicine, 103,* 983–988.

Wadden, T. A., & Stunkard, A. J. (1993). Psychosocial consequences of obesity and dieting: Research and clinical finding. In A. J. Stunkard & T. A. Wadden (Eds.), *Obesity: Theory and therapy* (pp. 163–177). New York: Raven Press.

23

Temperament, Mental Development, and Language in the Transition from Infancy to Early Childhood

Lorraine F. Kubicek
Robert N. Emde
Stephanie Schmitz

Specific temperamental characteristics such as shyness or activity level may affect children's experiences in ways that contribute to differential outcomes in language and cognitive development. For example, shy children may avoid initiating interactions with unfamiliar peers, thereby limiting the frequency and variety of their social and linguistic experiences. Very active children, on the other hand, may not focus on complex tasks long enough to learn how to complete them successfully.

In this chapter we explore relations between temperament and cognitive functioning in general, and language development in particular, during toddlerhood. Through behavioral genetics methods, we examine common genetic and environmental influences on these relations at each of four ages—14, 20, 24, and 36 months.

Numerous studies have reported positive correlations between standardized mental test scores in infancy and childhood and a number of different temperamental characteristics (see Matheny, 1989, for a review). In general, children who score higher on mental tests are described as being more attentive to and persistent on tasks, as more readily approaching unfamiliar persons, objects, and events, as more flexible in response to change, and more positive in mood. Of particular relevance is Matheny's (1989) longitudinal research from the Louisville Twin Study. Matheny assessed the relation between the Bayley Mental Development Index (MDI; Bayley, 1969) and three temperament factors identified from the rating scales of the Infant Behavior Record (IBR; Bayley, 1969)—affect-extraversion, task orientation, and activity, at 12, 18, and 24 months. He found that all of the correlations involving affect-extraversion and task orientation were positive and ranged from about .30 to .50. Correlations involving activity were positive and about .20 at 12 and 18 months only.

McCall and co-workers (1977) have suggested that the association between general cognitive ability and temperament may reflect, in part, an association between language ability and temperament. In fact, Matheny (1989) reported that when verbal and performance test items were evaluated separately for children in the Louisville Twin Study at ages 5–12 years, more correlations were found between verbal IQ and temperament. Unfortunately, comparable data are not available for these children when they were under 5 years of age because the measures used to assess general cognitive ability at the younger ages did not distinguish between verbal and performance IQ.

There are considerable individual differences in early language development that are attributable to differences in children's language-learning environments, to differences in the children themselves, as well as to the interaction of such variables (Bates et al., 1995; Fenson et al., 1994). Recently, temperament has been suggested as a contributing factor. For example, Wells (1986) noted that aspects of a child's temperament such as sociability or argumentativeness may affect how others interact with that child and, as a consequence, affect the amount and kind of information he or she learns about language through conversation. Bates et al. (1988) suggested that a child's strong need to be with or like others may enhance or facilitate a rote/holistic/imitative strategy for learning language. Both Wells and Bates et al. offered social explanations for a relation between temperament and language. Wells suggested that a child's temperament may affect how people interact with the child, whereas Bates et al. suggested that temperament may affect a child's style of interacting with others.

Bloom (1993), in contrast, offered a cognitive explanation which focuses on possible limitations in the young child's attentional capacities. She proposes that a temperament profile characterized by more time in neutral affect expression would facilitate early language learning by allowing for the reflective stance needed to construct the mental meanings for learning words. Conversely, the evaluative stance which underlies an emotion and its expression would preempt attention and the cognitive resources needed for learning words. Therefore, children who are temperamentally predisposed to more frequent emotional expression will learn words somewhat later.

Despite these proposals, to date, there has been little systematic study of the potential links between temperament characteristics and early language ability (with substantial numbers of young children). One exception is a recent, large-scale investigation by Slomkowski and colleagues (1992), which examined both contemporaneous and longitudinal relations between temperament and language in early and middle childhood. They found that affect-extraversion and task orientation (rated on the IBR) were significantly and positively correlated with both receptive and expressive language at age two (assessed on the Sequenced Inventory of Communication Development [SICD]; Hedrick et al., 1975). Moreover, affect-extraversion at age two made a unique contribution to individual differences in both receptive and expressive language at age three and to receptive language skills at age seven.

The main goals of this chapter are to (1) explore relations between aspects of temperament and cognitive functioning in general, and language development in particular, in a large sample at four ages during toddlerhood; (2) assess common genetic and environmental influences underlying these relations at each age; (3) determine whether these influences change during toddlerhood; and (4) explore whether children who are more advanced cognitively or linguistically are temperamentally different from those who are not.

This study builds on previous research in a number of important ways. First, links between temperament and language were assessed very early in language development, that is, either before or during the production of single words. The subjects followed by Slomkowski et al. (1992) were first assessed at 2 years, a time when many toddlers are already combining two or more words. Second, three of our measures were the same as those used by Matheny and Slomkowski et al.—namely, the Bayley MDI, IBR, and the SICD—which facilitate comparisons across studies. A parent-report measure of temperament, the Colorado Childhood Temperament Inventory (CCTI; Rowe & Plomin, 1977), and two observational measures of temperament, one for shyness observed in the home and the other for inhibition observed in the lab, were also added. The inclusion of these additional measures broadened our study of temperament, cognitive functioning, and language links to include temperament characteristics not evaluated in past research. Finally, the introduction of behavioral genetic analyses made it possible to explore common genetic and environmental influences on the relations between temperament and general cognitive functioning and/or language at each age.

METHODS

Subjects

Analyses were based on a maximum sample of 351 twin pairs, 190 monozygotic (MZ) and 161 dizygotic (DZ) twin pairs. They represent all twin pairs who were assessed in the home and laboratory at 14, 20, 24, and 36 months and had corresponding maternal report data at each age. Specific analyses may be based on a somewhat smaller sample due to noncompletion of procedures. Characteristics of the sample are described in detail in chapter 3.

Procedures

Home visits were scheduled at each mother's convenience for a time when her children were well rested and at their best. Each home visit was conducted by two female examiners and lasted approximately 2.5 hr. Procedures completed in the home were home entry, the Bayley Scales of Infant Development, the Stanford-Binet Form L-M, and the SICD.

Laboratory visits were conducted at the Institute for Behavioral Genetics on the University of Colorado Boulder campus, typically within 2 weeks of the home visit. Each lab visit was conducted by two female examiners and lasted approximately 2.5 hr. The one procedure relevant to these analyses that was conducted in the laboratory was behavioral inhibition at 14, 20, and 24 months only.

The father was asked to complete the Colorado Childhood Temperament Inventory for each twin during each home visit, and the mother was asked to do the same during each laboratory visit. Refer to chapter 3 for a more complete description of study procedures.

Measures

Measures within the domains of temperament, cognitive functioning, and language are described below.

Temperament

Colorado Childhood Temperament Inventory. At 14, 20, 24, and 36 months, both parents rated each child on a version of the CCTI (Rowe & Plomin, 1977) that had been modified to assess shyness as distinct from sociability (Buss & Plomin, 1984). Scales composed of five statements each assessed the temperament domains of activity, emotionality, persistence, shyness, sociability, and soothability. Parents used a 5-point scale (ranging from "not at all" to "a lot") to rate how much their child was like the statement. An example from the activity scale is "child is very energetic," and one from the sociability scale is "child finds people more stimulating than anything else." Because mothers' ratings were available for more of the sample than fathers' ratings were, only mothers' ratings were included in these analyses.

Bayley's Infant Behavior Record. At 14, 20, and 24 months, examiners used the IBR to rate each child's behavior during administration of the Bayley MDI (Bayley, 1969) and the SICD (Hedrick et al., 1975). The IBR items were aggregated on three temperament factors as suggested by Matheny (1980). They included affect-extraversion (as defined by social responsiveness, cooperativeness, fearfulness, and emotional tone), activity (as defined by activity, body motion, and energy), and task orientation (as defined by object orientation, goal directedness, and attention span). At 36 months, a modified version of the IBR was used to rate each child's behavior. Affect-extraversion was assessed during administration of the Stanford-Binet Form L-M (Terman & Merrill, 1973) and SICD as well as during the remainder of the home visit. Activity and task orientation were assessed during test administration only.

Home Entry. At each age, an observational measure of shyness was based on the child's initial reactions to the examiners' arrival in the home. Discrete behaviors such as reaction to an examiner's approach, the child's latency to approach a proffered toy, and the frequency of vocalization were recorded. In addition, global ratings of shyness and hesitation were completed by video-raters for each minute of the 5-min entry episode. Measures were averaged across each minute and then across the 5-min entry episode. A composite measure of shyness was based on an unrotated first principal component score derived from these ratings.

Behavioral Inhibition. At 14, 20, and 24 months, inhibited and uninhibited behaviors were assessed from behavioral reactions to unfamiliar events in the laboratory playroom. Videotapes of the laboratory playroom episode were coded for four latencies: (1) leaving the mother upon entry to the playroom, (2) approaching the toys, (3) approaching the stranger, and (4) approaching an unfamiliar object (e.g., furry monster at 14 months, tin can robot at 20 months, and remote control robot at 24 months). The total proportion of time the child spent proximal to the mother during the free play, stranger, and unfamiliar object episodes was also recorded. We computed an aggregate index (standard score) of the inhibited/uninhibted dimension by averaging the standard scores for each of the seven variables (four latencies and three time intervals when the child was proximal to the mother). This average standard score was treated as a continuous index of behavioral inhibition in an unfamiliar context.

Cognitive Functioning

At 14, 20 and 24 months, the Bayley Mental Scale from the Bayley Scales of Infant Development was administered, and the MDI was derived. At 36 months, the Stanford-Binet Form L-M was administered, and a full-scale IQ score was derived.

Language

The SICD (Hedrick et al., 1975) was administered at each age to provide a standard measure of language comprehension and production. Observational data were supplemented by maternal reports when necessary (see chapter 16). At 14, 20, and 24 months, expressive items focused on the child's spontaneous production of specific sounds and intonation patterns, picture naming, and use of words to comment, ask for things, and respond to questions. Imitation of speech, nonspeech sounds, and nonvocal motor acts were assessed as well. At these three ages, receptive items focused on identifying body parts and familiar objects, discriminating different environmental sounds, complying with simple requests, and comprehending the names of familiar people and objects, location prepositions, and big and little. A more advanced set of items was presented at 36 months. At this age, expressive items assessed the child's spontaneous production of why and how questions, use of plurals and conjunctions, counting, and responding to what, when, and where questions. Receptive items focused on the identification of colors, shapes, and textures and compliance with requests involving plurals, numbers, and multiple objects or actions.

RESULTS

Phenotypic Correlations

Maternal Reports

Phenotypic correlations between maternal reports of temperament (CCTI ratings) and general cognitive functioning (Bayley MDI or Stanford-Binet full-scale IQ) on the one hand, and language (SICD expressive and receptive) on the other hand, were either nonsignificant or low. They ranged from −.14 to .10 for general cognitive functioning (see table 23.1) and from −.21 to +.15 for language (see table 23.2). As table 23.1 illustrates, there were no significant correlations between CCTI ratings and Bayley performance at 14 months. The most consistent result was a significant negative correlation between CCTI shyness and general cognitive functioning at the three later ages. CCTI activity was significantly and negatively correlated with general cognitive functioning at 20 and 36 months, whereas CCTI persistence was significantly and positively correlated with general cognitive functioning at 24 and 36 months.

As table 23.2 shows, there was also a significant negative correlation between CCTI shyness and both expressive and receptive language at 20 months and expressive language at 24 months. CCTI activity was significantly and negatively correlated with receptive language at 20 months and both expressive and receptive language at 36 months. Moreover, CCTI sociability was significantly and positively correlated with expressive language at 14 months, both expressive and receptive language at 24 months, and receptive language at 36 months.

Table 23.1. Statistically significant phenotypic correlations of maternal ratings of temperament (Colorado Childhood Temperament Inventory) and general cognitive functioning

Measure/age	Emotionality	Activity	Shyness	Sociability	Persistence	Soothability
Bayley, 14 months						
Bayley, 20 months		−.12**	−.10**			
Bayley, 24 months	−.10**		−.10**	.09*	.10**	.09*
Binet, 36 months		−.14***	−.10*		.09*	

***p < .001; **p < .01; *p < .05.

Table 23.2. Statistically significant phenotypic correlations of maternal ratings of temperament (Colorado Childhood Temperament Inventory) and language (Sequenced Inventory of Communication Development; SICD)

SICD component/age	Emotionality	Activity	Shyness	Sociability	Persistence	Soothability
Expressive, 14 months				.08*		
Receptive, 14 months					.07*	
Expressive, 20 months			−.18***			
Receptive, 20 months		−.13**	−.11**			
Expressive, 24 months	−.11**		−.21***	.15***		
Receptive, 24 months				.14***		
Expressive, 36 months		−.14***				
Receptive, 36 months		−.10**		.08*		

***$p < .001$; **$p < .01$; *$p < .05$

A comparison of tables 23.1 and 23.2 indicates considerable overlap in the pattern of correlations involving general cognitive functioning and language with maternal reports of emotionality, activity, shyness, and sociability, especially at 20 and 24 months. This finding lends support to the suggestion by McCall et al. (1977) that some of the association between general cognitive functioning and temperament may reflect links between verbal ability and temperament.

Observational Measures

Phenotypic correlations involving observed IBR ratings of temperament were higher than those involving maternal reports of temperament, ranging from −.08 to .51 for general cognitive functioning (table 23.3) and from −.09 to .46 for language (table 23.4). As table 23.3 illustrates, both IBR affect-extraversion and task orientation were significantly and positively correlated with general cognitive functioning at 14, 20, 24, and 36 months. Activity level was significantly and negatively correlated with general cognitive functioning at 20 and 24 months. These correlations are comparable to those reported by Matheny

Table 23.3. Statistically significant phenotypic correlations of observational ratings of temperament (Infant Behavior Record) and general cognitive functioning

Measure/age	Affect-extraversion	Activity	Task orientation
Bayley, 14 months	.38***		.43***
Bayley, 20 months	.33***	−.09*	.33***
Bayley, 24 months	.44***	−.08*	.34***
Binet, 36 months	.23***		.51***

***$p < .001$; **$p < .01$; *$p < .05$

Table 23.4. Statistically significant phenotypic correlations of observational ratings of temperament (Infant Behavior Record) and language (Sequenced Inventory of Communication Development, SICD)

SICD component/age	Affect-extraversion	Activity	Task orientation
Expressive, 14 months	.24***		.19***
Receptive, 14 months	.17***		.12***
Expressive, 20 months	.33***		.23***
Receptive, 20 months	.31***		.21***
Expressive, 24 months	.46***		.21***
Receptive, 24 months	.38***	−.09*	.19***

***$p < .001$; **$p < .01$; *$p < .05$

(1989) for twins participating in the Louisville Twin Study at ages 12, 18, and 24 months.

As table 23.4 shows, both IBR affect-extraversion and task orientation were significantly and positively correlated with expressive and receptive language at 14, 20, and 24 months. In all cases, the magnitude of the correlations involving affect-extraversion were higher than those involving task orientation. Correlations at 24 months are comparable to those reported by Slomkowski et al. (1992) for a nontwin sample of the same age. Activity level was significantly and negatively correlated with receptive language only at 24 months. There were no significant correlations between IBR ratings of temperament and language at 36 months, a finding addressed later in the Discussion.

There was also considerable similarity in the pattern of correlations involving general cognitive functioning and language with observed IBR ratings of temperament at 14, 20, and 24 months, a finding that is consistent with the pattern noted earlier for correlations involving maternal reports of temperament.

Phenotypic correlations involving the observational measures of home shyness and behavioral inhibition with general cognitive functioning and language were either nonsignificant or low (−.07 or −.09) Home shyness was significantly and negatively correlated with general cognitive functioning at 36 months (−0.09) and expressive language at 20 (−0.09) and 24 months (−0.08). Behavioral inhibition was significantly and negatively correlated with expressive language at 20 months (−0.08).

Bivariate Modeling

As reported above, correlations involving measures of observed IBR affect-extraversion and task orientation were substantial and showed a relatively consistent pattern across age, in contrast to other correlations we found. To explore the nature of the correlations involving IBR affect-extraversion and task orientation more closely, we performed a series of bivariate analyses designed to

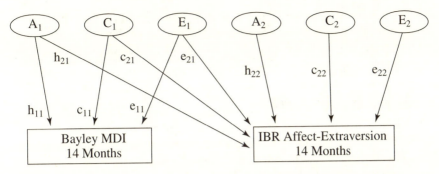

Figure 23.1. *ACE* model for correlation between Bayley Mental Development Index (MDI) and Infant Behavior Record (IBR) affect-extraversion at 14 months.

estimate the genetic and environmental influences underlying those correlations. These analyses used a model referred to as a Cholesky decomposition similar to those introduced in chapter 2 for the analysis of development. At each age, models were fitted separately for general cognitive functioning, expressive language, and receptive language with affect-extraversion and task orientation.

Figure 23.1 illustrates the full model for Bayley performance and observed affect-extraversion at 14 months. This model decomposes the phenotypic variance for ratings of Bayley MDI and IBR affect-extraversion into latent factors that are common to both (i.e., additive genetic influences, A_1; shared environmental influences, C_1; and nonshared environmental influences plus measurement error, E_1), as well as three factors that are unique to observed affect-extraversion (i.e., additive genetic, A_2, and shared, C_2, and nonshared, E_2, environmental influences).

The standardized path coefficients for the paths from the common factors to the phenotype for Bayley performance (h_{11}, c_{11}, e_{11}) indicate the relative magnitude of genetic and environmental influences on Bayley MDI at 14 months. The proportion of variance due to genetic or environmental influence (i.e., h^2, c^2, e^2) is obtained by squaring these path coefficients (available from the authors). The paths from the common factors to the phenotype for observed affect-extraversion (h_{21}, c_{21}, e_{21}) indicate the extent to which genetic and environmental influences are common to both Bayley performance and IBR affect-extraversion at 14 months. The paths from the three unique factors to the phenotype for observed affect-extraversion (h_{22}, c_{22}, e_{22}) represent genetic and environmental influences that are independent of those operating on Bayley performance.

Models were fitted using maximum-likelihood (ML) procedures in Mx, a structural equation modeling program (Neale, 1994). Chi-square statistics and their associated probabilities indicate how well the models fit the data (Heath et al., 1989). To test the significance of a particular path, submodels of the full decomposition were tested by constraining common paths to zero (first one at a time and then all together) and recalculating the likelihood ratio for the fit of the reduced model. The resulting change in χ^2 is the test statistic for the path excluded from the reduced model.

Results of the Bivariate Analyses

The models for IBR affect-extraversion and general cognitive functioning as indicated by the Bayley MDI show a significant influence of common non-shared environment at 14 and 20 months and a significant influence of common genetic factors at 24 months. As table 23.5 illustrates, the χ^2 difference between the full model and the submodel when the common path for non-shared environment (e_{21}) was excluded was significant at 14 and 20 months. Similarly, the χ^2 difference between the full model and the submodel when the common path for genetic influence (h_{21}) was excluded was significant at 24 months. The models for IBR affect-extraversion and expressive language (table 23.6) show a significant influence of common nonshared environment at 14, 20, and 24 months, as well as a significant influence of common genetic

Table 23.5. Model comparison: Infant Behavior Record affect-extraversion and general cognitive functioning

Model	χ^2	df	p	Models compared	χ^2 Diff	df	p
14 Months							
1. Full model	14.40	11	.21				
2. Drop h_{21}	14.41	12	.28	1 vs. 2	0.01	1	ns
3. Drop c_{21}	16.45	12	.17	1 vs. 3	2.05	1	ns
4. Drop e_{21}	62.94	12	.00	1 vs. 4	48.54	1	<.001
5. Drop all common paths	124.38	14	.00	1 vs. 5	109.98	3	<.001
20 Months							
1. Full model	7.63	11	.75				
2. Drop h_{21}	10.25	12	.59	1 vs. 2	2.62	1	ns
3. Drop c_{21}	8.01	12	.79	1 vs. 3	0.38	1	ns
4. Drop e_{21}	15.73	12	.20	1 vs. 4	8.10	1	<.01
5. Drop all common paths	64.26	14	.00	1 vs. 5	56.63	3	<.001
24 Months							
1. Full model	4.57	11	.95				
2. Drop h_{21}	12.47	12	.41	1 vs. 2	7.90	1	<.01
3. Drop c_{21}	5.90	12	.92	1 vs. 3	1.33	1	ns
4. Drop e_{21}	6.21	12	.91	1 vs. 4	1.64	1	ns
5. Drop all common paths	93.12	14	.00	1 vs. 5	88.55	3	<.001

Table 23.6. Model comparisons: Infant Behavior Record affect-extraversion and expressive language

Model	χ^2	df	p	Models compared	χ^2 Diff	df	p
14 Months							
1. Full model	28.33	11	.003				
2. Drop h_{21}	30.12	12	.003	1 vs. 2	1.79	1	ns
3. Drop c_{21}	28.34	12	.005	1 vs. 3	0.01	1	ns
4. Drop e_{21}	37.05	12	.000	1 vs. 4	8.72	1	<.01
5. Drop all common paths	73.04	14	.000	1 vs. 5	44.71	3	<.001
20 Months							
1. Full model	12.25	11	.35				
2. Drop h_{21}	14.78	12	.25	1 vs. 2	2.53	1	ns
3. Drop c_{21}	12.67	12	.39	1 vs. 3	0.42	1	ns
4. Drop e_{21}	20.46	12	.06	1 vs. 4	8.21	1	<.01
5. Drop all common paths	75.92	14	.00	1 vs. 5	63.67	3	<.001
24 Months							
1. Full model	9.86	11	.54				
2. Drop h_{21}	19.75	12	.07	1 vs. 2	9.89	1	<.01
3. Drop c_{21}	10.20	12	.60	1 vs. 3	0.34	1	ns
4. Drop e_{21}	17.51	12	.13	1 vs. 4	7.65	1	<.01
5. Drop all common paths	112.30	14	.00	1 vs. 5	102.44	3	<.001

factors at 24 months. In contrast, the only significant effect for IBR affect-extraversion and receptive language was for common shared environment at 24 months (χ^2_1 diff = 5.95, p <.05).

The models for IBR task orientation and general cognitive functioning (table 23.7) show a significant influence of common nonshared environment at 14, 24, and 36 months and a significant influence of common genetic factors at 24 months. As table 23.7 shows, the χ^2 difference between the full model and the submodel when the common path for nonshared environment (e_{21}) was excluded was significant at 14, 24, and 36 months. Similarly, the χ^2 difference between the full model and the submodel when the common path for genetic influence (h_{21}) was excluded was significant at 24 months. For IBR task orientation and expressive language, there was a significant influence of common shared environment at 14 months (χ^2_1 diff = 7.02, p < 01). For receptive language, there was a significant influence of common genetic factors at 20 months (χ^2_1 diff = 7.56, p <.01.

To summarize, genetic effects common to the different measures were minimal and primarily limited to relations at 24 months. Models indicated that there was a common genetic influence on IBR affect-extraversion and general

Table 23.7. Model comparisons: Infant Behavior Record task orientation and general cognitive functioning

Model	χ^2	df	p	Models compared	χ^2 Diff	df	p
14 Months							
1. Full model	15.41	11	.17				
2. Drop h_{21}	17.60	12	.13	1 vs. 2	2.19	1	ns
3. Drop c_{21}	15.44	12	.22	1 vs. 3	0.03	1	ns
4. Drop e_{21}	60.98	12	.00	1 vs. 4	45.57	1	<.001
5. Drop all common paths	171.45	14	.00	1 vs. 5	156.04	3	<.001
20 Months							
1. Full model	12.64	11	.32				
2. Drop h_{21}	14.81	12	.25	1 vs. 2	2.17	1	ns
3. Drop c_{21}	14.55	12	.27	1 vs. 3	1.91	1	ns
4. Drop e_{21}	16.30	12	.18	1 vs. 4	3.66	1	ns
5. Drop all common paths	68.49	14	.00	1 vs. 5	55.85	3	<.001
24 Months							
1. Full model	16.90	11	.11				
2. Drop h_{21}	23.75	12	.02	1 vs. 2	6.85	1	<.01
3. Drop c_{21}	16.94	12	.15	1 vs. 3	0.04	1	ns
4. Drop e_{21}	20.76	12	.05	1 vs. 4	3.86	1	<.05
5. Drop all common paths	87.33	14	.00	1 vs. 5	70.43	3	<.001
36 Months, Binet IQ							
1. Full model	3.52	11	.98				
2. Drop h_{21}	5.60	12	.94	1 vs. 2	2.08	1	ns
3. Drop c_{21}	5.63	12	.93	1 vs. 3	2.11	1	ns
4. Drop e_{21}	16.71	12	.16	1 vs. 4	13.19	1	<.001
5. Drop all common paths	109.41	14	.00	1 vs. 5	105.89	3	<.001

cognitive functioning and IBR affect-extraversion and expressive language at 24 months only. There was also a common genetic influence on IBR task orientation and general cognitive functioning at 24 months and on IBR task orientation and receptive language at 20 months.

Shared environmental effects common to the different measures were also minimal and showed no consistent pattern across measures at the different ages. Shared environment was a significant influence on the link between IBR affect-extraversion and receptive language at 24 months and between IBR task orientation and expressive language at 14 months.

Nonshared environmental effects common to the different measures were more prevalent than genetic and shared environment effects and were evident at all four ages. Nonshared environment was a significant influence on the links between IBR affect-extraversion and general cognitive functioning at 14 and 20 months and between IBR affect-extraversion and expressive language at 14, 20, and 24 months. Moreover, nonshared environment was a significant influence on IBR task orientation and general cognitive functioning at 14, 24, and 36 months.

DISCUSSION

The main goals of this study were to (1) explore relations between aspects of temperament and cognitive functioning in general, and language development in particular, at four ages during toddlerhood; (2) assess common genetic and environmental influences underlying these relations at each age; (3) determine whether these influences change during toddlerhood; and (4) explore whether children who are more advanced cognitively or linguistically are temperamentally different from those who are not.

Phenotypic Correlations

We found significant correlations between general cognitive functioning and observed affect-extraversion and task orientation at each age during the second year that were comparable to those reported by Matheny (1989) for the Louisville Twin Study using the same measures. Moreover, our results extend these findings to 3 years of age using related measures.

We also found significant correlations between receptive and expressive language and observed affect-extraversion and task orientation at each age during the second year that were comparable to those reported by Slomkowski et al. (1992) for 2 year olds in a nontwin sample using the same measures. Therefore, our results, which are based on a large sample of twins tested at multiple ages during the second year, extend the link between language and aspects of temperament to children as young as 14 months. Correlations involving affect-extraversion were higher than those involving task orientation at each age. Moreover, for affect-extraversion, the correlations increased with age for both receptive and expressive language, whereas for task orientation, the correlations were similar across ages.

Before moving on, we want to discuss our findings at 36 months. Recall that at this age, the Binet replaced the Bayley as the measure of general cognitive functioning, and items more appropriate for 36 month olds replaced those used at the younger ages on the IBR and the SICD. Although there were similarities in the content of the earlier and later items on the IBR, the earlier items on the SICD were completely different from the later ones. At 36 months, the SICD items focused exclusively on language abilities, whereas at the younger ages, they assessed more general communicative abilities and included both language and nonlanguage items. Moreover, IBR ratings of affect-extraversion were based on the entire home visit and were not limited to test administration as was true at 14, 20 and 24 months. For these reasons, it is somewhat difficult to interpret the differences that emerge at 36 months. Most striking among these was the absence of any significant correlations between either receptive or expressive language and affect-extraversion or task orientation. It remains an open question whether this is due to the change in measures and/or procedure or whether this represents a change in the relation between language and temperament at 36 months.

Behavioral Genetic Analyses

Models (behavioral genetic analyses) were fitted to estimate the genetic and environmental influences underlying the significant correlations between general cognitive functioning or language and aspects of observed temperament. Behavioral genetic analyses make it possible to separate the different influences that may affect these relations and to determine whether these influences change with age.

Influence of Genetic Factors

Common genetic effects were minimal and limited to relations at 20 and 24 months. Models indicated that there was a common genetic influence on IBR affect-extraversion and general cognitive functioning and IBR affect-extraversion and expressive language at 24 months only. There was also a common genetic influence on IBR task orientation and general cognitive functioning at 24 months and on IBR task orientation and receptive language at 20 months.

The second year of life represents a significant transition period which is associated with substantial developments in language, cognition, and temperament (e.g., Bates et al., 1987; Kagan, 1981; Piaget, 1952; Rothbart, 1989). Among these are the onset of grammar, an increase in the ability to reason symbolically and to engage in pretend play, and a greater differentiation and clarity in the expression of temperamental characteristics. We now know that genetic influences can change over time (Plomin, 1986; Plomin et al., 1993). Our findings suggest that as behavior in the domains of language, cognition, and temperament becomes more organized, differentiated, and complex toward the end of the second year, common genetic factors have more of an influence on the relations between these domains.

Influence of Shared Environment

Shared environmental effects common to the different measures were also minimal and showed no consistent pattern across measures or at the different ages. Shared environment was a significant influence on the link between IBR affect-extraversion and receptive language at 24 months and between IBR task orientation and expressive language at 14 months. Shared environmental factors are environmental influences that make children in the same family similar to one another. They may include general parental characteristics such as being responsive to children's comments or requests or specific parental behaviors such as encouraging quiet activities like reading or looking at books. Shared environmental influences on IQ and specific cognitive abilities are expected to decrease with age and become negligible during adolescence (Plomin & Daniels, 1987).

Influence of Nonshared Environment

Common nonshared environmental effects were more prevalent than genetic effects and were evident at all four ages. Nonshared environment was a significant influence on the links between IBR affect-extraversion and general cognitive functioning at 14 and 20 months and on IBR affect-extraversion and expressive language at 14, 20, and 24 months. Moreover, common nonshared environment was a significant influence on IBR task orientation and general cognitive functioning at 14, 24, and 36 months.

Nonshared environment is composed of measurement error and environmental influences that make children in the same family different from one another. These influences include idiosyncratic factors such as accidents or illnesses as well as more systematic factors such as nonshared prenatal environment, family structure, differential parental treatment, sibling interactions, and relationships outside the family. In contrast to shared environmental effects, these kinds of influences are likely to increase with age as each twin spends more time away from their family and apart from their cotwin due to the formation of separate friendships, placement in different classes at school, the pursuit of different interests and activities, and the like (Plomin & Daniels, 1987; Rowe & Plomin, 1981).

Our results, like those of earlier studies mentioned above, suggest links between aspects of temperament and general cognitive functioning as well as language. The question, however, is whether these links are simply an artifact of the testing situation or whether children who are more advanced linguistically or cognitively are temperamentally different from those who are less advanced. The first half of this question focuses on short-term links relating to differences in response to the specific demands of the testing situation, whereas the second half focuses on long-term links relating to differences in development. Each will be addressed in turn.

Short-term Links Relating to the Testing Situation

First let us consider how differences in a child's affective response to the testing situation might lead to links between temperament and general cognitive functioning or language. One possibility relates to practical issues regarding test administration and how a child's behavior might affect the examiner's ability to obtain a valid measure of best performance (Plomin et al., 1990).

Testing is obviously facilitated when the examiner can establish a comfortable rapport with the child. Clearly, some children are easier to engage and keep on task than others. If a child is attentive and expresses interest, readily responds to the examiner, and persists on the tasks at hand, it is more likely that measured performance will actually reflect the child's ability to do these tasks. If, on the other hand, a child stares blankly at the examiner or complains, is reluctant or refuses to respond at times, and does not make an consistent effort to do what is required, it is less likely that measured performance will actually reflect the child's ability to do these tasks.

A second possibility relating to the testing situation is less straightforward and concerns potential examiner bias. The IBR ratings of temperament were based either exclusively or in part (for affect-extraversion at 36 months only) on the child's behavior during administration of the Bayley or Binet and the SICD. Moreover, temperament was rated by the same examiner who administered these tests. As Haviland (1976) argued in her paper regarding the influential aspects of looking smart, we often rely on affect as a shorthand method for assessing intelligence. A child who is attentive and whose affect appears to be appropriate to ongoing events is more likely to be perceived as intelligent than a child who is inattentive and whose affect appears to be inappropriate to ongoing events. This perception might affect the examiner's expectations regarding the child's abilities and thereby influence efforts to elicit best performance. Following similar reasoning, it seems plausible that a child's cognitive or linguistic performance might affect an examiner's perception of the child's temperament. A child who performs well on these tasks might be perceived as having a more positive or agreeable temperament.

Long-term Links Relating to Temperament-by-Environment Interactions

The first half of our question focused on short-term links relating to differences in response to the testing situation. Now let us consider the second half of our question, which focuses on long-term links relating to individual differences in development. That is, how might links between aspects of temperament and general cognitive functioning or language reflect individual differences in development resulting, in part, from the interaction of a child's temperamental characteristics and his or her environment?

Children's experiences in the world have a profound effect on their development, and it is now recognized that a child's unique characteristics help shape these experiences (Lerner and Busch-Rossnagel, 1981; Sameroff, 1983; Scarr & McCartney, 1983; Thomas et al., 1963). This is especially true of children's social interactions, where parent and child mutually influence each other from birth (Bell, 1968). Because so much of a child's early experience occurs in this social context, individual characteristics that affect parent–child interactions are expected to have important consequences for development (Bornstein, 1995). Temperament is considered one of these characteristics

(Bates, 1987, 1989; Chess et al., 1956; Rothbart, 1989; Thomas et al., 1963, 1968).

It seems reasonable to assume that children who are temperamentally different will elicit different responses from their environment. Unfortunately, few research studies in this area involving toddlers have been conducted, and many of these have tended to focus on the relation between "difficult" temperament and parental control and teaching behavior (e.g., Daniels et al., 1984; Gordon, 1983; Lee & Bates, 1985; Maccoby et al., 1984; Matheny et al., 1987). Nevertheless, results from some of these studies suggest that there is a relation between "difficult" temperament and parental behavior. The work of Dunn and Plomin (1986) and Buss (1981), which focused on the temperamental characteristics of sociability and activity level, also suggests a link between child temperament and parental behavior. This is especially true when third variables are considered. The more recent appreciation of the importance of third variables such as nontemperamental characteristics of the child (e.g., age and sex), characteristics of the caregiver (e.g., sex, psychological health), and of the caregiving environment (e.g., social support, life stress, social class, and cultural affiliation) in mediating relations between child temperament characteristics and parenting behavior is likely to increase the number and kind of relations that are found (Bates, 1987; Sanson & Rothbart, 1995).

In this study, behavioral genetic analyses indicated that nonshared environment had a significant influence on the relation between affect-extraversion and general cognitive functioning, as well as on expressive language, during the second year. Based on these findings, we suggest that children who vary in this temperament dimension may not only elicit different responses from their environment, but, over time, both qualitative and quantitative differences in response to the children may lead to differential outcomes in linguistic or cognitive development. The interaction of child temperament characteristics and parenting behaviors has been proposed to explain developmental variation on other outcome measures such attachment, behavioral adjustment, school achievement, and self-esteem (Bates, 1989; Sanson & Rothbart, 1995). Clearly, the implications of a child's temperament for development will depend, to a large extent, on the "goodness of fit" between their temperament and their social partners' expectations and behaviors (Thomas et al., 1968). We expect that developmental outcome will be different for a sociable child whose numerous bids for attention are always responded to than for one whose bids are often ignored.

We have suggested that one way in which a child's temperament may affect development is by shaping or influencing his other social environment. However, there is another possibility. Children who are temperamentally different may also interpret and respond differently to the same environmental influences. Escalona's (1968) concept of "effective experience" and Wachs and Gruen's (1982) idea of "organismic specificity" represent early formulations of the notion that similar parenting can lead to different outcomes for children with different temperaments.

A child's temperament may affect nonsocial interactions as well. Behavioral genetic analyses indicated a significant effect of nonshared environment on the relation between task orientation and general cognitive functioning. Children who can focus for longer periods of time and are goal directed will likely score higher on a standardized test not only because they can adapt more easily to the test situation, but also because they may know more and be more competent due to greater opportunities to learn from their day-to-day experiences.

Evaluating Alternative Interpretations

A first step toward evaluating these alternative possibilities would be to use measures that assess temperament independently from cognitive functioning and language. A number of such measures were, in fact, included in this study, such as the CCTI, home shyness, and behavioral inhibition. All correlations involving these measures were much smaller than those involving the IBR. These smaller correlations may reflect the elimination of relations due to nonindependence of measurement discussed earlier. However, there are also differences among these temperament measures that may account for these smaller correlations as well. The most important difference is that these four measures assess different temperament dimensions so that direct comparisons are not possible. Neither the IBR nor the CCTI measure the kinds of behaviors evaluated by home shyness or behavioral inhibition. Although there is some overlap between the behaviors assessed by the three IBR domains, affect-extraversion, task persistence, and activity, and three of the six CCTI domains, sociability, persistence, and activity, respectively, their operational definitions are more different than they are alike. Moreover, the IBR evaluates child behavior that is somewhat specific to the testing situation, whereas the CCTI asks about child behavior in general. Secondly, the CCTI is a parent-report measure, whereas the others are observational measures. Both types of measures provide useful but different kinds of information about a child due to parent and experimenter differences in perspective and experience with that child. It is likely that there would only be moderate correlations between the parent and experimenter ratings even if they were using the same measure (Schmitz, 1994). Thus, these additional measures, though informative in their own right, cannot help in deciding between alternative explanations regarding these data.

FUTURE DIRECTIONS

As we discussed above, there are a number of methodological problems that need to be addressed in future studies designed to assess the relation between temperament and general cognitive functioning or language. The first concerns language assessment. In this study, the SICD was used to assess language ability. The SICD is a diagnostic tool designed to differentiate children who are developing normally from those who are not and in need of remediation. In all likelihood, this measure is not sensitive enough to evaluate the wide range of variability in early language that is characteristic of a normally developing population. Moreover, the SICD is not primarily a language measure per se, but rather a measure of general communicative ability which assesses pre-speech sounds, words, and motor acts. Both imitative and spontaneous expression is evaluated. It is interesting to speculate whether the link between language and temperament would be stronger if a more rigorous measure of language ability were used. The second problem that needs to be addressed is the assessment of temperament. We believe that it is important to include an observational measure of temperament that is separate from the measure of general cognitive functioning or language. Moreover, different raters need to be used for assessing behavior in each domain so that the evaluation of one domain, for example, temperament, does not influence the rater's evaluation of another, for example, general cognitive functioning. A final problem, and perhaps the most difficult to solve, is the need to find measures that can assess continuity of development within a domain while at the same time taking account of appropriate developmental change.

CONCLUSION

Despite the methodological limitations of this study, it is important to keep in mind that the results show a consistent pattern of correlations between observational measures of temperament and general cognitive functioning as well as language. Though the correlations are somewhat modest, they corroborate findings from previous studies with both twin and nontwin populations and support the contention that there are small but consistent relations between the different behavioral domains as early as 14 months of age. It remains for future studies to explore the reasons for these relations.

REFERENCES

Bates, J. E. (1987). Temperament in infancy. In J. D. Osofsky (Ed.), *Handbook of infant development* (pp. 1101–1149). New York: Wiley.

Bates, J. E. (1989). Applications of temperament concepts. In G. A. Kohnstamm, J. E. Bates, & M. K. Rothbart (Eds.), *Temperament in childhood* (pp. 321–355). Chichester, England: Wiley.

Bates, E., Bretherton, I., & Snyder, L. (1988). *From first words to grammar: Individual differences and dissociable mechanisms.* Cambridge: Cambridge University Press.

Bates, E., Dale, P. S., & Thal, D. (1995). Individual differences and their implications for theories of language development. In P. Fletcher & B. Mac Whinney (Eds.), *Handbook of child language* (pp. 96–151). Oxford: Blackwell.

Bates, E., O'Connell, B., & Shore, C. (1987). Language and communication in infancy. In J. D. Osofsky (Ed.), *Handbook of infant development* (pp. 149–203). New York: Wiley.
Bayley, N. (1969). *The Bayley Scales of Infant Development.* New York: Psychological Corporation.
Bell, R. Q. (1968). A reinterpretation of the direction of effects in studies of socialization. *Psychological Review, 75,* 81–95.
Bloom, L. (1993). *The transition from infancy to language.* New York: Cambridge University Press.
Bornstein, M. H. (1995). Parenting infants. In M. H. Bornstein (Ed.), *Handbook of parenting: Vol. 1. Children and parenting* (pp. 3–39). Hillsdale, NJ: Erlbaum.
Buss, D. M. (1981). Predicting parent-child interactions from children's activity level. *Developmental Psychology, 17,* 598–605.
Buss, A. H., & Plomin, R. (1984). *Temperament: Early developing personality traits.* Hillsdale, NJ: Erlbaum.
Cherny, S. S., Fulker, D. W., Emde, R. N., Robinson, J., Corley, R. P., Reznick, J. S., Plomin, R., & DeFries, J. C. (1994). A developmental-genetic analysis of continuity and change in the Bayley Mental Developmental Index from 14 to 24 months: The MacArthur Longitudinal Twin Study. *Psychological Science, 5*(6), 354–360.
Chess, S., Thomas, A., & Birch, H. G. (1965). *Your child is a person.* New York: Viking.
Daniels, D., Plomin, R., & Greenhalgh, J. (1984). Correlates of difficult temperament in infancy. *Child Development, 55,* 1184–1194.
Dunn, J., & Plomin, R. (1986). Determinants of maternal behavior towards 3-year-old siblings. *British Journal of Developmental Psychology, 4,* 127–137.
Escalona, S. (1968). *The roots of individuality.* Chicago: Aldine.
Fenson, L., Dale, P. S., Reznick, J. S., Bates, E., Thal, D. J., & Pethick, S. J. (1994). Variability in early communicative development. *Monographs of the Society for Research in Child Development, 59* (5, Serial No. 242).
Gordon, B. (1983). Maternal perception of child temperament and observed mother-child interaction. *Child Psychiatry and Human Development, 13,* 153–167.
Haviland, J. (1976). Looking smart: The relationship between affect and intelligence in infancy. In M. Lewis (Ed.), *Origins of intelligence: Infancy and early childhood* (pp. 353–377). New York: Plenum Press.
Heath, A. C., Neale, M. C., Hewitt, J. K., Eaves, L. J., & Fulker, D. W. (1989). Testing structural equation models for twin data using LISREL. *Behavior Genetics, 19,* 9–36.
Hedrick, D. L., Prather, E. M., & Tobin, A. R. (1975). *Sequenced Inventory of Communication Development.* Seattle: University of Washington Press.
Kagan, J. (1981). *The second year.* Cambridge, MA: Harvard University Press.
Lee, C., & Bates, J. (1985). Mother-child interaction at age two years and perceived difficult temperament. *Child Development, 56,* 1314–1326.
Lerner, R. M., & Busch-Rossnagel, N. A. (1981). Individuals as producers of their development: Conceptual and empirical bases. In R. M. Lerner & N. A. Busch-Rossnagel (Eds.), *Individuals as producers of their development: A life-span perspective* (pp. 1–36). New York: Academic Press.

Maccoby, E. E., Snow, M. E., & Jacklin, C. N. (1984). Children's dispositions and mother-child interaction at 12 and 18 months: A short-term longitudinal study. *Developmental Psychology, 20,* 459–472.

Matheny, A. P., Jr. (1980). Bayley's Infant Behavior Record: Behavioral components and twin analysis. *Child Development, 51,* 1157–1167.

Matheny, A. P., Jr. (1989). Temperament and cognition: Relations between temperament and mental test scores. In G. A. Kohnstamm, J. E. Bates & M. K. Rothbart (Eds.), *Temperament in childhood* (pp. 321–355). New York: Wiley.

Matheny, A. P. Jr., Wilson, R. S., & Thoben, A. (1987). Home and mother: Relations with infant temperament. *Developmental Psychology, 23,* 323–331.

McCall, R. B., Applebaum, M. I., & Hogarty, P. S. (1977). Developmental changes in mental performance. *Monographs of the Society for Research in Child Development, 38* (3, Serial No. 150).

Neale, M. C. (1994). *Mx: Statistical modeling* (2nd ed.). Richmond, VA: Department of Psychiatry, Medical College of Virginia.

Piaget, J. (1952). *The origins of intelligence in children.* New York: Norton.

Plomin, R. (1986). *Development, genetics, and psychology.* Hillsdale, NJ: Erlbaum.

Plomin, R., Campos, J., Corley, R., Emde, R. N., Fulker, D. W., Kagan, J., Reznick, J. S., Robinson, J., Zahn-Waxler, C., & DeFries, J. D. (1990). Individual differences during the second year of life: The MacArthur Longitudinal Twin Study. In J. Colombo & F. Fagen (Eds.), *Individual differences in infancy: Reliability, stability, prediction* (pp. 431–455). Hillsdale, NJ: Erlbaum.

Plomin, R., & Daniels, D. (1987). Why are children in the same family so different? *Behavioral and Brain Sciences, 10,* 1–16.

Plomin, R., Emde, R. N., Braungart, J. M., Campos, J., Corley, R., Fulker, D. W., Kagan, J., Reznick, J. S., Robinson, J., Zahn-Waxler, C., & DeFries, J. C. (1993). Genetic change and continuity from fourteen to twenty months: The MacArthur Longitudinal Twin Study. *Child Development, 64,* 1354–1376.

Rothbart, M. K. (1989). Temperament and development. In G. A. Kohnstamm, J. E. Bates & M. K. Rothbart (Eds.), *Temperament in childhood* (pp. 187–247). Chichester, England: Wiley.

Rowe, D. C., & Plomin, R. (1977). Temperament in early childhood. *Journal of Personality Assessment, 41,* 150–156.

Rowe, D. C., & Plomin, R. (1981). The importance of nonshared (E_1) environmental influences in behavorial development. *Developmental Psychology, 17*(3), 517–531.

Sameroff, A. J. (1983). Developmental systems: Contexts and evolution. In W. Kessen (Vol. Ed.) & P. H. Mussen (Series Ed.), *Handbook of child psychology: Vol. 1. History, theory, and methods* (pp. 237–294). New York: Wiley.

Sanson, A., & Rothbart, M. K. (1995). Child temperament and parenting. In M. H. Bornstein (Ed.), *Handbook of parenting: Vol. 1. Children and parenting* (pp. 299–321). Hillsdale, NJ: Erlbaum.

Scarr, S., & McCartney, K. (1983). How people make their own environments: A theory of genotype-environmental effects. *Child Development, 54,* 424–435.

Schmitz, S. (1994). Personality and temperament. In J. C. DeFries, R. Plomin,

& D. W. Fulker (Eds.), *Nature and nurture in middle childhood* (pp. 120–140). Cambridge, MA: Blackwell.

Slomkowski, C. L., Nelson, K., Dunn, J., & Plomin, R. (1992). Temperament and language: Relations from toddlerhood to middle childhood. *Developmental Psychology, 28,* 1090–1095.

Terman, L. M., & Merrill, M. A. (1973). *Stanford-Binet Intelligence Scale.* Boston: Houghton-Mifflin.

Thomas, A., Chess, S., & Birch, H. G. (1968). *Temperament and behavior disorders in children.* New York: New York University Press.

Thomas, A., Chess, S., Birch, H. G., Hertzig, M. E., & Korn, S. (1963). *Behavioral individuality in early childhood.* New York: New York University Press.

Wachs, T. D., & Gruen, G. (1982). *Early experience and human development.* New York: Plenum Press.

Wells, G. (1986). *The meaning makers: Children learning language and using language to learn.* London: Hodder & Stoughton.

24

Early Predictors of Problem Behavior at Age Four

Stephanie Schmitz
David W. Fulker
Robert N. Emde
Carolyn Zahn-Waxler

This chapter examines the antecedents of problem behavior in the MacArthur Longitudinal Twin Study, as indicated by the Child Behavior Checklist (CBCL/ 4-18; Achenbach, 1991) at age four. Specifically, we examined several antecedent characteristics of the child and the measured environment during the second and third years that may impose a risk for the later development of problem behavior. Because the assessment of problem behavior also depends on the rater, mother's and father's CBCL ratings were used separately in relationship to the child's earlier assessments. There is growing interest in clinical and developmental psychology in the antecedents and correlates of problem behavior in young children. This chapter covers a number of associations previously reported in the literature and examines them with a young twin sample.

Campbell (1991) noted that the definition of normal and abnormal behaviors in children does not imply specific problem behaviors but rather relates to the dimensions of their frequency, intensity, chronicity, constellation, and social context. Thus, the definition of disorder in young children should include patterns of symptoms that persist over time and situations, patterns that are likely to impede a child's ability to negotiate developmental tasks involving the self, the family, and, as the children get older, peers.

The CBCL is the most developed, empirically derived behavior rating scale for children and adolescents between the ages of 4 and 18 years (Barkley, 1988). First published in 1983 by Achenbach and Edelbrock and revised in 1991 by the first author, the CBCL aims for psychometric rigor. The reported good correspondence between CBCL scales and clinical diagnoses indicate it to be of clinical usefulness.

There are several studies that document the correspondence of the CBCL to psychiatric diagnostic measures, mainly the DSM-III-R (American Psychiatric Association, 1987). For the externalizing scale, Campbell and Ewing (1990) reported that the the summary externalizing symptom count from the DISC-P (Diagnostic Interview Schedule for Children—Parent version; Costello et al., 1982) yielded high correlations; similar results were obtained for the internalizing dimension. Biederman et al. (1993), in their sample of boys with attention deficit-hyperactivity disorder and control boys, aged 6–17 years, used both categorical diagnoses and dimensional measurements (quantitative deviation from normal, rather than discrete clinical entity). The authors reported excellent convergence between the CBCL scales and structured interview data.

The CBCL is appealing due to several additional advantages, including, for example, that of a large item pool, a broad range of symptom scales, generally good reliabilities, and norms that reflect both age and sex differences in the prevalence and patterns of symptoms. Using continuous scales, an advantage of the CBCL is that it can be used not only for clinical samples but also for meaningfully characterizing children displaying a normal range of behaviors.

STABILITY OF PROBLEM BEHAVIOR IN YOUNG CHILDREN

The question of whether problem behavior at a young age, such as at age four, is just a transitional problem or whether it really is a precursor of problem behavior at later ages, has variously been addressed in the literature (Campbell, 1991; Richman et al., 1982); generally, a continuity has been shown. Researchers set out to show that problem behavior in children is not of a transient nature but relatively stable; that is, finding predictive variables would be expected. Campbell and Ewing (1990) reported that 67% of the hard-to-manage preschoolers in their sample who showed clinically significant problems at age six met DSM-III-R criteria for externalizing disorders at age nine.

In their large-scale epidemiological study on behavior problems, Richman et al. (1982) found that 15% of 3 year olds showed mild problems, while an additional 7% showed moderate to severe problems. Stevenson and Richman (1978) and Stevenson et al. (1985) found that behavior problems at age three strongly predicted behavioral deviance at age eight, as well as a low score on a language measure relating to later problem behavior.

Children's problem behaviors in longitudinal studies were shown to be moderately stable over a 1 year follow-up period; initial symptom levels, maternal self-reported depression, and negative maternal control predicted follow-up ratings of externalizing problems (Campbell et al., 1991). Although only those children who showed persisting problems at age six met DSM-III criteria (American Psychiatric Association, 1980) for an externalizing disorder at age nine, earlier child behavior still was a good predictor, particularly when assessed through maternal ratings (Campbell & Ewing, 1990; Campbell et al., 1986). Although behavior problems may continue, they may not be expressed the same way at different ages (e.g., externalizing at a younger age and internalizing at a later; Campbell, 1991). Correlations for particular aspects of behavior for up to 7-year time periods were moderate, particularly those regarding more withdrawn behavior, whereas those for total problem scores were larger (Achenbach, 1992; Schmitz et al., 1995).

Achenbach et al. (1995) reported on a national normative sample of initially 4–12 year olds a mean correlation of .42 for all scales of the CBCL over a period of 6 years, with highest continuities for the total problem score, externalizing, aggression, and attention problems. The variance accounted for at the later age point varied between 18% (somatic complaints) and 45% (aggression), with similar rates in boys and girls. The strongest predictor for each syndrome was that particular syndrome at the previous age point, rather than the remaining variables (socioeconomic status, ethnicity, geographic location with the USA), both in combination with other predictors and when looking only at the unique contributions.

Factors contributing to behavior problems in children fall into several domains, namely those of (1) the child (such as nature of conduct problems, age, sex, race, temperament, and problem-solving abilities); (2) the environment (such as life events, particularly negative ones, and peers), and (3) the family (such as composition, parental behavior, parenting/child-rearing skills, personal and marital distress, and socioeconomic status [SES]; Campbell, 1995). As these factors may have different influences as antecedents for different aspects of problem behavior, we review them below.

CHILD CHARACTERISTICS AS ANTECEDENTS

Temperament has been proposed as an antecedent of problem behavior and "difficult temperament" in particular. Thomas and Chess (1982), in the New York Longitudinal Study (NYLS), showed associations with later adjustment scores from age three onward. All aspects of temperament as measured by the questionnaires developed by Carey and colleagues, which are based on the nine NYLS temperament categories, have been shown to be under significant heritable influences (see Cyphers et al., 1990), regardless of the child's age.

Campbell (1991) noted that maternal reports of difficult temperament in infancy, primarily irritability and fussiness, predicted later ratings of behavior problems at preschool and kindergarten age. Regarding reported behavior problems, internalizing problems, such as anxiety and fearfulness, were less likely to persist than externalizing problems, characterized by aggression, noncompliance, and poor impulse control. Using a longitudinal sample of adolescents who were identified as being extremely (un)inhibited as toddlers, Schwartz and co-workers (1996) reported group differences regarding externalizing but not internalizing problems.

Rende (1993) examined correlations between different aspects of temperament in infancy and early childhood, as measured with the Emotionality Activity, Sociability, Impulsivily (EASI)/Colorado Childhood Temperament Inventory questionnaire (Buss & Plomin, 1984; Rowe & Plomin, 1977), with problem behavior when the children were 7 years old. The emotionality aspect of temperament was most consistently related to problem behavior at later ages, while there were only sporadic relations for activity and sociability. In boys, high emotionality in infancy and early childhood was associated with later anxiety/depression, the ratings at each age group explaining 4% of the variance. For girls, high emotionality and low sociability predicted anxiety/depression, with 9% of the variance explained from ratings in infancy and 8% in early childhood. No link was detected between any of the EASI dimensions and delinquent behavior.

Using CBCL data from this sample when the children were 5 years of age, Zahn-Waxler et al. (1996) related internalization of control, as indicated by frustration tolerance and good impulse control at earlier ages, to a lesser amount of externalizing problems in teacher-rated behavior.

Prematurity at birth has been mentioned as a possible factor influencing temperament and has therefore been associated with problem behavior. Oberklaid and colleagues (1991), addressing problems due to prematurity at birth, reported no significant differences on individual dimensions or clinical categories of temperament between premature and full-term infants. This makes it unlikely that our sample of twins, who are more likely to be born prematurely, would show more problem behavior than singletons.

The relationship between cognitive variables and temperament has been addressed by Maziade et al. (1987). In their study, children who were more temperamentally difficult displayed higher IQs, particularly in middle and upper socioeconomic classes. Campbell et al. (1982), in contrast, reported that the children in the parent-referred problem group had a lower IQ, but they were unable to replicate this with a larger sample. Thus, the mediating role of cognitive factors on temperament and problem behavior in young children is unclear.

In an attempt to maximize predictive power, Oberklaid et al. (1993) reported that the combination of overall ratings of temperament and sex were a better predictor of problem behavior than difficult temperament alone. For the preschool years, however, sex differences regarding problem behavior seem to be less important (Earls, 1987). Campbell (1995), in her review of the literature, states that although Prior et al. (1992) found a higher incidence of externalizing problem behavior in boys, other studies have revealed relatively trivial sex differences for preschoolers.

FAMILY AND ENVIRONMENTAL FACTORS AS ANTECEDENTS

Problem behavior in children can be either a reflection of, or is at least influenced by, parental problems, or family factors predicting later conduct problems. In the Canadian sample of Maziade et al. (1989), however, neither indicators of SES nor communication style added significantly to the explained variance of behavior problems.

Richman (1977) showed that factors other than characteristics of the child are predictive of the child's behavior problems—for example, a strained marital relationship between the parents. Lancaster et al. (1989) showed that maternal characteristics were good predictors of children's behavior ratings. Children of depressed mothers showed more behavioral problems (Fergusson et al., 1993; Wolkind & De Salis, 1982); however, the direction of causation is unclear in this case (Spiker et al. 1992).

Lancaster et al. (1989) argued that maternal ratings, even though biased, are the most viable means to know about a preschooler's behavior over an extended period of time. Earls (1980) reported that fathers underestimate behavior problems in their children, which might either be due to fathers having less information about their children, the children behaving differently for each parent, mother/father characteristics influencing their perceptions differently, or a combination of all of these factors.

Williams et al. (1990) reported that, among other characteristics, the marital status of the parents distinguished between children with and without emotional disorder. Meyer et al. (1993) modified these observations by showing that divorce led to increased conduct disorder in girls but not in boys, further stressing the importance of looking at the combination of risk factors.

One of the few studies addressing the relationship between marital data and children's problem behavior with a genetically informative sample is that of Braungart-Rieker et al. (1995). They reported that interparental conflict was associated with a greater incidence of childhood adjustment problems, with boys being at a greater risk for externalizing problems. Because in their sample mother and father ratings on the Family Environment Scale (FES; Moos & Moos, 1981) correlated highly (between .37 to .71), and the correlations between the three time points (when the children were 1, 3, and 5 years old) were also substantial (ranging from .60 to .85 for mother ratings and from .56 to .81 father ratings), FES scores were averaged across time and parents as a way of data reduction. However, even though parental FES correlate highly with each other, there are differences in how these scales correlate with parental ratings of their children's behavior. Schmitz and Fulker (1996) showed that different aspects of the family environment were associated with parental ratings of children's problem behavior.

We used the MacArthur Longitudinal Twin Study study to test (1) the associations between temperament and various other domains, expecting significant correlations with temperament; (2) whether these associations differ for boys and girls, assuming that sex differences do not yet show up at this age; and (3) whether certain associations are specific to the rater—that is, a parent or a tester.

METHODS

Sample

The sample consisted of all those pairs of twins for whom either their mother, father, or both had completed the CBCL for 4 to 18 year olds (Achenbach, 1991; Achenbach & Edelbrock, 1983) at age 4 years. These assessments were then related to data that are available for these children from earlier ages. Maternal CBCL ratings were available for 234 individual monozygotic (MZ) twins (107 female and 127 male) and 215 individual dizygotic (DZ) twins (81 female and 134 male). Paternal CBCL ratings are available on 176 MZ twins (80 female and 96 male) and 161 DZ twins (66 female and 95 male). The lower sample size for paternal ratings reflects the decreased cooperation of fathers over the duration of the study.

Measures

Problem behavior was first assessed with the CBCL (see above) at age four. This questionnaire has widely been used in both research (see introductory chapters) and as a screening instrument in clinical practices (e.g., Garrison & Earls, 1985). The broadband groupings of internalizing and externalizing as well as the total problems score were used as indicators of problem behaviors. Due to the differences in assessment by mothers versus fathers, separate ratings by each parent were used.

The following measures were used as potential antecedents of later problem behavior; for a more detailed description of the measures, see chapter 2.

Temperament

Questionnaire measures for the temperament domain at 14, 20, 24, and 36 months included a modified version of the Colorado Childhood Temperament Inventory (CCTI; Buss & Plomin, 1984; Rowe & Plomin, 1977) and the Toddler Temperament Survey (TTS; Fullard et al., 1984), which were both completed by the twins' parents. Apart from items that are summed to form scales, the TTS also asks questions regarding the difficulty of the child's temperament. At the first three age-points, the tester-rated Infant Behavior Record (IBR; Bayley, 1969) was also used. At age 3, a modified version of the IBR was used, with more age-appropriate items (Plomin et al., 1988a). Further observational measures included an unrotated first principal component score as a composite measure of shyness, derived from video ratings of the children's reactions during the first 5 min of both the home and the lab visit. The first 5 min followed a standard procedure developed by Plomin and Rowe (1979). Video ratings were also used for behavioral inhibition (Kagan et al., 1988; Robinson et al., 1992). This aspect of temperament is described by averaging z-scores for for seven measures (e.g., the latency to leave the mother after the child entered a playroom in the lab). Aggression and prosocial behavior ratings were compiled from video ratings of the free-play session at the end of each home visit (Zahn-Waxler et al., 1992).

Emotion

At 14 and 36 months, the twins' mother completed the Differential Emotions Scale (DES; Izad, 1972; modified from Fuenzalida et al., 1981), a measure to assess the experience of the child's discrete emotions. From video recordings of the home visit, the child's hedonic tone (Easterbrooks & Emde, 1983) was assessed both during cognitive testing and during the unstructured free play at 14, 20 and 24 months. Empathic response rating scales developed by Zahn-Waxler and colleagues (Zahn-Waxler et al., 1979, 1992) were used to create a composite rating of the child's reaction to various empathy probes during the home and the lab visit at 24 and 36 months.

Language and Cognition

Standardized measures for testing at 14, 20, and 24 months included the Mental Development Index (MDI) of the Bayley Scales of Infant Development (Bayley, 1969), an index of general intelligence, and the Sequenced Inventory of Communication Development (SICD; Hedrick et al., 1975), a cumulative assessment of expressive and receptive language development. An abbreviated age-appropriate form of the SICD was used at 36 months, along with the Stanford-Binet Intelligence Scale Form L-M (Terman & Merrill, 1973) and the Peabody Picture Vocabulary Test (PPVT, Form L; Dunn & Dunn, 1981).

Parental Measures

Because it has been suggested in the literature that problem behavior in children can be either a reflection of, or at least influenced by, parental problems, or that family factors predict later conduct problems (Loeber & Stouthammer-Loeber, 1986), a number of relevant measures were requested from the parents at the 14 and 36-month home visit. These included the Family Environment Scale (FES; Moos & Moss, 1981) in a shorter version yielding eight scales of five items each (Plomin et al., 1988b), the Dyadic Adjustment Scale (DAS; Spanier, 1976) to assess satisfaction with the marital relationship, and the Eysenck Personality Inventory (EPI; Eysenck & Eysenck, 1964) to assess the parental personality dimensions of extraversion and neuroticism.

RESULTS

Descriptive Data on the CBCL at Age Four

Table 24.1 gives the means and standard deviations on the CBCL broadband groupings as well as the total problem score, separately for boys and girls at

Table 24.1. Means and standard deviations on each Child Behavior Checklist (CBCL) broadband grouping and comparison to norm sample

CBCL scale	MZ	DZ	Norms[a]
Boys			
Total, mothers	21.85 ± 12.98	25.31 ± 18.37	24.2 ± 15.6
Total, fathers	22.92 ± 12.67	21.35 ± 12.92	
Externalizing, mothers	10.20 ± 6.73	10.91 ± 8.27	9.8 ± 7.0
Externalizing, fathers	10.27 ± 6.25	9.38 ± 6.64	
Internalizing, mothers	2.83 ± 2.58	4.52 ± 4.95	5.5 ± 4.7
Internalizing, fathers	2.61 ± 2.33	3.20 ± 3.20	
Girls			
Total, mothers	19.80 ± 12.49	22.65 ± 15.96	23.1 ± 15.5
Total, fathers	22.83 ± 16.17	18.89 ± 13.80	
Externalizing, mothers	7.87 ± 6.31	8.77 ± 6.33	8.2 ± 6.1
Externalizing, fathers	8.88 ± 7.05	7.56 ± 6.09	
Internalizing, mothers	3.84 ± 3.08	4.26 ± 4.41	6.3 ± 5.5
Internalizing, fathers	3.96 ± 3.79	2.98 ± 3.07	

[a] Norms for this age group, regardless of sex of rater (Achenbach, 1991). See text for sample sizes.

age four. For comparison, means from the 1991 norm samples are listed. Both parents rated the twins within the normal range, with mothers usually reporting slightly higher means than fathers, as previously been reported in the literature (Campbell, 1991; Earls, 1980).

Sex differences regarding parental ratings on the CBCL were apparent for maternal ratings regarding the total problem score, attention, delinquent, aggression, and externalizing, such that boys' scores were higher than girls' scores. Paternal ratings were affected by the child's sex regarding somatic complaints, attention, aggression, and externalizing such that the mean for boys was greater than that for girls. These differences, however, explained only between 1.0 and 2.7% of the variance. Even though these mean differences between the sexes in parent-rated behaviors exist, this does not preclude that the same etiological factors can be operating.

Phenotypic Correlations between Potential Predictors and CBCL

Temperament

The CCTI scales associated with problem behavior showed predictive consistency across time, in that mothers' assessment of the children's emotionality at all four ages correlated significantly with mothers' ratings of internalizing, externalizing and total problem score at age four. Additionally, the CCTI scale

Table 24.2. Significant correlations between parental Colorado Childhood
Temperament Inventory and Child Behavior Checklist

	14 Months	20 Months	24 Months	36 Months
Emotionality, maternal ratings				
Internalizing	.20***	.18***	.26***	.22***
Externalizing	.19***	.22***	.22***	.30***
Total problems	.20***	.20***	.25***	.28***
Emotionality, paternal ratings				
Internalizing		.17**	.13*	.17*
Externalizing	.12*	.20***	.28***	.20**
Total problems		.200***	.25***	.22**
Shyness, maternal ratings				
Internalizing	.22***	.16***	.26***	.25***
Externalizing				
Total problems	.13**		.13**	.11*
Shyness, paternal ratings				
Internalizing		.21***	.24***	.23**
Externalizing				
Total problems		.12*	.19**	

Number of observations, 380–426.
***p < .001; **p < .01; *p < .05.

shyness was predictive of mother-assessed internalizing scores in both sexes,
again, at all four age points. Table 24.2 presents some correlations of parent-
rated temperament and later problem behavior.

In addition to the correlations reported in table 24.2, soothability at 20 and 24
months as well as persistence at 36 months were predictors of maternally assessed externalizing scores for both sexes. Similar to maternal ratings, the father's
impression of the child's shyness at 20, 24, and 36 months correlated significantly
with internalizing at 4 years. Some sex differences existed in that the father's assessment of daughters' emotionality across the four age points was significantly
correlated with his ratings of their internalizing, externalizing and total problem
scores at age four but less consistently so for the ratings of sons.

Although some paternal ratings of temperament were consistently correlated
with later behavior ratings, maternal ratings showed a greater degree of consistency across time. Our results confirm those of Rende (1993), who reported
a significant correlation between emotionality and later problem behavior.

Mother's assessment of the children's temperament at 14 months correlated
significantly, for both boys and girls, with later problem behavior (total prob-

lems score of the CBCL) with respect to adaptability, intensity, and mood; in addition, approach-withdrawal and mood correlated with the internalizing score at 4 years. Adaptability, intensity, and mood correlated with externalizing scores; however, none of the associations at later age points was found to be either consistent or statistically significant. None of the father's TTS scale ratings that correlated with father's assessment of problem behavior was the same for boys and girls. The "difficult temperament" score did not show consistent associations with aspects of problem behavior either.

Observer-rated behavior, as indicated by the IBR, showed only occasional and inconsistent correlations with later problem behavior, such that activity ratings at 20 and 36 months correlated negatively ($r = -.13$ and $r = -.18$, respectively) with internalizing scores. Similarly, measures of observer-rated shyness correlated at low levels and inconsistently with parent-rated problem behaviors at the later age; for example, shyness as assessed during the 20- and 24-month home visits correlated to .11 and .10, respectively, with later rated internalizing scores.

Emotion

The DES records how often the mother has seen her child express a range of feelings. Maternal ratings of the child's emotions correlated with all three aspects of problem behavior assessed with the CBCL, with mainly emotions of sadness and anger correlating positively with the CBCL problem behavior scales (.10 to .30) and enjoyment correlating negatively ($-.11$ to $-.22$) with them. The DES scales which correlated with the father's CBCL ratings for both boys and girls were enjoyment at 36 months with the total problem score ($-.14$), and shyness/shame and the total negative score at 36 months with internalizing (.17 and .19, respectively). Although these the correlations between experienced emotions and problem behavior are of a low level (between .10 and .30), they nevertheless show the consistent association of the perceived anger and sadness with all aspects of maternal rated problem behavior.

Empathic response ratings at 24 months were consistently correlated with later problem behavior for girls but not for boys. Girls who displayed indifferent behavior during the empathic probes were rated as displaying more problem behavior, both in the internalizing and externalizing domain, at age four, with correlations being of a similar magnitude for both maternal and paternal ratings (r ranging from .20 to .35). Moreover, empathic concern was related to the mother's assessment of all three aspects of problem behavior at age four (r ranging from $-.19$ to $-.24$), while hypothesis testing at age two during empathic probes correlated significantly with the father's assessment of total behavior problems and internalizing scores at age four for girls (r of .25 and .30, respectively).

Empathic response ratings at 36 months did not correlate with problem behavior at later ages, showing that levels of early empathic development are not consistently correlated to later expressions of problem behavior. Zahn-Waxler et al. (1995) had previously reported no correlations between early empathy and problem behavior.

Language and Cognition

Language and cognition were predictive of girls' problem behavior, both when assessed by the mother and the father. Generally speaking, the lower the girls' MDI and the less advanced their expressive and receptive language development, the more items were endorsed by their parents regarding internalizing, externalizing and total problem score, with correlations ranging from $r = -.15$ (MDI at 14 months and maternally rated externalizing) to $-.28$ (PPVT and paternally rated externalizing). While it seemed that language was associated with parent-rated problem behavior in girls, there were no significant relations for boys.

Parental Measures

For both mothers and fathers, their impression of the family environment, assessed by the FES scales, and that of the marital relationship, as assessed by the DAS, correlated with ratings of later problem behavior. However, particular aspects of the FES associated with later CBCL differ by the parents (see also Schmitz, 1996; Schmitz & Fulker, 1996). For maternal ratings, the more emotional aspects of the family environment seem to be more closely related to the assessment of their children's problem behavior, as evidenced by the associations with scales such as cohesion (r ranging from $-.13$ to $-.17$) and conflict (r ranging from .17 to .24). Mothers who experienced their families as less cohesive and having more conflict tended to rate more problem behavior in their children later. For paternal ratings, the more structural aspects of family life seemed to be more important, as shown by the correlations with activities (r ranging from $-.18$ to $-.20$) and family organization (r ranging from $-.19$ to $-.30$). Fathers, who were under the impression that family members participated in few activities together or that responsibilities within the family were not clear, were more prone to rate higher levels of problem behavior in their children. Thus, parents' ratings of their children's behavior seemed to be related to how they experienced the family as a whole. Conversely, the children's behavior might influence the way a family functions. The DAS total score correlated between $-.11$ and $-.25$ with aspects of problem behavior, meaning parents who are satisfied with their marital relationship reported less problem behaviors in their children.

Maternal neuroticism scores correlated with all three dimensions of the CBCL for girls (mother's and father's ratings; r ranging from .15 to .29) but correlated slightly less so for boys. The correlations between father's neuroticism and extraversion and children's problem behavior were less consistent. The relationship between parent's personality and the problem behavior ratings in their children could be due to several factors: the expression of problem behavior is mediated through personality factors and the child inherited the parent's personality (heritability estimates from meta-analyses are about .30 for the major personality domains; Loehlin, 1992), or it could be that more neurotic parents see more problem behavior in their children.

Behavior Genetic Models

It is not within the limits of this chapter to explore the underlying etiologies of all these observed correlations. To demonstrate the ability of behavior genetic designs to address these questions with genetically informative data, two questions are treated more extensively: the nature of the association between temperament and later problem behavior, since CCTI scales were consistently correlated with CBCL measured problem behavior, and a possible model regarding parental (dis)agreement in rating their children's behavior.

Covariation between Temperament and Problem Behavior

The CCTI-assessed temperament was fairly stable during early childhood, particularly from one age point to the next. These correlations were similar for boys and girls. Maternal ratings of all domains assessed by the CCTI for boys and girls correlate highly, between .44 (from 14 to 36 months) and .62 (from 20 to 24 months) for emotionality, and between .43 (from 14 to 36 months) and .70 (from 20 to 24 months) for shyness. Not surprisingly, the range included higher values for adjacent ages, ranging from .54 to .70 for both. These correlations indicate that either the children's temperament stays relatively stable over the period from 14 to 36 months, or that mothers experience their children as the same, or both. Correlations for paternal ratings showed a similar pattern and were of similar magnitude.

As shown in table 24.2, maternal ratings of children's temperament, as assessed by the CCTI, and children's problem behavior, as assessed by the CBCL, correlated substantially. Applying a bivariate Cholesky decomposition (see chapter 2), the magnitude of genetic and environmental influences on these domains as well as their associations were estimated. Comparing models in which these parameters were estimated separately for boys and girls with those that were applied to the whole sample led to the conclusion that no sex-specific estimation procedures were needed.

As already reported in the literature (Plomin, et al., 1991; Schmitz, 1994; see also chapter 6, this volume), individual variation in aspects of maternal-rated temperament is due to genetic and nonshared environmental influences but not due to shared environmental influences. Although several approaches are possible to model these covariations, a full *ACE* model was used for the following analyses, that is, additive genetic (*A*) and shared (*C*) as well as nonshared (*E*) environmental influences are hypothesized. For a slightly different model leading to similar results, see Schmitz et al. (1999).

Table 24.3. Components of variance for the Colorado Childhood Temperament Inventory (CCTI), based on a number of bivariate analyses of the CCTI and the Child Behavior Checklist

Measure	14 Months	20 Months	24 Months	36 Months
Emotionality				
h^2	.33*	.47*	.23*	.22*
c^2	.02	.00	.06	.03
e^2	.65	.52	.71	.75
Shyness				
h^2	.23*	.34*	.33*	.43*
c^2	.00	.02	.00	.01
e^2	.77	.64	.67	.56

h^2 denotes variance due to genetic influences, c^2 variance due to shared environmental influences, and e^2 variance due to nonshared environmental influences, including measurement error.
*$p < .05$.

Tables 24.3–24.5 present the results of the bivariate behavior genetic analyses of CCTI and CBCL—that is, the results as they were obtained from the joint analyses of CCTI and CBCL scales. Table 24.3 shows the components of variance in two aspects of temperament in the current sample, emotionality and shyness. Table 24.4 shows the components of variance for the CBCL broadband groupings at age four, again showing the estimates from the bivariate analyses, and table 24.5 shows the genetic and environmental correlations between temperament at the preceding four age points and problem behavior at age four. Due to the bivariate analyses which take into account the covariance between the two measures, parameter estimates in tables 24.3 and 24.4 may vary from those arrived at with univariate analyses (e.g., see chapter 6, this volume). All behavior genetic analyses were conducted with the structural equation modeling package Mx (Neale, 1994).

Table 24.4. Components of variance for the Child Behavior Checklist (CBCL), based on a number of bivariate analyses of the Colorado Childhood Temperament Inventory and CBCL

Measure	Internalizing	Externalizing	Total problems
h^2	.26*	.30*	.13*
c^2	.42*	.51*	.72*
e^2	.32	.19	.15

h^2 denotes variance due to genetic influences, c^2 variance due to shared environmental influences, and e^2 variance due to nonshared environmental influences, including measurement error.
*$p < .05$.

Table 24.5. Genetic and environmental correlations between Colorado Childhood Temperament Inventory and Child Behavior Checklist, based on bivariate analyses

Correlation	14 Months	20 Months	24 Months	36 Months
Emotionality and internalizing				
rG	.57	.57*	.29	.59
rC	1.00	.00	1.00	1.00
rE	.01	.08	.18*	.01
Emotionality and externalizing				
rG	.53(*)	.66*	.29	.46
rC	.00	1.00	1.00*	1.00(*)
rE	.08	.05	.02	.18(*)
Emotionality and total score				
rG	.75	1.00*	.16	.55
rC	1.00*	.00	1.00*	1.00*
rE	.09	.15	.14	.14
Shyness and internalizing				
rG	1.00*	.29	.53	.96*
rC	.00	.00	.00	1.00
rE	.07	.07	.33*	.18*

rG stands for the genetic correlation, rC for the shared environmental correlation, and rE for the nonshared environmental correlation.
*p < .05; (*)p < .10.

Table 24.3 illustrates that heritable influences were statistically significant at each age point for both emotionality and shyness, while c^2 estimates were either zero or not significantly different from zero. The magnitude of non-shared environmental influences is substantial, meaning that factors that make children in the same family different from each other are important. Problem behavior in children of this age, in contrast, seems to be due to genetic as well as shared and nonshared environmental influences, as shown in table 24.4.

Both genetic as well as shared environmental influences could not be excluded from the model without a significant deterioration of fit. Although shared environmental factors account for 42% and 51% of the variation in internalizing and externalizing at 4 years of age, respectively, their impact on the total problem score is even larger, accounting for more than 70% of the variation. This high estimate of c^2 might be due to the fact that the ratings were completed by the same person or that parents of young children might be reluctant to report problems (e.g., because they believe they will grow out of it). These estimates of shared environmental influences are higher than those reported in the literature, for example, by Edelbrock, et al. (1995) for school-aged children. However, heritable influences seem to increase over time, while shared environmental ones decrease (Loehlin, 1992; Schmitz et al., 1995; Silberg et al., 1994), and the current sample is considerably younger than samples cited in the literature.

As shown in table 24.5, the observed covariation between CCTI at younger ages and CBCL at age four is mainly due to common genetic influences, as well as some small shared and nonshared environmental influences at the later age points. Genetic influences and those specific to each twin contribute to the observed moderate covariation between temperament and problem behavior at the different age points. Because shared environmental influences play a role only in the etiology of problem behavior at this young age but do not contribute significantly toward individual differences in temperament, they generally do not contribute toward the covariation between these two domains. Even though the same shared environmental factors may influence temperament and problem behavior, leading to an r_c of 1.0, as is, for example, the case for emotionally and internalizing at 14 months, this seemingly high correlation is nonsignificant because the shared environment does not contribute significantly to individual variation in emotionality (see table 24.3; c^2 estimate of .02). These shared environmental influences on problem behavior are further explored through behavior analyses of rater bias.

Parental Agreement in Ratings

Throughout the literature as well as this book (chapter 7), the issue of low parental agreement in rating children's behavior has been discussed. The agreement between parental ratings at age four on the CBCL was moderate, as was to be expected from the literature (Achenbach et al., 1987). Table 24.6 shows the parental agreements separately by the four sex and zygosity groups.

Curiously, parents of DZ twins agree more on their children's behavior ratings than do parents of MZ twins. Reasons for this discrepancy need to be explored further and might be related to differences in how parents of MZ twins emphasize behavioral contrasts in order not to see their children as a unit. Saudino and Cherny (chapter 7) addressed the question of possible me-

Table 24.6. Correlations between maternal and paternal Child Behavior Checklist ratings at age four

Scale	MZ		DZ	
	Boys	Girls	Boys	Girls
Total score	.13	.27*	.55***	.50***
Externalizing	.17	.35***	.50***	.49***
Internalizing	.18	.17	.49***	.51***

***p < .001; *p < .05.

diators of these discrepancies by partitioning the observed parental correlations into components of genetic and environmental origin, using CCTI ratings. Applying this approach, a Cholesky decomposition, to the CBCL data (Schmitz, 1996; Schmitz & Fulker, 1996), parental ratings showed a high genetic correlation (usually unity). The low shared environmental correlations, ranging from .20 for total problems to .24 for internalizing, point toward different aspects of the child's behavior being rated, either because mothers and fathers experience their children in different situations, and/or they experience their children differently in general.

Another approach involves the so-called rater bias model (Hewitt et al., 1992, used CBCI/4–18 data; van den Oord et al., 1996, applied the model CBCL/2–3 data). This approach partitions the observed variance in the CBCL broadband groupings into different sources: those due to parental perception or situational specificity (subsumed under "bias" in this model), those due to measurement unreliability, and true trait variance (e.g., internalizing or externalizing). Latent genetic and environmental factors are assumed to influence this true trait variance. By allowing this variance to have a different impact on maternal and paternal ratings, the amount of bias can differ for each parent.

Parameters in this model did not need to be estimated separately for boys and girls for this sample. The heritability estimate of the traits by this form of analysis increased over those reported in table 24.4, as is often seen in rater bias models (Hewitt et al., 1992). This is due to shared environmental influences being separated into those due to parental bias and additional ones influencing the true variance. Although all shared environmental influences for externalizing are explained through parental reporting bias, there are additional shared environmental influences on internalizing and the total problem score (34% and 30% of the latent trait variance, respectively). For the total problem score and externalizing, the true trait variance had more of an impact on paternal ratings, and thus the analyses showed less bias in those ratings, whereas maternal rating bias contributed a larger percentage toward the observed variance. The reverse situation was observed for internalizing: paternal ratings showed more rating bias, and the impact of the true trait variance was smaller than for maternal ratings (detailed results available from the authors upon request).

DISCUSSION

As had already been shown by Thomas and Chess in 1982, correlations between difficult–easy temperament scores during the first 2 years of life show no significant correlations with later adjustment scores. However, from age three onward, these temperament scores do correlate moderately (r values in the .30 range) with adjustment ratings in their studies (Thomas & Chess, 1982).

Parameter estimates regarding genetic influences on problem behavior in the bivariate analyses (table 24.6) are lower than those reported by Edelbrock et al. (1995) for school-aged children. Edelbrock and colleagues reported heritable influences to be larger than shared environmental ones; for internalizing they estimated an h^2 of .42 and a c^2 of .25; for externalizing the corresponding estimates were .45 and .28. These discrepancies support the theory that genetic influences on personality, and presumably related domains such as problem behavior, increase with age (see Loehlin, 1992). At younger ages, therefore, the latent shared environmental factors, such as those indicated by the FES, seem to exert a greater influence.

Sanson et al. (1991) reported that single risk factors, including difficult temperament, resulted in only modest increases in the prevalence of later maladjustment. However, certain combinations of risk factors were associated with markedly increased prevalence rates; for example, difficult temperament, when associated with any of the three biological factors of male sex, perinatal stress, or being born prematurely, increased the risk of negative outcome. In the current sample, sex differences were not detected in the association between aspects of temperament and problem behavior.

Difficult temperament, long thought to be a precursor of problem behavior (Rende, 1993; Thomas & Chess, 1982), did not show the expected correlations in a number of longitudinal studies (e.g., Katz & Gottman, 1993). In the current sample of 4-year-old twins, the correlations between problem behavior and the difficult temperament construct were either nonsignificant or inconsistent.

Generally, when in the literature maternal characteristics were taken into account, they provided greater contributions to the maternal perceptions of the child, rather than the observed child behavior per se. The reported correlations between maternal ratings on the CCTI and the DES with those on the CBCL, coupled with the lack of significant correlations between observer rated temperament and later problem behavior, seem to support this point of view.

The above results raise questions about the sources of parental behavior or perceptions of the child that are relevant to later individual differences in CBCL ratings. Some of those possibilities have been discussed in the literature. Plomin and Bergeman (1991) hypothesized that parental behavior, as assessed on measures of the home environment, might be due to characteristics of the parents as well as those of the children. Using a twin sample, they reported significant heritability estimates for the FES scales of expressiveness, cultural orientation, organization, and control. Within-pair twin correlations indicated significant genetic influences for the warmth dimension (relationship), but no genetic influences were found for the control/system maintenance dimension. The relationship between estimates of the shared environment as well as parental rating bias and measures of the family environment in this sample, as assessed by the FES and the DAS, was statistically significant (Schmitz, 1996; Schmitz & Fulker, 1996).

A more complex design also could take other parental characteristics into consideration. Even though Plomin et al. (1988b) reported that correlations of FES scales with personality variables were low, aspects of the parental personality might be mediating the association between the ratings of the family environment and of children's problem behavior.

Although Garrison and Earls (1985) postulated that parents might agree less when rating internalizing aspects of behavior, parental correlations in the current study are similar for externalizing and internalizing, particularly when rating their DZ twin children. This not only applies to the phenotypic correlation (table 24.6) but also to the genetic and environmental correlations, which were similar for both broadband groupings (Schmitz, 1996; Schmitz & Fulker, 1996).

In extending a rater bias model to two time points, Schmitz and Fulker (1995) showed that the continuity in child behavior ratings is partly due to continuity in the genetic and environmental influences on behavior, but also partly due to continuity in the bias on the part of the rater. Although maternal bias was correlated across time in a similar magnitude for boys and girls, the paternal bias correlation was lower for boys than for girls. The association between early childhood parameters and later problem behavior obviously warrants more detailed studies, particularly with respect to the normal range of behavior and genetically informative samples.

CONCLUSIONS AND SUGGESTIONS FOR FURTHER STUDY

Considering the many associations with problem behavior reported in this chapter, the following results stand out:

- Individual differences in problem behavior at age four are correlated with a number of domains earlier in life. Correlations with maternally rated temperament and emotions during the child's second and third years are are the most consistent ones. Correlations with paternal ratings showed less consistency, and observer rated behavior of the child was not predictive.
- Both genetic and shared environmental influences contributed to individual differences in problem behavior at age four. Parental rating bias and/or situational specificity made up a large proportion of the shared environment. Once this was taken into account, the results showed that genetic influences were stronger for externalizing than for internalizing.
- Generally, no sex differences were found with respect to the prediction of individual differences in problem behavior at age four. The only exceptions to this were the correlations between earlier cognitive measures and empathic responses with later problem behaviors for girls. Although this seems to imply that general cognitive ability and language development may mediate the association with problem behaviors for girls, this finding is only suggestive and requires still further investigation.

- Questions of situational and/or relationship specificity also need to be addressed in further studies, since the results reported in this chapter differ according to who assessed the child. In other words, we need to know more about how much it matters who assesses a child and why this is so.
- Finally, future research needs to address questions about patterns of profiles of behavior. Some research reported in the literature has demonstrated a small increase in predictability when several variables were combined; thus, future research could aggregate a number of variables. Moreover, the MacArthur Longitudinal Twin Study sample on which our analyses are based is a low-risk, nonclinical sample. Future research involving groups of children showing extreme early behaviors might point to more consistent patterns of associations with later problem behaviors.

REFERENCES

Achenbach, T. M. (1991). *Manual for the Child Behavior Checklist/4–18 and 1991 Profile*. Burlington: Department of Psychiatry, University of Vermont.

Achenbach, T. M. (1992). *Manual for the Child Behavior Checklist/2–3 and 1992 Profile*. Burlington: Department of Psychiatry, University of Vermont.

Achenbach, T. M., & Edelbrock, C. (1983). *Manual for the Child Behavior Checklist and Profile*. Burlington: Department of Psychiatry, University of Vermont.

Achenbach, T. M., Howell, C. T., McConaughy, S. H., & Stanger, C. (1995). Six-year predictors of problems in a national sample of children and youth: I Cross-informant syndromes. *Journal of the American Academy of Child and Adolescent Psychiatry, 34*(3), 336–347.

Achenbach, T. M., McConaughy, S. H., & Howell, C. T. (1987). Child/adolescent behavioral and emotional problems: Implications of cross-informant correlations for situational specificity. *Psychological Bulletin, 101,* 213–232.

American Psychiatric Association (1980). *Diagnostic and statistical manual of mental disorders* (3rd ed). Washington, DC: American Psychiatric Association.

American Psychiatric Association (1987). *Diagnostic and statistical manual of mental disorders* (3rd ed. revised). Washington, DC: American Psychiatric Association.

Barkley, R. A. (1988). Child behavior rating scales and checklists. In M. Rutter, A. H. Tuma, & I. S. Lann (Eds.), *Assessment and diagnosis in child psychopathology* (pp. 113–155). New York: Guilford Press.

Bayley, N. (1969). *Manual for the Bayley Scales of Infant Development*. New York: Psychological Corporation.

Biederman, J., Faraone, S. V., Doyle, A., Lehman, B. K., Kraus, I., Perrin, J., & Tsuang, M. T. (1993). Convergence of the Child Behavior Checklist with structured interview-based psychiatric diagnoses of ADHD children. *Journal of Child Psychology and Psychiatry, 34*(7), 1241–1251.

Braungart-Rieker, J., Rende, R. D., Plomin, R., DeFries, J. C., & Fulker, D. W. (1995). Genetic mediation of longitudinal associations between family environment and childhood behavior problems. *Development and Psychopathology, 7,* 233–245.

Buss, A. H., & Plomin, R. (1984). *Temperament: Early developing personality traits.* Hillsdale, NJ: Erlbaum.

Campbell, S. B. (1991). Longitudinal studies of active and aggressive preschoolers: Individual differences in early behavior and in outcome. In D. Cicchetti & S. L. Toth (Eds.), *Internalizing and externalizing expressions of dysfunction* (pp. 57–89). Hillsdale, NJ: Erlbaum.

Campbell, S. B. (1995). Behavior problems in preschool children: A review of recent research. *Journal of Child Psychology and Psychiatry, 36,* 113–149.

Campbell, S. B., & Ewing, L. J. (1990). Follow-up of hard-to-manage preschoolers: Adjustment at age 9 and predictors of continuing symptoms. *Journal of Child Psychology and Psychiatry, 31,* 871–889.

Campbell, S. B., Ewing, L. J., Breaux, A. M., & Szumowski, E. K. (1986). Parent-referred problem three-year-olds: Follow-up at school entry. *Journal of Child Psychology and Psychiatry, 27,* 473–488.

Campbell, S. B., March, C. L., Pierce, E. W., Ewing, L. J., & Szumowski, E. K. (1991). Hard-to-manage preschool boys: Family context and the stability of externalizing behavior. *Journal of Abnormal Child Psychology, 19,* 301–318.

Campbell, S. B., Szumowski, E. K., Ewing, L. J., Gluck, D., & Breaux, A. M. (1982). A multidimensional assessment of parent-identified behavior problem toddlers. *Journal of Abnormal Child Psychology, 10*(4), 569–591.

Costello, A., Edelbrock, C., Kalas, R., Kessler, M. D., & Klarie, S. H. (1982). *The NIMH Diagnostic Interview Schedule for Children (DISC).* Pittsburgh: Department of Psychiatry, University of Pittsburgh.

Cyphers, L. H., Phillips, K., Fulker, D. W., & Mrazek, D. A. (1990). Twin temperament during the transition from infancy to early childhood. *Journal of the American Academy of Child and Adolescent Psychiatry, 29,* 392–397.

Dunn, L. M., & Dunn, L. M. (1981). *Peabody Picture Vocabulary Test—Revised.* Circle Pines, MN: American Guidance Service.

Earls, F. (1980). The prevalence of behavior problems in 3-year-old children: Comparison of the reports of mothers and fathers. *Journal of the American Academy of Child Psychiatry, 19,* 439–452.

Earls, F. (1987). Sex differences in psychiatric disorders: Origins and developmental influences. *Psychiatric Development, 5,* 1–23.

Easterbrooks, A., & Emde, R. N. (1983). Hedonic tone and overall mood scales. Unpublished manuscript.

Edelbrock, C., Rende, R., Plomin, R., & Thompson, L. A. (1995). A twin study of competence and problem behavior in childhood and early adolescence. *Journal of Child Psychology and Psychiatry, 36,* 775–785.

Eysenck, H. J., & Eysenck, S. B. G. (1994). *Manual of the Eysenck Personality Inventory.* London: University Press.

Fergusson, D. M., Lynskey, M. T., & Horwood, L. J. (1993). The effect of maternal depression on maternal ratings of child behavior. *Journal of Abnormal Child Psychology, 21,* 245–269.

Fuenzalida, C., Emde, R. N., Pannabecker, B. J., & Stenberg, C. (1981). Validation of the differential emotions scale in 613 mothers. *Motivation and Emotion, 5,* 37–45.

Fullard, W., McDevitt, S. C., & Carey, W. B. (1984). Assessing temperament in one-to three-year-old children. *Journal of Pediatric Psychology, 9,* 205–217.

Garrison, W. T., & Earls, F. (1985). The Child Behavior Checklist as a screening instrument for young children. *Journal of the American Academy of Child Psychiatry, 24,* 76–80.

Hedrick, D. L., Prather, E. M., & Tobin, A. R. (1975). *Sequenced Inventory of Communication Development.* Seattle: University of Washington Press.

Hewitt, J. K., Silberg, J. L., Neale, M. C., Eaves, L. J., & Erickson, M. (1992). The analysis of parental ratings of children's behavior using LISREL. *Behavior Genetics, 22,* 293–317.

Izad, C. E. (1972). *Patterns of emotion.* New York: Academic Press.

Kagan, J., Reznick, J. S., & Snidman, N. (1988). Biological bases of childhood shyness. *Science, 240,* 167–171.

Katz, L. F., & Gottman, J. M. (1993). Patterns of marital conflict predict children's internalizing and externalizing behavior. *Developmental Psychology, 29,* 940–950.

Lancaster, S., Prior, M., & Adler, R. (1989). Child behavior ratings: The influence of maternal characteristics and child temperament. *Journal of Child Psychology and Psychiatry, 30,* 137–149.

Loeber, R., & Stouthammer-Loeber, M. (1986). Family factors as correlates and predictors of juvenile conduct problems and delinquency. In M. Tonry & N. Morris (Eds.), *Crime and justice,* Vol. 7 (pp. 29–149). Chicago: University of Chicago Press.

Loehlin, J. C. (1992). *Genes and environment in personality development.* Newbury Park, CA: Sage.

Maziade, M., Côté, R., Bernier, H., Boutin, P., & Thivierge, J. (1989). Signifance of extreme temperament in infancy for clinical status in pre-school years II: Patterns of temperament change and implications for the appearance of disorders. *British Journal of Psychiatry, 154,* 544–551.

Maziade, M., Côté, R., Boutin, P., Bernier, H., & Thivierge, J. (1987). Temperament and intellectual development: A longitudinal study from infancy to four years. *American Journal of Psychiatry, 144,* 144–150.

Meyer, J., Eaves, L., Silberg, J., Hewitt, J., Erickson, M., & Neale, M. (1993). Contextual effects on adolescent behavioral development: The Virginia school-age twin study. *Abstracts of the 60th Anniversary Meeting of the Society for Research in Child Development, 9,* 147.

Moos, R. H., & Moos, B. S. (1981). *Family Environment Scale manual.* Palo Alto, CA: Consulting Psychologists Press.

Neale, M. C. (1994). *Mx: Statistical modeling* (2nd ed.). Richmond, VA: Department of Psychiatry, Medical College of Virginia.

Oberklaid, F., Sanson, A., Pedlow, R., & Prior, M. (1993). Predicting preschool behavior problems from temperament and other variables in infancy. *Pediatrics, 91,* 113–120.

Oberklaid, F., Sewell, J., Sanson, A., & Prior, M. (1991). Temperament and behavior of preterm infants: A six-year follow-up. *Pediatrics, 87*(6), 854–861.

Plomin, R., & Bergeman, C. S. (1991). The nature of nurture: Genetic influence on "environmental measures." *Behavioral and Brain Sciences, 14,* 373–427.

Plomin, R., Coon, H., Carey, G., DeFries, J. C., & Fulker, D. W. (1991). Parent-offspring and sibling adoption analyses of parental ratings of temperament in infancy and childhood. *Journal of Personality, 59,* 705–732.

Plomin, R., DeFries, J. C., & Fulker, D. W. (1988a). *Nature and nurture during infancy and early childhood.* Cambridge: Cambridge University Press.

Plomin, R., McClearn, G. E., Pedersen, N. L., Nesselroade, J. R., & Bergeman, C. S. (1988b). Genetic influence on childhood family environment perceived retrospectively from the last half of the life span. *Developmental Psychology, 24*(5), 738–745.

Plomin, R., & Rowe, D. C. (1979). Genetic and environmental etiology of social behavior in infancy. *Developmental Psychology, 15,* 62–72.

Prior, M., Smart, D., Sanson, A., Pedlow, R., & Oberklaid, F. (1992). Transient versus stable behavior problems in a normative sample: Infancy to school age. *Journal of Pediatric Psychology, 17*(4), 423–443.

Rende, R. D. (1993). Longitudinal relations between temperament traits and behavioral syndromes in middle childhood. *Journal of the American Academy for Child and Adolescent Psychiatry, 32*(2), 287–290.

Richman, N. (1977). Behaviour problems in pre-school children: Family and social factors. *British Journal of Psychiatry, 131,* 523–527.

Richman, N., Stevenson, J. E., & Graham, P. (1982). *Preschool to school: A behavioural study.* London: Academic Press.

Robinson, J. L., Kagan, J., Reznick, J. S., & Corley, R. (1992). The heritability of inhibited and uninhibited behavior: A twin study. *Developmental Psychology, 28,* 1030–1037.

Rowe, D. C., & Plomin, R. (1977). Temperament in early childhood. *Journal of Personality Assessment, 41,* 150–156.

Sanson, A., Oberklaid, F., Pedlow, R., & Prior, M. (1991). Risk indicators: Assessment of infancy predictors of pre-school behavioural maladjustment. *Journal of Child Psychology and Psychiatry, 32*(4), 609–626.

Schmitz, S. (1994). Temperament and personality. In J. C. DeFries, R. Plomin, & D. W. Fulker (Eds.), *Nature and nurture during middle childhood* (pp. 120–140). Oxford: Blackwell.

Schmitz, S. (1996). Familial correlates of children's problem behavior in a longitudinal twin sample. Unpublished doctoral dissertation, University of Colorado, Boulder.

Schmitz, S., & Fulker, D. W. (1995). Continuity due to which factors? An extension to the rater bias model. *Behavior Genetics, 25,* 287.

Schmitz, S., & Fulker, D. W. (1996). Family environment and children's problem behavior. *Behavior Genetics, 26,* 596.

Schmitz, S., Fulker, D. W., & Mrazek, D. A. (1995). Problem behavior in early and middle childhood: An initial behavior genetic analysis. *Journal of Child Psychology and Psychiatry, 36,* 1443–1458.

Schmitz, S., Fulker, D. W., Plomin, R., Zahn-Waxler, C., Emde, R. N., & DeFries, J. C. (1999). Temperament and problem behavior during early childhood. *International Journal of Behavioral Development, 23*(2), 333–355.

Schwartz, C. E., Snidman, N., & Kagan, J. (1996). Early childhood temperament as a determinant of externalizing behavior in adolescence. *Development and Psychopathology, 8,* 527–537.

Silberg, J. L., Erickson, M. T., Meyer, J. M., Eaves, L. J., Rutter, M. L., & Hewitt, J. K. (1994). The application of structural equation modelling to maternal ratings of twins' behavioral and emotional problems. *Journal of Clinical and Consulting Psychology, 62,* 510–521.

Spanier, G. B. (1976). Measuring dyadic adjustment: New scales for assessing the quality of marriage and similar dyads. *Journal of Marriage and the Family, 38,* 15–28.

Spiker, D., Kraemer, H. C., Constantine, N. A., & Bryant, D. (1992). Reliability and validity of behavior problem checklists as measures of stable traits in low birth weight, premature preschoolers. *Child Development, 63,* 1481–1496.

Stevenson, J., & Richman, N. (1978). Behavior, language, and development in three-year-old children. *Journal of Autism and Child Schizophrenia, 8,* 299–313.

Stevenson, J., Richman, N., & Graham, P. (1985). Behavior problems and language abilities at three years and behavioural deviance at eight. *Journal of Child Psychology and Psychiatry, 26*(2), 215–230.

Terman, L. M., & Merrill, M. A. (1973). *Stanford-Binet Intelligence Scale: 1972 norms edition.* Boston: Houghton-Mifflin.

Thomas, A., & Chess, S. (1982). Temperament and follow-up to adulthood. In R. Porter & G. M. Collins (Eds.), *Temperamental differences in infants and young children* (pp. 168–175). London: Pittmann.

van den Oord, E. J. C. G., Verhulst, F. C., & Boosma, D. I. (1996). A genetic study of maternal and paternal ratings of problem behavior in three-year-old twins. *Journal of Abnormal Psychology, 105*(3), 349–357.

Williams, S., Anderson, J., McGee, R., & Silva, P. A. (1990). Risk factors for behavioral and emotional disorder in preadolescent children. *Journal of the American Academy of Child and Adolescent Psychiatry, 29*(3), 413–419.

Wolkind, S. N., & De Salis, W. (1982). Infant temperament, maternal mental state and child behavioural problems. In R. Porter & G. M. Collins (Eds.), *Temperamental differences in infants and young children* (pp. 221–239). London: Pittman.

Zahn-Waxler, C., Cole, P. M., Welsh, J. D., & Fox, N. A. (1995). Psychophysiological correlates of empathy and prosocial behaviors in preschool children with behavior problems. *Development and Psychopathology, 7,* 27–48.

Zahn-Waxler, C., Radke-Yarrow, M., & King, R. A. (1979). Child-rearing and children's initiations toward victims of distress. *Child Development, 50,* 319–330.

Zahn-Waxler, C., Robinson, J. L., & Emde, R. N. (1992). The development of empathy in twins. *Developmental Psychology, 28,* 1038–1047.

Zahn-Waxler, C., Schmitz, S., Fulker, D., Robinson, J. L., & Emde, R. (1996). Behavior problems in five-year-old MZ and DZ twins: Estimates of genetic and shared environmental influences. *Development and Psychopathology, 8,* 103–122.

Part VI

Major Themes and Conclusions

25

An Experiment in Collaborative Science

Robert Plomin
Robert N. Emde
John K. Hewitt
Jerome Kagan
John C. DeFries

This book celebrates the success of an experiment in collaborative science. The MacArthur Longitudinal Twin Study brought together some key developmental psychologists and behavioral geneticists and gave them the opportunity to discuss the difficult issues of individual differences and nature and nurture, not just in the abstract, but in the concrete context of planning a collaborative study that integrated their diverse interests and expertise. Although there was friction along the way, the sparks produced by this collaborative culture were in the end channeled into producing a unique multivariate longitudinal twin study that focused on development during the critical but neglected second year of life.

This concluding chapter begins by carrying forward the previous part's theme of cross-domain integration by systematically examining relations within and between the major domains of temperament, emotion, and cognition. The rest of the chapter highlights some of the new findings that have emerged.

RELATIONS WITHIN AND BETWEEN DOMAINS IN THE SECOND YEAR OF LIFE

The MacArthur Longitudinal Twin Study was designed as a multivariate project that focused on three areas of considerable contemporary interest in developmental psychology: temperament, emotion, and cognition. Indeed, the multivariate nature of the project was the driving force behind the collaboration because the co-investigators could investigate their own areas of special interest in the context of and in comparison to the other domains. However, the first report of the project at 14 months (Plomin et al., 1990) suggested a surprising degree of independence within as well as between domains. For example, within the domain of temperament, shyness and sociability were relatively independent. Within the emotion domain, positive and negative hedonic tone correlated only slightly, and empathy was uncorrelated with other measures in the emotion domain. Only within the cognitive domain was moderate overlap seen, for example, between mental development and communication development. Across domains, there was even less overlap. For example, measures of temperament showed no significant associations with measures of emotion or cognition. One of the few cross-domain correlations involved positive hedonic tone and mental and communication development, although the correlations were only about .20.

The emerging conclusion that behavioral development is multifaceted both within and between domains continues to be supported from the multivariate analyses that appear in this volume. For example, in chapter 23, Kubicek et al. report that negligible associations were found between the domains of temperament and cognition, except for temperament assessed during the cognitive testing situation. Other measures such as parent ratings and observations of shyness and behavioral inhibition showed little association with measures of cognitive and language development. Tighter links might be expected for the domains of temperament and emotion, but temperament and anger showed few associations (see chapter 10). The authors were especially struck by the lack of links within the domain of anger. For example, the restraint measure of anger did not correlate above .10 with any of the other measures of anger such as anger outbursts. Differentiation is also the rule rather than the exception within the domain of temperament (see chapter 8).

This differentiation between and within domains is even more apparent in terms of analyses that are not reported in previous chapters. The multivariate design stimulated many attempts to analyze links between and within domains, but most of these analyses ended up on the "cutting-room floor" rather than surviving to the printed page because the lack of correlations seemed like a negative result, especially considering the statistical power of the large sample to detect such correlations. However, the multifaceted nature of behavioral development is not a negative finding. It is among the most important messages from this research. In addition, some of the significant correlations that have been found within and between domains, although modest, are of considerable interest.

The purpose of this chapter is to present the results of a systematic analysis of correlations between and within the temperament, emotion, and cognition domains of the MacArthur Longidutinal Twin Study during the second year of life (14, 20, and 24 months). The sample size now is nearly 800 individuals, four times larger than the one used for our original preliminary report (Plomin et al., 1990). We also include twice as many measures as in our original report, measures described in our early *Child Development* papers (Emde et al., 1992; Plomin et al., 1993). These include general factor scores (unrotated first principal component) derived from observational measures of behavioral inhibition, shyness in the home and in the laboratory, empathy, and anger. Positive and negative hedonic tone were assessed by time-sampled ratings of the children's strongest positive and negative affect from videotape recordings of the children during administration of the Bayley test and during free play. Testers rated children's temperament (activity, task orientation) and emotion (affect) using the Infant Behavior Record during administration of the Bayley mental scales and again during the non-Bayley portion of the home visit. In addition to these more objective measures, we examined results for parental ratings of temperament. Parents were administered the Colorado Childhood Temperament Inventory to rate their children's temperament (activity, persistence, shyness, sociability) and emotionality (emotionality, soothability); mothers' and fathers' ratings were averaged. Positive and negative composites were derived from maternal ratings of the expression of discrete emotions using the Differential Emotions Scale (DES). The cognitive domain was assessed using the Bayley Mental Development Index (MDI), the expressive and receptive scores of the Sequenced Inventory of Communication Development, and experimental measures of memory for locations, sorting, and word comprehension. We also included average heart rate and heart rate variability in these analyses. The same measures were obtained at 14, 20, and 24 months, with the exception that at 24 months shyness in the laboratory and sorting were not assessed.

Because the sample size is so large, very small correlations are significant. For this reason, we only present correlations significant at the $p < .001$ level, but even these correlations can be as small as .12, which explains $< .015$ of the variance and warrants modesty in interpretation. Using a stringent criterion for significance greatly attenuates the problem of false positives when examining so many correlations. Because we examined about 500 correlations, use of a p value of .001 means that we would expect no significant correlations by chance alone. Moreover, the power provided by this large sample makes it possible to accept the null hypothesis that two measures are not correlated, at least with an effect size $>2\%$.

We begin by examining relations within domains at 14 months, followed by analyses between domains at 14 months. Next, developmental changes in these multivariate relations at 20 and 24 months are briefly described. Finally, we consider continuity and change during the second year.

Relations within Domains at 14 Months

For each variable, the first row in table 25.1 lists significant ($p < .001$) correlations within the domain of temperament at 14 months. Some of the correlations were presented separately by gender in chapter 8. Behavioral inhibition correlates most with measures of shyness. Behavioral inhibition correlates more strongly with shyness observed during the children's entry in the labo-

Table 25.1. Correlations within the temperament domain at 14, 20, and 24 months

				IBR		CCTI			
Measure/age	Behavioral inhibition	Home shyness	Lab shyness	Activity	Task orientation	Activity	Persis-tence	Shyness	Soci-ability
Behavioral inhibition									
14 Months		.25	.36			−.13		.30	
20 Months		.32	.29	−.18		−.14		.34	
24 Months		.27						.25	
Home shyness									
14 Months				.34	−.14	−.12			.37
20 Months			.37	−.19				.35	
24 Months				−.18		−14		.39	
Lab shyness[a]									
14 Months								.29	
20 Months				−.16				.25	
IBR activity									
14 Months						.19	−.13		
20 Months						.22	−.14	−.18	
24 Months								−.16	
IBR task orientation									
14 Months									
20 Months							.18		
24 Months									
CCTI activity									
14 Months								−.31	.28
20 Months								−.25	.33
24 Months							−.14	−.30	.34
CCTI persistence									
14 Months									−.16
20 Months									−.28
24 Months									−.23
CCTI shyness									
14 Months									−.37
20 Months									−.38
24 Months									−.39
CCTI sociability									
14 Months									

All correlations are significant at $p < .001$. CCTI, Colorado Childhood Temperament Inventory; IBR, Infant Behavior Record.
[a]Lab shyness was not assessed at 24 months.

ratory (.36) than with shyness observed during the examiner's entry into the home (.25). Indeed, behavioral inhibition correlates as strongly with these measures of shyness as the home and laboratory measures correlate with each other (.34). Observed behavioral inhibition is also correlated almost as highly (.30) with shyness as rated by parents on the CCTI. These findings suggest that shyness is a major component of behavioral inhibition.

Behavioral inhibition also correlates negatively with parental ratings of activity ($-.13$) and with examiner ratings of activity ($-.10$, $p = .006$, not shown in table 25.1). Although activity is not usually discussed in relation to behavioral inhibition, it makes sense that more active children are seen as somewhat less inhibited.

The observational shyness variables showed a similar pattern of relationship to parental ratings of shyness and, negatively, to activity as rated by the examiner and by the parent. The observational measures of shyness and parental ratings of sociability were not correlated. Moreover, the parental ratings of shyness and sociability were only moderately correlated ($-.37$). This might seem surprising given the common assumption that shyness is synonymous with low sociability. However, sociability is the tendency to affiliate with others and to prefer being with others rather than being alone. Shyness, on the other hand, refers to reactions to strangers. When assessed independently, shyness and sociability are only modestly correlated (e.g., $-.30$) in adults (Buss & Plomin, 1984), and we find a similar low level of correlation in children as rated by their parents.

Examiner IBR ratings of activity correlate only modestly (.19) with parental ratings of activity on the CCTI, although the IBR rating has more to do with fidgeting than with overall energy output, which is what the CCTI attempts to assess. IBR activity also correlates ($-.13$) with parental ratings of CCTI persistence, even though IBR activity does not correlate with IBR task orientation. Parental CCTI ratings show some modest overlap across temperaments, for example, between activity and shyness ($-.31$) and between activity and sociability (.28).

Table 25.2 presents correlations within the emotion domain. Most notable is the lack of association between positive and negative emotions. Videotape observations of positive and negative hedonic tone correlate only $-.11$ (not significant at $p < .001$ and thus not shown in table 25.2). Maternal DES ratings of positive and negative emotional expressions show a slight positive correlation (.15). This finding supports the emerging consensus that positive and negative aspects of emotion are not merely opposite ends of a single continuum.

The only substantial correlation ($-.50$) in table 25.2 is between videotape ratings of negative hedonic tone and examiner IBR ratings of affect (which essentially assesses happiness). However, although substantial, this correlation seems likely to be at least in part due to the fact that the videotape raters and the examiners rated some of the same samples of behavior. Negative hedonic tone was also modestly correlated (.23) with the experimental measure of an-

Table 25.2. Correlations within the emotion domain at 14, 20, and 24 months

| | Hedonic tone | | | | | CCTI | | DES | |
Measure/age	Positive	Negative	Empathy	Anger	IBR affect	Emotion-ality	Sooth-ability	Positive	Negative
Hedonic tone positive									
14 Months					.29				
20 Months					.38				
24 Months					.35				
Hedonic tone negative									
14 Months			.23		−.50	.14			
20 Months					−.29	.15			
24 Months					−.38				
Empathy									
14 Months									
Anger response to restraint									
14 Months					−.30	.15	−.14		
20 Months					−.33		−.13		
24 Months					−.27				
IBR affect									
14 Months						−.15			−.15
20 Months									−.13
24 Months									−.17
CCTI emotionality									
14 Months							−.48		.42
20 Months							−.52		.43
24 Months							−.46		.39
CCTI soothability									
14 Months								.14	−.32
20 Months								.20	−.31
24 Months								.22	−.29
DES positive									
14 Months									.15
DES negative									
14 Months									

All correlations are significant at $p < .001$. IBR, Infant Behavior Record; CCTI, Colorado Childhood Temperament Inventory; DES, Differential Emotions Scale.

ger, as was the IBR rating of affect (−.30). Videotape observations of hedonic tone do not correlate with maternal DES ratings of emotional expression.

The biggest surprise is that the global measure of empathy does not correlate with any of the other variables in the emotion domain, suggesting that empathy, as we have assessed it, is an independent dimension. Of course, other measures such as measures of the private experience of empathy, might yield different results.

Maternal CCTI ratings of emotionality correlate modestly with videotape ratings of negative hedonic tone (.14), anger (.15), and tester ratings of IBR affect (−.15). Maternal CCTI ratings of emotionality also correlate with maternal DES ratings of negative emotional expression (.42) and with maternal CCTI ratings of soothability (−.48).

Correlations within the domain of cognition are shown in table 25.3. The Bayley MDI correlated slightly but significantly with memory for locations (.14), but not with sorting or word comprehension. The highest intercorrelations were between the Bayley and SICD expressive (.42) and receptive (.43), which were almost as high as the correlation between SICD expression and receptive scales (.46). The experimental word comprehension task correlated with the SICD expressive score (.19), but not with the receptive score.

Despite the natural tendency to focus on positive findings, it should be emphasized that most of the correlations within each domain do not reach significance. Moreover, the strongest correlations are generally in the .3–.4 range. In other words, these results suggest that behavioral development is quite differentiated within the major domains of temperament, emotion, and cognition at 14 months.

Relations across Domains at 14 Months

Given the surprising degree of independence of dimensions within each domain, correlations across domains seemed unlikely. However, some pairs of domains (especially temperament and emotion) yield proportionately as many significant correlations as there are within domains. Other domains (especially temperament and cognition) are independent.

Table 25.4 lists significant correlations that emerged between temperament and emotion. Behavioral inhibition was correlated with anger (.22) and DES negative expression (.14); shyness are rated from videotapes in the home and in the laboratory showed similar patterns of correlations. The IBR task orientation measure of temperament yielded an unexpected correlation with positive hedonic tone (.22), IBR affect (.35), and DES positive emotional expression (.14), suggesting perhaps that tester ratings on IBR task orientation are influenced by positive emotionality. Parental ratings of CCTI temperament yielded a smattering of significant correlations. CCTI shyness correlated with the most scales, including observed anger (.26), IBR affect (−.19), CCTI emotionality (.28) and soothability (−.29), and DES negative (.36).

Table 25.3. Correlations within the cognitive domain at 14, 20, and 24 months

Measure/age	MDI	Memory for locations	Sorting	Word comprehension	SICD Expressive	SICD Receptive
MDI						
14 Months		.14			.42	.43
20 Months		.15		.30	.59	.65
24 Months		.21		.32	.63	.68
Memory for locations						
14 Months						.14
20 Months					.15	.15
24 Months					.15	.22
Sorting[a]						
14 Months						
Word comprehension						
14 Months						.19
20 Months					.21	.38
24 Months					.23	.29
SICD expressive						
14 Months						.46
20 Months						.61
24 Months						.59
SICD receptive						
14 Months						

All correlations are significant at p < .001. MDI, Bayley Mental Development Index; SICD, Sequenced Inventory of Communication Development.
[a] The sorting task was not included at 24 months.

Table 25.5 shows significant correlations between temperament and cognition. Only 3 of the 54 correlations were significant. All three involved examiner ratings of IBR task orientation, which correlated with the Bayley Mental Development Index (.43) and with SICD receptive (.20) and expressive (.12) scores.

Table 25.6 indicates a few more significant correlations between emotion and cognition. Positive hedonic tone is related to the Bayley (.20) and to SICD expressive (.18) and receptive (.13) measures. Negative hedonic tone is also related to the Bayley (−.19) but not to the SICD communication measures. IBR

Table 25.4. Correlations between temperament and emotion domains at 14, 20, and 24 months

Measure/age	Hedonic tone				IBR affect	CCTI		DES	
	Positive	Negative	Empathy	Anger		Emotion-ality	Sooth-ability	Positive	Negative
Behavioral inhibition									
14 Months				.22					.14
20 Months					−.16	.14	−.16		.20
24 Months				.22	−.22	.13	−.13		
Home shyness									
14 Months				.20	−.17				.14
20 Months				.29	−.22				.16
24 Months				.25	−.23		−.15		
Lab shyness[a]									
14 Months	−.16			.16		.13			.19
20 Months				.17	−.19				.16
IBR activity									
14 Months		.14							
20 Months		.19							
IBR task orientation									
14 Months	.22				.35			.14	
20 Months	.24				.34				
24 Months		−.16			.30				
CCTI activity									
14 Months						−.13	.17	.14	−.14
20 Months						−.15			−.17
24 Months						−.15			−.15
CCTI persistence									
14 Months						−.19	.23		
20 Months						−.22	.18		
24 Months						−.14	.24		
CCTI shyness									
14 Months				.26	−.19	.28	−.29		.36
20 Months			−.17	.24	−.30	.23	−.27	−.18	.38
24 Months			−.17	.26	−.22	.22	−.30	−.14	.37
CCTI sociability									
14 Months							.16		.15
24 Months			.17				.18		

All correlations are significant at $p < .001$. IBR, Infant Behavior Record; CCTI, Colorado Childhood Temperament Inventory; DES, Differential Emotions Scale.

[a] Lab shyness was not assessed at 24 months.

Table 25.5. Correlations between temperament and cognition domains at 14, 20, and 24 months

Measure/age	MDI	Memory for locations	Sorting[a]	Word Comprehension	SICD	
					Expressive	Receptive
IBR task orientation						
14 Months	.43				.20	.12
20 Months	.33	.14	.15		.23	.22
24 Months	.34			.13	.21	.19
CCTI persistence						
24 Months	.13	.16				
CCTI shyness						
20 Months					−.20	−.15
24 Months					−.21	

All correlations are significant at *p* <.001. MDI, Bayley Mental Development Index; IBR, Infant Behavior Record; CCTI, Colorado Childhood Temperament Inventory; SICD, Sequenced Inventory of Communication Development.
[a] Sorting was not assessed at 24 months.

affect shows the strongest correlation with the Bayley (.38) and also correlates with SICD expressive (.23) and receptive (.17) communication. The DES positive expression but not negative expression correlates with SICD expressive (.19) and receptive (.20) communication. The only correlation between empathy and any other measure occurred with SICD expressive (.18) and receptive (.21) communication.

We also examined correlations between heart rate mean and variability and the three domains of behavioral development. Only one significant correlation emerged, an expected negative correlation (−.13) between heart rate variability and behavioral inhibition. The expected positive correlation between heart rate mean and behavioral inhibition was not significant at our *p*<.001 threshold (r=.10, p=.008)

Although some of these correlations were not expected, such as the correlations between shyness and anger, between IBR affect and Bayley scores and between empathy and communication development, none is counterintuitive. As with any such correlations, disentangling the direction of effects will require more work. For example, empathy might lead children to pay more attention to other people and thus develop communication skills at an earlier age, or children who are more advanced in language at 14 months might have more of a basis for empathic responding. Adding the longitudinal dimension to these relations can shed some light on their origins. Despite the few interesting correlations across domains, the overall pattern of differentiated devel-

Table 25.6. Correlations between emotion and cognition at 14, 20, and 24 months

Measure/age	MDI	Sorting[a]	Word Comprehension	SICD Expressive	Receptive
Hedonic tone positive					
14 Months	.20			.18	.13
20 Months	.17			.22	
24 Months	.16			.22	.21
Hedonic tone negative					
14 Months	−.19				
20 Months				−.15	−.16
24 Months	−.24			−.17	−.20
Empathy					
14 Months				.18	.21
20 Months	.21			.22	.30
24 Months	.18			.23	.15
Anger response to restraint					
24 Months				−.16	
IBR affect					
14 Months	.38			.23	.17
20 Months	.33	.15		.33	.31
24 Months	.44			.47	.39
DES positive					
14 Months				.19	.20
DES negative					
24 Months					−.14

All correlations are significant at $p < .001$. MDI, Bayley Mental Development Index; SICD, Sequenced Inventory of Communication Development; DES, Differential Emotions Scale.
[a] Sorting was not assessed at 24 months.

opment is apparent from the blanks (indicating nonsignificance) tables 25.4–25.6.

Relations at 20 and 24 Months

How do the relations within and between domains change during the second year of life? Significant correlations at 20 months are shown for each variable in tables 25.1–25.3. The results are surprisingly similar to those described for 14 months. For example, looking at the 14-month correlations .20 in table 25.1 (temperament), without exception, the same significant correlations are found at 20 and 24 months. There were no substantial developmental changes in the patterns of phenotypic relations between and within domains from 14 to 24 months.

Table 25.2 (emotion) suggests two developmental changes involving negative hedonic tone. One developmental change is marginal: Negative hedonic tone correlates with the anger measure at 14 months (.23) but not at the other ages; however, the correlation just misses significance both at 20 months (.12) and 24 months (.12). More interesting is the change in the correlation between negative hedonic tone and IBR affect from −.50 at 14 months to −.29 at 20 months and −.38 at 24 months. One of many possible interpretations is that children become somewhat more differentiated in their emotional responding across situations.

Table 25.3 (cognition) shows a consistent pattern of higher correlations at 20 and 24 months as compared to 14 months. This might reflect greater validity of assessment and greater involvement of language skills towards the end of the second year. Similar results emerged at 20 and 24 months across domains, as seen in tables 25.4–25.6. The orderliness of these results from 14 to 24 months speaks well for the reliability of data collection in the MacArthur study and for the power of its large sample size.

Longitudinal Relations across Ages

We expected 20 and 24 months to be a major transitional period that would encompass many changes such as language, sense of self, empathy, and more complex emotional communication. This period of change was bracketed on either side by assessments at 14 and 36 months, which we anticipated would be ones of relative consolidation rather than change.

Our preliminary analyses at 14 and 20 months found some continuity in the face of considerable change from 14 to 20 months based on a sample only one quarter as large and with only half as many variables as the present analyses (Plomin et al., 1990). Significant ($p < .001$) longitudinal correlations within the temperament, emotion, and cognition domains, respectively, are listed in tables 25.7–25.9. The full matrix of significant correlations is included. For example, in table 25.7, above the diagonal, behavioral inhibition at 14 months is correlated with home shyness at 20 months (.17); below the diagonal, home shyness at 14 months is correlated with behavioral inhibition at 20 months (.27). Few of these cross-lag longitudinal correlations are significantly different; correlations must differ on the order of .00 versus .15 to be significantly different at $p < .01$, even with samples as large as ours.

Homotypic continuities are listed as the diagonals. As indicated in other chapters in this volume, observed measures of temperament and emotion show much more change than continuity, with correlations typically in the .20–.30

Table 25.7. Longitudinal correlations within the temperament domain at 14–20 months, 20–24 months, and 14–24 months

				IBR		CCTI			
Measure/age	Behavioral inhibition	Home shyness	Lab shyness	Activity	Task orientation	Activity	Persistence	Shyness	Sociability
Behavioral inhibition									
14–20 Months	.30	.17	.21	−.23				.23	
20–24 Months	.27	.37				−.16		.33	
14–24 Months		.19		−.16				.16	
Home shyness									
14–20 Months	.27	.25	.13	−.17				.25	
20–24 Months		.20	.42						.35
14–24 Months		.28						.24	
Lab shyness									
14–20 Months	.25	.25	.31	−.16				.26	
20–24 Months	.17	.33						.25	
14–24 Months		.26						.22	
IBR activity									
14–20 Months	−.14	−.12		.21		.16			
20–24 Months		−.15		.22		.18	−.20	−.14	
14–24 Months				.23					
IBR task orientation									
14–20 Months					.18	.16			
20–24 Months					.26				
14–24 Months					.25	.14			
CCTI Activity									
14–20 Months	−.19			.16		.65	−.18	−.22	.20
20–24 Months						.69	−.17	−.25	.29
14–24 Months						.59		−.23	.20
CCTI persistence									
14–20 Months							.51		−.15
20–24 Months						−.14	.60		−.25
14–24 Months							.45		
CCTI shyness									
14–20 Months	.21	.26		−.18		−.28		.62	−.26
20–24 Months	.16	.29				−.21		.68	−.25
14–24 Months		.25		−.14		−.26		.52	−.24

(*continued*)

Table 25.7. Continued

				IBR		CCTI			
Measure/age	Behavioral inhibition	Home shyness	Lab shyness	Activity	Task orientation	Activity	Persistence	Shyness	Sociability
CCTI sociability									
14–20 Months						.18		−.22	.47
20–24 Months						.24	−.22	−.24	.59
14–24 Months						.15	−.15	−.25	.44

All correlations are significant at $p < .001$. IBR, Infant Behavior Record; CCTI, Colorado Childhood Temperament Inventory.
[a] Lab shyness was not assessed at 24 months.

Table 25.8. Longitudinal correlations within the emotion domain at 14–20 months, 20–24 months, and 14–24 months

	Hedonic tone				IBR	CCTI		DES	
Measure/age	Positive	Negative	Empathy	Anger	affect	Emotion-ality	Sooth-ability	Positive	Negative
Hedonic tone positive									
14–20 Months	.32				.20				
20–24 Months	.36				.23			.18	
14–24 Months	.25		.15		.15				
Hedonic tone negative									
14–20 Months		.19							
20–24 Months	−.23	.20		.17	−.17				
14–24 Months		.18		.17	−.14				
Empathy									
14–20 Months	.16		.25						
20–24 Months			.31						
14–24 Months			.24						
Anger response to restraint									
14–20 Months				.33	−.18				
20–24 Months				.40	−.17				
14–24 Months				.26					
IBR affect									
14–20 Months	.14	−.23		−.18	.26				−.15
20–24 Months	.21	−.19		−.25	.33				
14–24 Months	.22	−.17			.27		.15		−.15

(continued)

Table 25.8. Continued

| Measure/age | Hedonic tone | | Empathy | Anger | IBR affect | CCTI | | DES | |
	Positive	Negative				Emotion-ality	Sooth-ability	Positive	Negative
CCTI emotionality									
14–20 Months						.56	−.31		.30
20–24 Months						.60	−.34		.32
14–24 Months						.49	−.33		.26
CCTI soothability									
14–20 Months					−.38		.50		−.22
20–24 Months							.53	.17	−.23
14–24 Months							.44		−.24
DES positive									
14–20 Months							.20	.66	
20–24 Months							.20	.67	
14–24 Months			.16				.21	.58	
DES negative									
14–20 Months						.28	−.19		.61
20–24 Months						.32	−.27		.68
14–24 Months						.25	−.21		.61

All correlations are significant at $p < .001$. IBR, Infant Behavior Record; CCTI, Colorado Childhood Temperament Inventory, DES, Differential Emotions Scale.

range. Parental ratings are much more stable, in the .50–.60 range. The Bayley Mental Development Index and expressive and receptive language are moderately stable, especially from 20 to 24 months, when the correlations are in the .60's.

Heterotypic continuities within domains are largely as expected given the pattern of correlations at each age (tables 25.1–25.3) and stabilities (table 25.7–25.9). For example, the correlations among behavioral inhibition, observed shyness, and parental ratings of shyness, discussed in relation to table 25.1, are seen longitudinally, especially from 20 to 24 months (table 25.7). Although it could be a chance result, behavioral inhibition at 20 months correlates significantly more highly with home shyness at 24 months (.37) than shyness at 20 months correlates with inhibition at 24 months (.20). One possible interpretation is that inhibition at 20 months is causally related to later shyness.

Concerning emotion (table 25.8), the largest correlation in table 25.2 involved videotaped negative hedonic tone and tester ratings of affect. However, we suggested that the correlation might be due to the fact that videotape raters

Table 25.9. Longitudinal correlations within the cognitive domain at 14–20 months, 20–24 months, and 14–24 months

Measure/age	MDI	Memory for locations	Sorting	Word comprehension	SICD Expressive	SICD Receptive
MDI						
14–20 Months	.48		.18	.23	.35	.39
20–24 Months	.67	.17		.28	.52	.56
14–24 Months	.43	.14		.19	.30	.37
Memory for locations						
14–20 Months					.16	.15
20–24 Months	.20	.17			.18	
14–24 Months	.17				.19	
Sorting						
14–20 Months			.19			
Word comprehension						
20–24 Months	.32			.24	.26	
SICD expressive						
14–20 Months	.28				.42	.35
20–24 Months	.51			.18	.64	.46
14–24 Months	.25				.34	.25
SICD receptive						
14–20 Months	.43			.28	.40	.54
20–24 Months	.59	.14		.30	.55	.67
14–24 Months	.44			.22	.34	.46

All correlations are significant at $p < .001$. MDI, Bayley Mental Development Index; SICD, Sequenced Inventory of Communication Development.
ª Sorting was not assessed at 24 months.

and examiners rated some of the same samples of behavior during a testing session. This hypothesis is supported by the results in table 25.8, which show only modest longitudinal correlations between these two variables.

Cognition (table 25.9) shows the largest correlations, especially for the Bayley MDI and the expressive and receptive scales of the SICD. Expressive and receptive SICD scores are closely linked to MDI scores longitudinally. One significant difference is that 24-month MDI scores are predicted significantly better by receptive SICD scores at 14 months ($r = .44$) than by expressive SICD scores at 14 months ($r = .25$). The cross-lag correlations between expressive and receptive language cosistently support the hypothesis that early receptive SICD scores are more predictive of later expressive SICD scores than vice versa, but these differences fall short of statistical significance.

Because correlations across domains within age were modest (tables 25.4–25.6), we do not present comparable cross-domains correlations longitudinally. The longitudinal cross-domain results held few surprises. One minor exception concerns the correlation between MDI and IBR task orientation (.43, .33, and .34 at 14, 20, and 24 months, respectively). These contemporaneous correlations might involve an artifact in that testers might rate children as higher on task orientation if they performed better on the MDI. However, rather than task orientation reflecting MDI, the longitudinal correlations suggest the reverse. Task orientation at 14 months predicts MDI at 20 months significantly better than MDI at 14 months predicts task orientation at 20 months (.26 versus .10). The same trend exists from 14 to 24 months, but the difference is not significant (.20 versus .14). As shown in table 25.6, IBR affect also correlated substantially with MDI (.38, .33, and .44, at 14, 20, and 24 months, respectively), but the longitudinal cross-lag correlations were similar.

An intriguing cross-domain correlation, the only cross-domain correlation involving empathy, was a correlation with SICD communication, with correlations in the .20's at 14, 20, and 24 months (table 25.6). The cross-lag longitudinal correlations from 14 to 24 months indicate that 14-month empathy predicts 24-month receptive language better than 14-month receptive language predicts 24-month empathy (.27 versus .13). A similar pattern was found for 20 and 24 months (.28 versus .12). This developmental priority of empathy over receptive language is interesting even if it is not causal.

In this first half of the chapter, we have systematically examined relations within and between the domains of temperament, emotion, and cognition during the second year of life. Our general conclusion is that development is multifaceted, within as well as between domains. The second half of the chapter continues the theme of cross-domain integration and presents the main findings of the book from our interdisciplinary perspective.

HERITABILITY AND ENVIRONMENTAL COMPONENTS OF VARIANCE

An important advantage of studying individual differences is that behavioral genetic designs make it possible to move beyond description to explanation of the genetic and environmental origins of these individual differences. These designs can be used to go beyond asking whether and how much genetic and environmental factors are responsible for individual differences. However, it is a reasonable first step toward understanding the origins of individual differences to ask these rudimentary questions.

Table 25.10. Summary of genetic and environmental results

Variable	Heritability	Shared environment
Behavioral inhibition	++	0
Shyness (home)	+	++
IBR temperament (tester ratings)	++	0
Parental ratings of temperament	++	0
Empathy	++	0
Cheerfulness	+	+
Anger response to restraint	++	++
DES anger	++	++
General cognitive ability	++	++
SICD expressive language	++	++
SICD receptive language	+	+++
Height	+++	0
Weight	+++	0

IBR, Infant Behavior Record; DES, Differential Emotions Scale; SICD, Sequenced Inventory of Communication Development. 0 = negligible influence (<10% of the variance), + = modest influence (10–30%); ++ = moderate influence (30–50%); +++ = strong influence (> 50%).

Consider empathy, for example. The MacArthur Longitudinal Twin Study is the first behavioral genetic study of this important domain other than self-report studies of adults. In general, empathy yields genetic results similar to measures of temperament (see chapter 11). That is, most components of empathy and a composite measure show moderate genetic influence, except, interestingly, at 20 months. Also similar to findings for temperament (except shyness), there is no evidence for shared environmental influence, which seems especially surprising for empathy. Unlike findings for temperament, maternal reports of prosocial acts indicate substantial shared environmental influence but not genetic influence. These latter results might indicate that mothers are not able to differentiate their two children's empathic responses. For temperament, mothers seem to exaggerate differences between their children, especially for fraternal twins.

Another example of a novel measure is observed anger-related reactions to restraint (see chapter 10). For this measure, genetic influence and shared family environmental influence are suggested in equal measure at 14, 20, and 24 months. Maternal reports of anger expression showed genetic influence at all three ages and shared environmental influence, especially at 24 months.

Table 25.10 summarizes the results of some of the genetic analyses presented in this book. This summary makes the important point that no simple explanation, like "it's all genetic," can capture the complexity of development during the second year. Although most variables are moderately heritable, some show less genetic influence and some show more. For example, shyness observed in the home appears to show less genetic influence than behavioral inhibition, even though these variables are related. Receptive language shows less genetic influence than expressive language, which is as heritable as general cognitive ability. Some variables show greater heritability. The most striking example is weight, which shows heritabilities greater than 80% from 14 through 36 months. One chapter examined whether heritability differed for boys and girls, but no significant differences were found for language (see chapter 16).

Because heritability is usually moderate at most, these same data indicate that environmental factors are important. In all cases, environmental influences unique to the individual contribute to phenotypic variance. For anthropometric and cognitive variables, these nonshared environmental effects are generally smaller than those influencing behaviors in the domains of temperament and emotion, where nonshared environments may be the predominant influence. To some extent, the nonshared environment includes what is often referred to as unreliability of measurement, but of course this is really another way of saying that there is situation-specific and occasion-specific individual variance. For many aspects of behavior, much of what takes place when 2-year-old infants are observed in a given situation is largely unpredictable, whether from knowledge of their genotypes or from family circumstances or parental behavior. This unpredictability may be more marked in childhood than in later life.

In addition to nonshared environment, the results in the second year of life show much more evidence for shared environmental influence than is seen later in life (Plomin et al., 1994). For example, shared environmental influence is reported for shyness, cheerfulness, anger, general cognitive ability, and language. Shared environment appears to be especially important for receptive language. However, little evidence of shared environmental influence is found for behavioral inhibition, tester and parent ratings of temperament, empathy, and height and weight.

BEYOND HERITABILITY AND COMPONENTS OF VARIANCE

Combining developmental and genetic research strategies lights the way for exploring questions that go beyond estimating heritability and environmental components of variance. Major categories of questions include development, multivariate relations, and the interplay between nature and nurture.

Development

Longitudinal genetic data make it possible to chart developmental changes and continuities in genetic and environmental influences. Several interesting results emerged from comparisons across ages. For example, empathy shows significant genetic influence at 14, 24, and 36 months but not at 20 months. Another example is that genetic influence tends to increase for cognitive ability

Table 25.11. Summary of genetic and environmental longitudinal results

Variable	Genetic		Nonshared environment		Shared environment	
	Continuity	Change	Continuity	Change	Continuity	Change
Behavioral inhibition	Yes	Yes	No	Yes	—[a]	—[a]
Shyness (home)	Yes	No	No	Yes	Yes	Yes
IBR temperament (tester ratings)	Yes	No	No	Yes	—[a]	—[a]
Parental ratings of temperament	Yes	No	Yes	Yes	—[a]	—[a]
Empathy	Yes	No	No	Yes	—[a]	—[a]
Cheerfulness	Yes	Yes	No	Yes	Yes	Yes
Anger response to restraint	Yes	No	No	Yes	Yes	Yes
DES anger	Yes	Yes	No	Yes	Yes	No
General cognitive ability	Yes	Yes	No	Yes	Yes	No
SICD expressive language	Yes	Yes	No	Yes	Yes	No
SICD receptive language	Yes	No	No	Yes	Yes	Yes
Height	Yes	Yes	No	Yes	—[a]	—[a]
Weight	Yes	No	No	Yes	—[a]	—[a]

IBR, Infant Behavior Record; DES, Differential Emotions Scale; SICD, Sequenced Inventory of Communication Development.
[a] No significant shared environment.

during the second year. Shared environmental influence for receptive language decreases during the second year.

Such age differences can be addressed using cross-sectional data. More interesting are longitudinal genetic analyses of age-to-age change and continuity. In general, as summarized in table 25.11, continuity is largely accounted for by genetic factors and nonshared environment accounts for change. A good example of longitudinal genetic analysis involves observed temperament (see chapter 7). Genetic factors contribute almost entirely to age-to-age stability in temperament from 14 to 36 months for observed shyness and for tester ratings of temperament. Nonshared environment is responsible for change. This finding that genetics contributes to continuity and that nonshared environment contributes to change mirrors results for the few extant longitudinal twin studies of personality in adulthood (McGue et al., 1993).

There are many exceptions to the general finding of genetic continuity and nonshared environmental change. For example, for behavioral inhibition, genetic factors contribute more to change than to continuity. Results showing genetic change are sometimes interpreted to mean that new genetic influences come into play at later ages. This would not necessarily mean that genes are literally turned on or off. It could mean that genes that have certain effects at one age have different effects at later ages when the organism in which they operate has changed so dramatically in terms of cognition and emotion. Regardless of the specific mechanism of genetic change, these quantitative genetic results pointing to age-to-age genetic change imply that, if a specific gene were found that is associated with behavioral inhibition at one age, it would not necessarily be associated with behavioral inhibition at another age. This finding also implies that in order to capture genetic influence on behavioral inhibition in infancy (e.g., in order to predict later behavior), multiple measurement occasions are required. When shared environmental influence was found, it often contributed to both change and continuity. For example, shared environment contributes to change as well as to continuity for observed shyness, the only temperament variable that shows shared environmental influence.

Parental ratings of temperament yielded an unusual result (see chapter 6). Like observed temperament measures, genetic factors largely contribute to stability for parental ratings. However, parental ratings showed significant nonshared environmental contributions to continuity. This stability of nonshared environment for parental ratings might be due to stable biases in parental ratings but, if this were the explanation, the biases would have to be specific to individual twins or reflect a stable tendency to contrast the twins in a particular way. Another possibility is that there are individual environmental events that begin to shape temperament even in early infancy, and that these temperamental characteristics can be detected by parents who frequently interact with their children but not by nonparental observers who base their assessments on a limited sample of behavior.

For emotion, the usual pattern of results (genetic continuity and nonshared environmental change) was found for empathy and for anger responses to restraint. In addition, shared environment contributed to change as well as continuity for observed anger expression. Some genetic change as well as continuity was found for cheerfulness.

For cognitive development, some hints of genetic change from age to age were found, although most genetic influence contributes to continuity. Significant new genetic variation appears to be expressed at 24 and 36 months. A reasonable hypothesis is that this new genetic variation might be due to the increasing presence of language on measures of mental development toward the end of the second year. For general cognitive ability, shared environmental influences are involved in continuity, not change. That is, no new shared environmental influences emerge at 20 or 24 months. This suggests the presence of a monolithic and static influence such as socioeconomic status. As usual, nonshared environmental effects largely account for change from age to age.

Expressive language shows a similar developmental pattern to general cognitive ability. However, expressive language shows greater genetic change than any other trait. New genetic variation appears at each age from 14 to 36 months that is greater than genetic variation in common across ages. Receptive language as assessed by the SCID shows a different developmental pattern. Unlike other variables in the cognitive domain, genetic factors contribute entirely to continuity, and shared environmental factors contribute more to change than to continuity across the four ages. Although receptive language ability is moderately correlated with expressive language and with general cognitive ability, it seems to be giving us many clues that it is different etiologically. For example, receptive language is less heritable from 14 to 24 months, and it shows high but declining shared environmental influence from 14 to 36 months. However, we will need at least a few more clues to understand its message fully.

Finally, another developmental pattern was observed for height and weight (see chapter 22). New genetic influences affect height between 14 and 20 months, but genetic factors are stable thereafter. In contrast, genetic influences on weight are stable from 14 to 36 months, with no indication of the emergence of new genetic influences.

Again, the data are telling us that no simple story does justice to the interesting complexity of developmental patterns of genetic and environmental influence. Although the general theme is that genetic factors contribute to continuity and nonshared environment contributes to change, nearly all possible patterns of genetic and environmental influence on change and continuity were seen.

Multivariate Relations

The relative independence of domains and of traits within domains, discussed earlier, stifles the potential that the MacArthur Longitudinal Twin Study offers for multivariate genetic analysis of covariance. In univariate genetic analysis, if there is no variance, there can be no genetic analysis of variance (e.g., for species-universal traits such as bipedalism or binocular vision). Similarly, in multivariate genetic analysis, there can be no genetic analysis of covariance if there is no covariance to analyze. For example, in chapter 22, Chambers et al. examined the relationship between body types and temperament, but the phenotypic correlations were too low for meaningful multivariate genetic analyses.

Part IV, Cognition, provides several interesting multivariate genetic results. As explained in the preface to that part (chapter 13), current developmental theory of modularity assumes that development proceeds from molarity to modularity. However, multivariate genetic analyses of cognitive development suggest the opposite. For example, genetic overlap between experimental word comprehension and memory tasks and the Bayley measure of general mental development increases during the second year (see chapter 14). Shared environmental influences overlap completely among these measures at all three ages.

Even more striking is the finding that, at 14 months, genetic effects on expressive and receptive language as assessed by the SICD are independent of genetic effects on the Bayley (see chapter 16). However, by 36 months, the situation is reversed, providing strong evidence for a developmental trend from modularity to molarity. In other words, a period of integration and interconnectedness may follow an earlier period of differentiation and relative independence. That is, a common genetic factor affects expressive and receptive language and scores on the Bayley, although there is also some genetic variation unique to receptive language. Shared environmental effects show a similar trend, beginning at 14 months as specific factors and losing their specificity by 36 months. Results at 20 and 24 months show intermediate transitional patterns.

Relations between temperament and cognition (chapter 23) also hinted at increasing genetic overlap from 20 to 24 months, but these genetic links between temperament and cognition emerged only for temperament assessed during the cognitive testing situation. In chapter 24, Schmitz et al. used multivariate genetic analysis to show that maternal ratings of temperament and emotion in infancy predict parental ratings of behavior problems at 4 years, primarily for genetic reasons.

A novel example of going beyond heritability is the application of multivariate genetic analysis of the covariance between maternal and paternal ratings (chapter 6). Mothers and fathers correlate about .50 on average in their temperament ratings of their children. Despite the high apparent heritability for both maternal and paternal ratings, a bivariate genetic analysis indicated that genetic factors do not contribute nearly as much to the correlation between maternal and paternal ratings as does nonshared environment. These multivariate genetic results suggest that mother–father agreement may be due to contrast biases held in common by mothers and fathers which magnify differences between their twins.

Another type of multivariate analysis considers relations between dimensions of normal variation and disorders, the normal and the abnormal. Although it is important to understand the etiology of the full range of individual differences, the clinical end of behavioral dimensions is the problem end—for example, shy children, unempathic children, and language-delayed children. Behavioral genetic analyses can, of course, be applied to samples selected on the basis of some cut-off or diagnosis. Psychiatric geneticists routinely study diagnosed cases and compare concordances, for example, for identical and fraternal twins. However, a technique called DF extremes analysis (see chapter 5) makes it possible to investigate the genetic and environmental links between dimensions and disorders. This technique assesses the extent to which disorders are the quantitative extreme of the same genetic and environmental factors responsible for variation throughout the distribution.

Chapter 5 focused on the analysis of extremes using observational measures of temperament. The results conform to an emerging pattern of results from other genetic analyses of this type: The genetic and environmental origins of the extremes of dimensions are not qualitatively different from the origins of the rest of the dimension. In other words, the extremes of dimensions appear to be the quantitative extremes of the same genetic and environmental factors that affect individual differences throughout the dimension. Similar results were found for analyses of low cognitive ability not reported in this book (Petrill et al., 1997). These results suggest strong genetic and environmental links between the normal and abnormal. More concretely, this emerging pattern of results implies that when genes are identified that are associated, for example, with risk for being highly inhibited, the same genes will account for genetic variation in inhibition throughout the dimension. Of course, results could dif-

fer for even more extreme children than those studied in the MacArthur Longitudinal Twin Study sample (Deater-Deckard et al., 1997). As chapter 5 indicates, such analyses address issues of great importance for understanding
the etiological links between the normal and abnormal.

Interplay between Nature and Nurture

Three of the most important findings that have emerged from behavioral genetic research involve the environment rather than genes. The MacArthur Longitudinal Twin Study offers additional documentation of all three points. The
first finding is that genetic research provides the best evidence that we have
for the importance of the environment. That is, although genetic studies such
as ours consistently point to strong genetic influence, most of the variance is
not explained by genetic factors. The second finding is the importance of nonshared environment. That is, to a substantial extent, salient environmental factors are those that make children growing up in the same family different from
one another. The third finding is at the heart of the interplay between nature
and nurture. Ostensible measures of the environment (e.g. parenting) often
paradoxically show genetic influence when treated as dependent variables in
genetic designs. This finding is an example of genotype–environment correlation, suggesting that parents respond to genetic dispositions of their children.

Although we have emphasized findings in relation to genetic influences, the
MacArthur Longitudinal Twin Study is no exception to the rule that genetic
research provides strong evidence of environmental influence. Heritabilities
are less than 50% for all behavioral traits reported in this book, which means
that nongenetic factors account for most of the variance. Although it is not
news to most developmentalists that the environment is important, as the mo-
mentum of genetic research pushes the pendulum of fashion toward the side
of nature, this point is critical to maintain a balanced perspective.

Concerning nonshared environment, we have emphasized the appearance of
shared environmental influence because it is surprisingly powerful for several
traits in infancy. However, nonshared environment always accounts for a substantial amount of variance. The possibility remains that nonshared environment is largely due to idiosyncracies of measurement, which can be less char-
itably called error of measurement. This is the conclusion in several of the
chapters concerning the cognition domain.

However, chapter 18 considers maternal perceptions of differential treatment
as assessed by a semistructured interview in which mothers were asked directly about their differential treatment. Modest stability was found for maternal perceptions of differential treatment from 14 to 36 months, especially for
differential discipline. The most interesting result was that differential treatment was reported to be greater for fraternal twins than for identical twins,
again, especially for discipline. This is consistent with the hypothesis that
mothers respond to genetic differences between their children. In the past,
such results would have been interpreted as a violation of the equal environ-

ments assumption of the twin method. However, other designs such as adoption designs consistently find that measures of parenting show genetic influence (Plomin et al., 1994). If parents respond to genetic differences between their children, we would expect that identical twins would be treated more similarly than fraternal twins. It is helpful to think about specific genes. If genes were identified that were associated with children's discipline-relevant behavior, we would expect that those genes would also be associated with parental discipline. Clearly, this would be a genetic effect, not a violation of the equal environments assumption of the twin method.

Another specific measure of the environment that is especially relevant to twins is perinatal factors. The relations between a dozen perinatal variables and development were examined in (chapter 21). Birth weight was the strongest correlate of development, yielding significant correlations with cognitive development at 14, 20, 24, and 36 months of age. Rather than assuming that the relation between birth weight and cognitive development is environmental in origin, this assumption was tested using differences within pairs of identical twins. Differences in birth weight within pairs of identical twins are not related to their differences in cognitive ability, which suggests that the environmental hypothesis may not be correct. That is, genetic factors might contribute both to birth weight differences and differences in cognitive development. To test this hypothesis, a multivariate genetic analysis is needed.

CONCLUSIONS

The general message of the MacArthur Longitudinal Twin Study is that there is no general message in terms of the genetic and environmental origins of individual differences in the second year of life. For those who want simple answers, this will be seen as a disappointing result. However, we view the complexity of these findings as an exciting conclusion that we hope will stimulate more research focused on specific traits.

Why is receptive language less heritable than expressive language? Why is there so much shared environmental influence for shyness and receptive language? Why is there so little shared environmental influence for behavioral inhibition, tester ratings of temperament, and empathy? Why do genetic factors contribute to change for behavioral inhibition and for expressive language? What are the processes responsible for the extensive genetic overlap between general cognitive ability and expressive language? What are the shared environmental factors that are constant in their effect on general cognitive development and expressive language during the second year? What are the shared environmental factors that contribute to change from age to age for receptive language?

Although the MacArthur Longitudinal Twin Study has taken major steps toward advancing our knowledge of individual differences in development during the transition from infancy to early childhood, there is a long way to go. Three related directions for research are most exciting: identifying specific environmental mechanisms, identifying specific genes, and identifying specific correlations and interactions between genes and environment. Specific genes responsible for heritability of complex behavioral traits are beginning to be identified (Hamer, 1997; Hamer & Copeland, 1998). These are not simple single-gene effects but rather genes that operate in multiple-gene and multiple-environment systems, often called quantitative trait loci (QTLs) or suscepbility genes. Once specific genes are found that are related to behavior, it is relatively easy and inexpensive to use these genes with any sample (not just twins or adoptees) to ask developmental and multivariate questions about the interface between nature and nurture (Plomin & Rutter, 1998).

This experiment in collaborative science has been a clear success in producing a whole that is much greater than the sum of its parts. The MacArthur Longitudinal Twin Study is an ongoing project with assessment continuing through middle childhood, so its best is yet to come.

REFERENCES

Buss, A. H., & Plomin, R. (1984). *Temperament: Early developing personality traits*. Hillsdale, NJ: Erlbaum.

Deater-Deckard, K., Reiss, D., Hetherington, E. M., & Plomin, R. (1977). Dimensions and disorders of adolescent adjustment: A quantitative genetic analysis of unselected samples and selected extremes. *Journal of Child Psychology and Psychiatry, 38,* 515–525.

Emde, R. N., Plomin, R., Robinson, J., Reznick, J. S., Campos, J., Corley, R., DeFries, J. C., Fulker, D. W., Kagan, J., & Zahn-Waxler, C. (1992). Temperament, emotion, and cognition at 14 months: The MacArthur Longitudinal Twin Study. *Child Development, 63,* 1437–1455.

Hamer, D. (1997). The search for personality genes: Adventures of a molecular biologist. *Current Directions in Psychological Science, 6,* 111–114.

Hamer, D., & Copeland, P. (1998). *Living with our genes*. New York: Doubleday.

McGue, M., Bacon, S., & Lykken, D. T. (1993). Personality stability and change in early adulthood: A behavioral genetic analysis. *Developmental Psychology, 29,* 96–109.

Petrill, S. A., Saudino, K. J., Cherny, S. S., Emde, R. N., Hewitt, J. K., Kagan, J., & Plomin, R. (1997). Exploring the genetic etiology of low general cognitive ability from 14 to 36 months. *Developmental Psychology, 33,* 544–548.

Plomin, R., Campos, J., Corley, R., Emde, R. N., Fulker, D. W., Kagan, J., Reznick, J. S., Robinson, J., Zahn-Waxler, D., & DeFries, J. C. (1990). Individual differences during the second year of life: The MacArthur Longitudinal Twin Study. In J. Colombo & J. Fagen (Eds.), *Individual differences in infancy: Reliability, stability, and predictability* (pp. 431–455). Hillsdale, NJ: Erlbaum.

Plomin, R., Chipuer, J. M., & Neiderhiser, J. M. (1994). Behavioral genetic evidence for the importance of nonshared environment. In E. M. Hetherington, D. Reiss, & R. Plomin (Eds.), *Separate social worlds of siblings: The impact of nonshared environment on development* (pp. 1–31). Hillsdale, NJ: Erl-baum.

Plomin, R., Emde, R. N., Braungart, J. M., Campos, J., Corley, R., Fulker, D. W., Kagan, J., Reznick, J. S., Robinson, J., Zahn-Waxler, C., & DeFries, J. C. (1993). Genetic change and continuity from fourteen to twenty months: The MacArthur Longitudinal Twin Study. *Child Development, 64,* 1354–1376.

Plomin, R., & Rutter, M. (1998). Child development, molecular genetics, and what to do with genes once they are found. *Child Development, 69,* 1223–1242.

Index

activity
  assessment, 57, 91, 310
  continuity and change, 107, 367–368
  contrast effects, 78
  correlations with other measures of temperament, 307–308, 312–314, 358–359
  extreme group familiality, 62
  extreme group heritability, 62
  extreme group means, 60–61
  genetic influence, 105–106
  heritability, 62, 103
  model fitting, 100
  twin correlations, 93–94
adaptive functions of emotions, 124
affect-extraversion
  assessment, 57, 91, 310
  common influences with general cognitive ability, 316
  continuity and change, 99, 105, 368–369
  correlations with cognitive measures, 308–309, 313–314, 364–365

  correlations with temperament, 361–363, 366
  extreme group contrast effects, 64
  extreme group familiality, 64
  extreme group heritability, 64–65
  extreme group and individual differences compared, 65
  extreme group means, 63–64
  extreme group nonadditive genetic influence, 64
  extreme group nonshared environment, 65
  extreme group shared environment, 65
  heritability, 64–65, 103
  model fitting, 102, 316
  twin correlations, 93–94
affective responses, 321–322
age-stability
  emotions, 131–132, 135–138
  summary of results, 374
  temperament, 76–80, 90, 94–95
aggression, 259
Akaike Information Criterion, 296
altruism, 142, 144, 156